SOCIETY FOR EXPERIMENTAL BIOLOGY
*SEMINAR SERIES*: 36

# BIOMECHANICS IN EVOLUTION

# SOCIETY FOR EXPERIMENTAL BIOLOGY SEMINAR SERIES

A series of multi-author volumes developed from seminars held by the Society for Experimental Biology. Each volume serves not only as an introductory review of a specific topic, but also introduces the reader to experimental evidence to support the theories and principles discussed, and points the way to new research.

2. Effects of pollutants on aquatic organisms. *Edited by A.P.M. Lockwood*
6. Neurones without impulses: their significance for vertebrate and invertebrate systems. *Edited by A. Roberts and B.M.H. Bush*
8. Stomatal physiology. *Edited by P.G. Jarvis and T.A. Mansfield*
10. The cell cycle. *Edited by P.C.L. John*
11. Effects of disease on the physiology of the growing plant. *Edited by P.G. Ayres*
12. Biology of the chemotactic response. *Edited by J.M. Lackie and P.C. Williamson*
14. Biological timekeeping. *Edited by J. Brady*
15. The nucleolus. *Edited by E.G. Jordan and C.A. Cullis*
16. Gills. *Edited by D.F. Houlihan, J.C. Rankin and T.J. Shuttleworth*
17. Cellular acclimatisation to environmental change. *Edited by A.R. Cossins and P. Sheterline*
19. Storage carbohydrates in vascular plants. *Edited by D.H. Lewis*
20. The physiology and biochemistry of plant respiration. *Edited by J.M. Palmer*
21. Chloroplast biogenesis. *Edited by R.J. Ellis*
23. The biosynthesis and metabolism of plant hormones. *Edited by A. Crozier and J.R. Hillman*
24. Coordination of motor behaviour. *Edited by B.M.H. Bush and F. Clarac*
25. Cell ageing and cell death. *Edited by I. Davies and D.C. Sigee*
26. The cell division cycle in plants. *Edited by J.A. Bryant and D. Francis*
27. Control of leaf growth. *Edited by N.R. Baker, W.J. Davies and C. Ong*
28. Biochemistry of plant cell walls. *Edited by C.T. Brett and J.R. Hillman*
29. Immunology in plant science. *Edited by T.L. Wang*
30. Root development and function. *Edited by P.J. Gregory, J.V. Lake and D.A. Rose*
31. Plant canopies: their growth, form and function. *Edited by G. Russell, B. Marshall and P.G. Jarvis*
32. Developmental mutants in higher plants. *Edited by H. Thomas and D. Grierson*
33. Neurohormones in invertebrates. *Edited by M. Thorndyke and G. Goldsworthy*
34. Acid toxicity and aquatic animals. *Edited by R. Morris, E.W. Tayler, D.J.A. Brown and J.A. Brown*
35. Division and segregation of organelles. *Edited by S.A. Boffey and D. Lloyd*
36. Biomechanics in evolution. *Edited by J.M.V. Rayner and R.J. Wootton*
37. Techniques in comparative respiratory physiology: An experimental approach. *Edited by C.R. Bridges and P.J. Butler*
38. Herbicides and plant metabolism. *Edited by A.D. Dodge*
39. Plants under stress. *Edited by H.G. Jones, T.J. Flowers and M.B. Jones*
40. *In situ* hybridisation: application to developmental biology and medicine. *Edited by N. Harris and D.G. Wilkinson*
41. Physiological strategies for gas exchange and metabolism. *Edited by A.J. Woakes, M.K. Grieshaber and C.R. Bridges*
42. Compartmentation of plant metabolism in non-photosynthetic tissues. *Edited by M.J. Emes*
43. Plant Growth: interactions with nutrition and environment. *Edited by J.R. Porter and D.W. Lawlor*
44. Feeding and the texture of foods. *Edited by J.F.V. Vincent and P.J. Lillford*
45. Endocytosis, exocytosis and vesicle traffic in plants. *Edited by C.R. Hawes, J.O.D. Coleman and D.E. Evans*

# BIOMECHANICS IN EVOLUTION


*Edited by*

**J. M. V. Rayner**

*Department of Zoology, University of Bristol*

*and*

**R. J. Wootton**

*Department of Biological Sciences, University of Exeter*



*The right of the University of Cambridge to print and sell all manner of books was granted by Henry VIII in 1534. The University has printed and published continuously since 1584.*

CAMBRIDGE UNIVERSITY PRESS

*Cambridge*

*New York Port Chester Melbourne Sydney*

Published by the Press Syndicate of the University of Cambridge
The Pitt Building, Trumpington Street, Cambridge CB2 1RP
40 West 20th Street, New York, NY 10011-4211, USA
10 Stamford Road, Oakleigh, Victoria 3166, Australia



First published 1991

Printed in Great Britain at the University Press, Cambridge

*A catalogue record of this book is available from the British Library*

*Library of Congress cataloguing in publication data*

Biomechanics and evolution / edited by J. M. V. Rayner and R. J. Wootton. p. cm. – (Society for Experimental Biology seminar series; 36)
Includes index.
ISBN 0–521–34421–2
1. Biomechanics. 2. Evolution (Biology) I. Rayner, J. M. V., 1953– II. Wootton, R. J. (Robin Jeremy), 1936– . III. Series: Seminar series (Society for Experimental Biology (Great Britain)); 36.
QH513.B56 1991
575 – dc20 91–4122 CIP

ISBN 0 521 34421 2 hardback

TAG

# CONTENTS


*List of contributors* vii

*Preface* xi

Ken Machin, 1924–1988
*R. McNeill Alexander and Charles P. Ellington* xv

Biomechanics and evolution: integrating physical and historical biology in the study of complex systems 1
*George V. Lauder*

The mechanical design of fossil plants 21
*Julian F. V. Vincent and George Jeronimidis*

Biomechanics of locomotion in fossil arthropods 37
*Derek E. G. Briggs, John E. Dalingwater and Paul A. Selden*

Cephalopod locomotor design and evolution: the constraints of jet propulsion 57
*John A. Chamberlain, Jr*

Biomechanics and the origin of insect flight 99
*Robin J. Wootton and Charles P. Ellington*

The hydrodynamics of cephalaspids 113
*Sarah J. Bunker and Kenneth E. Machin*

The evolution of underwater flight and the locomotion of plesiosaurs 131
*Jürgen Riess and E. Frey*

Pterosaurs: were they functional birds or functional bats? 145
*Kevin Padian*

Feeding mechanisms in some small herbivorous dinosaurs: processes and patterns 161
*David B. Norman and David B. Weishampel*

Avian flight evolution and the problem of *Archaeopteryx* 183
*Jeremy M. V. Rayner*

Functional interpretation of locomotory adaptations during equid evolution 213
*Jeffrey J. Thomason*

The evolution and mechanical design of horns and antlers 229
*Andrew C. Kitchener*

Characteristics and advantages of human bipedalism 255
*R. McNeil Alexander*

*Index* 267

# CONTRIBUTORS

*R. McNeill Alexander*
Department of Pure and Applied Biology, Baines Wing, University of Leeds, Leeds LS2 9JT, UK

*Derek E. G. Briggs*
Department of Geology, University of Bristol, Wills Building, Bristol BS8 1RJ, UK

*Sarah J. Bunker*
Department of Biological Sciences, University of Exeter, Hatherly Laboratories, Prince of Wales Road, Exeter EX4 4PS, UK

*John A. Chamberlain Jr*
Department of Geology, Brooklyn College, The City University of New York, Brooklyn, New York 11210, USA

*John E. Dalingwater*
Department of Environmental Biology, Williamson Building, The University, Manchester M13 9PL, UK

*Charles P. Ellington*
Department of Zoology, University of Cambridge, Downing Street, Cambridge CB2 3EJ, UK

*E. Frey*
Institut und Museum für Geologie und Paläontologie, Universität Tübingen, Sigwartstraße 10, D-7400 Tübingen, Germany
*Present address*: Staatliches Museum für Naturkunde, Erbprinzenstraße 13 (Friedrichsplatz), D-7500 Karlsruhe, Germany

*George Jeronomidis*
Biomechanics Group, Department of Engineering, University of Reading, Whiteknights, P.O. Box 225, Reading RG6 2AY, UK

*Andrew C. Kitchener*
Department of Pure and Applied Zoology, University of Reading, Whiteknights, P.O. Box 228, Reading RG6 2AJ, UK
*Present address*: Royal Museum of Scotland, Chambers Street, Edinburgh EH1 1JF, UK

*George V. Lauder*
Department of Developmental and Cell Biology, School of Biological Sciences, University of California, Irvine, California 92717, USA

*Kenneth E. Machin*
Department of Zoology, University of Cambridge, Downing Street, Cambridge CB2 3EJ, UK (*deceased*)

*David B. Norman*
Department of Zoology and University Museum, University of Oxford, South Parks Road, Oxford OX1 3PW, UK
*Present address*: Palaeontology Section, Nature Conservancy Council, Northminster House, Peterborough PE1 1UA, UK

*Kevin Padian*
Department of Integrative Biology, University of California, Berkeley, California 94720, U.S.A.

*Jeremy M. V. Rayner*
Department of Zoology, University of Bristol, Woodland Road, Bristol BS8 1UG, UK

*Jürgen Riess*
Institut und Museum für Geologie und Paläontologie, Universität Tübingen, Sigwartstraße 10, D-7400 Tübingen, Germany
*Present address*: Department of Biology, Southwest Texas State University, San Marcos, Texas 78666, USA

*Paul A. Selden*
Department of Extra-Mural Studies, The University, Manchester M13 9PL, UK

*Jeffrey J. Thomason*
College of Osteopathic Medicine and, Department of Zoological and Biomedical Sciences, Ohio University, Athens, Ohio 45701, USA

*Julian F. V. Vincent*
Department of Pure and Applied Zoology, University of Reading, Whiteknights, P.O. Box 228, Reading RG6 2AJ, UK

*David B. Weishampel*
Department of Anatomy and Cell Biology, School of Medicine, Johns Hopkins University, 725 North Wolfe Street, Baltimore, Maryland 21205, USA

*Robin J. Wootton*
Department of Biological Sciences, University of Exeter, Hatherly Laboratories, Prince of Wales Road, Exeter EX4 4PS, UK

# PREFACE

One of the first questions asked by the discoverer of a new fossil is how the organism in question worked. Among the oldest themes in the study of palaeontology has been the reconstruction of extinct plants and animals, in relation to how they functioned mechanically, and to where and how they may have lived. Biomechanics – the study of the function of an organism and of its component parts – has become a central tool in such research. More recently, a rigorous approach to the problems of evolution has led to the concept of constraints on the processes of adaptation and survival, and among the most important constraints are those imposed by the morphology of the organism, by the properties of its tissues, and by the mechanical environment in which it evolved and lived. Biomechanicists interested in the functioning of modern organisms not unnaturally ask how the systems they are studying evolved. As well as being a subject of considerable intrinsic interest, the question of the evolution of mechanical specializations is often crucial to interpreting their overall biological significance.

These three independent, but closely interlinked, approaches to the problems of biomechanics in evolution are illustrated in the chapters in this volume. The contributors are drawn from a range of backgrounds, and each takes a different starting point and a different goal. Some are concerned with understanding fossil structures or the mechanics and movement of extinct organisms, others with the mechanical processes underlying the appearance of major animal groups or of characteristic behaviour patterns, and yet others address problems in the interpretation of the fossil record. But all concur in their belief that biomechanics forms one of the most important tools in their research.

The chapters in this volume were originally presented as contributions to a joint seminar of the Society for Experimental Biology Biomechanics Group and the Palaeontological Association, which took place at the University of Manchester in January 1987. This was the first meeting, we believe, entirely devoted to the application of biomechanical principles and techniques to palaeontology and evolution. Circumstances have delayed publication, and several chapters have already been cited incorrectly as being 'in press' for earlier years. All contributions have now where necessary been updated by the authors to November 1990.

Dr Robin Wootton joined Dr Jeremy Rayner as co-editor in September 1990 to assist with the closing stages of publication. The editors wish to thank those contributors and others who assisted with the refereeing of chapters in this volume.

It was with great sadness that the editors and contributors to this volume learnt of the death of Kenneth E. Machin in 1988. Ken was a friend, teacher, inspiration and colleague to many of the contributors to this volume. The book is dedicated to his memory.

J.M.V.R. & R.J.W.
*December* 1990

# Ken Machin 1924–1988

The Department of Zoology at the University of Cambridge earned a tremendous reputation in the 1950s and 1960s for research on animal mechanics and biophysics. Exciting experimental investigations and ingenious equipment were the hallmarks of a dynamic group of researchers who quickly advanced their fields beyond recognition. The names from that era still invoke a deep respect: Richard Bainbridge, Ralph Brown, James Gray, Hans Lissmann, Donald Parry, John Pringle, Mark Pryor, Victor Rothschild – and Ken Machin.

After wartime work on the development of radar, Ken joined the Cambridge team that invented radio-astronomy. He was the first radio astronomer to program a computer, the leviathan Edsac I, for his research. Through the Gliding Club he met John Pringle, who quickly appreciated Ken's talents and ingenuity. The Zoology Department clearly needed a resident physicist adaptable enough to help in many lines of zoological research, and in 1953 it enticed Ken away from the radio-astronomers, much to their loss.

One of us (R.McN.A.) was Ken's first research student, in the mid 1950s. Ken seemed to be involved in all the many exciting things happening in the Department. With Pringle, he designed the elegant experiments which showed that the fibrillar flight muscle of insects would drive a mass-spring system at its resonance frequency. Lissmann had discovered the weak electric sense of some tropical fishes and, with Ken, was training one to make even finer discriminations between objects of different conductivities. Meanwhile, Ken theorized about how it might work, and competed with the fish by building electronic devices that would do the same job. He was continually being consulted about the design of apparatus, and about equipment that was causing trouble; for example, the great Bainbridge fish wheel that kept failing under the enormous torques needed to accelerate it at the same rate as the fish swimming inside. He was also drawn by Rothschild and Gray into the mechanics of flagellar locomotion. He published a paper concluding that the active elements must be distributed along the flagellum, that it could not simply be wagged by a motor at the base, before the sliding filament hypothesis was ever extended to cilia and flagella.

The other of us (C.P.E.) was Ken's last research student, in the mid 1970s. The intervening years had taken their toll of the dynamism and personnel of the original group. Ken, still very active, but with less stimulation in research, devoted more of his energies to teaching and to invaluable practical contributions; he helped to create an impressive television studio for the teaching laboratories, was very involved in the design and construction of a new wing, and supervised the re-wiring of the Department. Queens' College, Cambridge, also derived great benefit from his practical expertise for their building programme, in addition to his more pastoral role as Senior Tutor. He helped a number of researchers over those years, including Torkel Weis-Fogh, and he is gratefully acknowledged in many significant publications. But despite these research contributions he missed the vital spirit and camaraderie of an enthusiastic laboratory team. The trends of modern research with off-the-shelf equipment had started and were anathema to Ken; if you cannot buy equipment to do the research, then investigate something else! In more recent years, the new generation of insect flight research increasingly captured his interest, and we began to build some marvellous equipment in the grand old tradition. Almost certainly it was the equipment and the experimental problems that really caught his attention, and not insect flight per se. But it was that characteristic which made Ken so invaluable to many researchers over the decades: he happily collaborated on anything which posed interesting problems in physics or the design of equipment, and which allowed him to exercise his remarkable ingenuity. As Sir James Beament wrote: 'That was Ken's chosen role: he did not seek problems – he collected them, because everyone brought them to him, and it was his delight to solve them for others.'

Those that knew Ken were pleased to learn that he was again spending a great deal of time in the laboratory. He resigned as Senior Tutor and planned a new research programme on fibrillar flight muscle, the subject of his finest research back in the 1950s. The new equipment was designed and construction had started when he died suddenly. It was a great loss to those of us indebted to Ken, and we will sorely miss his unstinting helpfulness, his rapier-like physical insight, and his engaging laugh in the enjoyment of science. For others, this volume is a fitting tribute to a linchpin of biomechanics.

Neill Alexander
Charlie Ellington

GEORGE V. LAUDER

# Biomechanics and evolution: integrating physical and historical biology in the study of complex systems

One of the key problems in the study of biomechanics and evolution is that posed by Darwin: how do complex structures and functions evolve? If the form and function of parts of organisms are tightly integrated, then how easy is it for complex structures to be modified in evolution? Two approaches, not mutually exclusive, have been applied to the problem of complexity. First, biomechanical or biophysical analysis has been applied to organisms to elucidate the relation between environmental demands and the design of morphological systems. Such analyses contribute both an understanding of the design constraints on organismal form and explicit mechanical or mathematical models of organismal function. Second, historical, phylogenetic analysis has allowed the definition of evolutionary sequences of structural and functional change in organisms. In this chapter, two possible research programmes that integrate historical and biophysical analysis are presented. The first addresses the question of the relations among evolutionary change at different levels of biological organization: how independent is evolutionary change in form, function and behaviour? The second research programme uses biomechanical models to provide a causal theory that forms the basis for interpreting historically correlated changes in form, function and behaviour. By integrating physical and historical approaches to complex organismal design, we will be able to resolve the problem of complexity raised by Darwin.

## Introduction

A central fact confronting biologists and engineers as they attempt to understand organismal form and function is the apparent complexity of animal and plant design. Even the simplest structure is composed of many materials, may have more than one function, may be used by the organism in many ways, and has an evolutionary history which is not accessible to direct experimental analysis. More complex structures pose even greater problems as we attempt to provide scientific explanations for both the current diversity of biological form and function and for the historical patterns and processes that have produced this diversity. Complex morphological systems, because of their many parts, the intricacy of the interactions among the parts, and the difficulty of isolating individual components for study, have long been both a source of frustration for biologists and a magnet to which we are drawn in an attempt to apply the latest techniques and analytical tools. If we can understand the relation between form, function and the historical processes that have produced current diversity in a few complex morphological systems, then we will have made significant progress in our attempts to understand the causal basis of organismal design. This is an ultimate goal of many studies of biomechanics and evolution, the topic of this volume.

The 'problem of complexity' was well known to Darwin, who felt that complex morphological systems posed a special difficulty for his theory of evolution by natural selection. In a chapter outlining possible difficulties with his theory, Darwin (1859, chapter VI, pp. 186–187) discussed the eye, noting that it is difficult to imagine how such a complex structure could have been formed by a series of small structural and functional modifications. Indeed, it *is* difficult to imagine how the eye 'with all its inimitable contrivances for adjusting the focus to different distances, for admitting different amounts of light, and for the correction of spherical and chromatic aberration' (Darwin 1859, p. 186) could be modified in stages and how individual parts could be changed without disrupting the function of the entire structure.

A key concept underlying the problem of complexity, both as seen by Darwin and in modern formulations, is that organisms are highly integrated systems. How is it possible to change just one component of a structurally and functionally integrated unit? It is easy to imagine that altering one part of a complex structure could render the whole non-functional or at least of greatly reduced efficiency. The more tightly integrated are all the structures and functions of a complex system, the more difficult it should be to modify any single component.

The idea that organisms are highly integrated systems in which modification of one small feature has consequences for all the other structures and functions has been manifested many times in the history of biology (Russell 1916). Cuvier, for example, felt that all morphological features were so highly correlated and interrelated that by finding just one small part of a new fossil animal he could predict the form of the remaining parts. Russell (1916, p. 35) summarizes Cuvier's views by commenting that 'Cuvier's famous principle of correlation, the cornerstone of his work, is simply the practical application to the facts of structure of the principle of functional adaptedness. By the principle of correlation, from one part of an animal, given sufficient knowledge of the structure of its like, you can in a general way construct the whole.' Quoting from Cuvier (1826, p. 310), Russell continues, 'This must necessarily be so: for all the organs of an animal form a single system, the parts of which hang together, and act and react upon one another; and no modifications can appear in one part without bringing about corresponding modifications in all the rest.' More recently, similar views have been expressed by Taylor & Weibel (1981) and Weibel (1984), who advocate an extreme view of optimal design and intercorrelation among parts in the design of the mammalian respiratory system: 'in a functional system as complex as the one we are considering there is important interaction and cross-influence from one part to the other' (Weibel 1984, p. 60). These authors view all levels of organismal design, from cells to tissues and organs, as being coadapted so that changes at any one level necessarily entail alterations in structure and function at other levels.

This chapter addresses the problem of complexity in the context of studies of biomechanics and evolution. If organisms are in fact highly integrated systems in which any one part is difficult to alter without changing all the others, then it is indeed difficult to imagine how historical changes could take place in complex morphological systems. On the other hand, if there is considerable flexibility in complex systems such that characters and individual components of complex designs can be changed independently of other parts, then it is easier to envisage how changes in a complex structure could be achieved. In essence, this suggests that a central question in biomechanics and evolution today is: how many degrees of freedom does any complex morphological system possess? If there are few degrees of freedom, then complex designs should be relatively stable through time and changes in parts highly correlated with each other. If there are many degrees of freedom, then any one part of a complex design should be alterable without affecting other components.

## Physical and historical biology

Over the last twenty years two main approaches have been taken toward understanding complex organismal designs. The first is the biomechanical or biophysical approach (physical biology), while the second is the historical or evolutionary approach (historical biology). The two avenues are by no means mutually exclusive, but they have not often been applied by the same investigator to any one morphological system.

Physical biology is well exemplified by several chapters in this volume and by the works of Wainwright *et al.* (1976), Alexander (1968, 1971), Currey (1970), and the classic volume of Thompson (1917). The main goals of this approach are to define extrinsic constraints on animal design by precisely formulating both the physical and mechanical properties of biological materials and of the environment experienced by organisms. By precisely specifying in mathematical terms the demands of the environment on biological systems, and by then examining the mechanical design of organisms quantitatively, it is possible to determine the extent to which organismal design accurately matches environmental demands. Particularly good examples of this type of analysis include studies of animal flight (Rayner 1981, 1985; see also Wootton & Ellington, Riess & Frey, Padian and Rayner, this volume).

Historical biology, on the other hand, has had as a major goal the definition of precise historical sequences of structural and functional change (Lauder 1981, 1982*a*). This goes beyond the inference of functional attributes of fossils based on living species and biomechanical models in an attempt to document and explain historical sequences of change in an explicitly phylogenetic framework. Thus, not only are historical biologists interested in how a particular animal (either living or fossil) worked, but also in the exact historical sequence by which a particular morphological design was constructed.

A simple hypothetical example to illustrate an historical approach to organismal design is shown in figure 1. A hypothetical complex morphological system (shown at the top) is composed of a number of bones (indicated by circles), ligaments (broken lines), and muscles (black lines). This is the configuration possessed by species A. The phylogenetic relationships of species A and several closely related species (B to G) are shown underneath. After a phylogeny for these species has been obtained, the morphological configuration of each species is determined (by dissection, for example). Mapped onto the phylogeny is the level at which several of the morphological links of the complex design above evolved. Thus, species A, B, C, D and E all share the presence of muscle 4, while species G and F lack it. Muscle 4 is then considered to have evolved between nodes 4 and 3 and its occurrence diagnoses species A to E as forming a natural (monophyletic) lineage. Muscle 4 and ligament 6 form one biomechanical pathway (i) between bones c and d, while muscles 10, 9 and ligament 7 form a second (ii) biomechanical pathway (figure 1*a*). In species A, and only in species A, bone d can be moved by either (or both) of biomechanical pathways i and ii.

The phylogenetic analysis (figure 1*b*) indicates that the biomechanical

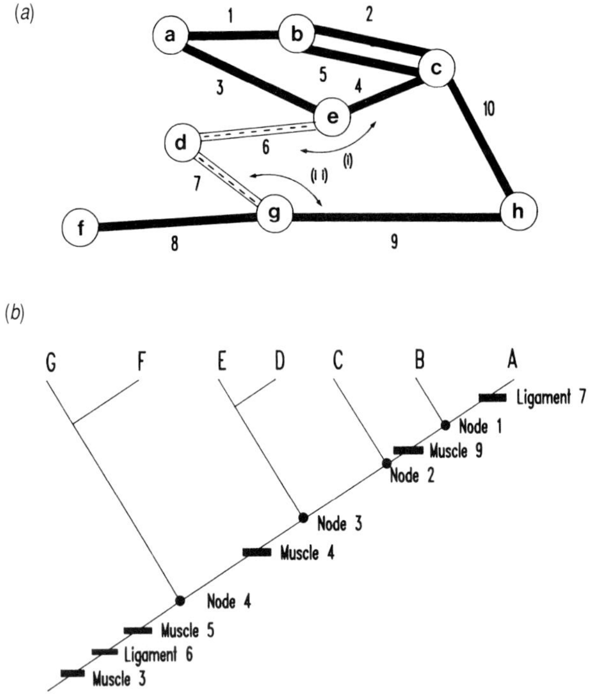


Figure 1. Schematic diagram representing (*a*) a hypothetical complex morphological system and (*b*) the historical sequence by which it was built up. Bones are represented by circles labelled with lower-case letters. Ligaments and muscles are numbered and are indicated by broken lines and black lines, respectively. Two pathways, numbered (i) and (ii), indicate two biomechanical linkage systems for moving bone d. (*b*) The phylogenetic relationships of seven species (A to G) with the historical origin of several components of the complex morphological system of (*a*) indicated. The diagram in (*a*) shows the morphology of species A in (*b*).

control of bone d evolved by a two stage process. First, pathway i arose, becoming complete by node 3 with the addition of muscle 4. All species at or above node 3 have the ability of moving bone d by pathway i. Then, pathway ii evolved, the first step being the addition of muscle 9 to the design, and the second, final, step being the addition of ligament 7 after node 1.

This type of analysis is fundamentally important in studies of biomechanics and evolution for several reasons. First, and most basically, it demonstrates the exact historical sequence by which the morphological design in species A was constructed. Second, it places considerable constraints on the types of allowable explanation for why the morphological design possessed by any of these species evolved. For example, any hypothesis that species C is adapted to function in a particular environment because of its morphology must also account for the fact that species D and E share the same morphology. Thus, the environment that is historically correlated with the single pathway i in species C is actually that environment present at node 3 on the phylogeny. If species A experiences the same environmental forces as species B to E, then it has simply retained a primitive ecological pattern that is not correlated with the evolution of two morphological novelties (muscle 9 and ligament 7). By focusing our biomechanical models and analyses on only one species and its present day environment, we may mistake a current correlation between form and environment for a causal relationship. The historical, phylogenetic analysis thus allows us to reject many proposed correlations as spurious, but does not itself enable us to establish a causal connection. How this can be done is discussed in more detail below.

Some attempts have been made to integrate both analyses of historical patterns and biophysical attributes of organismal design (see, for example, Kingsolver & Koehl (1985), Gans (1960, 1968), Lauder (1983, 1986), Liem (1973), Lombard & Wake (1986) and Schaefer & Lauder (1986)). But, in general, there has been little effort to develop case studies explicitly (an area where this symposium will be particularly valuable) and conceptual methodologies for integrating historical and biophysical approaches to organismal design.

In this chapter, I will outline two areas in which integrating physical and historical approaches could make significant contributions to our understanding of the problem of complexity and to the biomechanics and evolution of organismal design.

The first area involves the problem of complexity: how tightly integrated are organisms? Are changes in one component of a complex system necessarily reflected in other parts? Are changes at one level of biological organization reflected at other levels? A major contribution toward resolving the problem of complexity will be to determine the extent of integration in organismal design.

The second area concerns the conceptual role that biomechanical models and analyses play in analysing historical patterns to biomechanical design. Can biomechanical models be used as a causal link between historical correlations in patterns of structural, functional and behavioural/ecological change? If so, such biomechanical analyses will play a crucial role in conjunction with historical analysis in helping us to avoid the pitfalls of a purely correlational approach to the evolution of design.

These two areas will now be considered in turn, with the general aim of illuminating the relation between biomechanical and evolutionary analysis.

Table 1. *Levels of analysis of complex morphological systems*

| Level of analysis | Structures/functions to be studied |
|---|---|
| 1 Behavioural | Movements and movement sequences |
| 2 Peripheral morphology | Anatomy of the musculo-skeletal system |
| 3 Motor pattern | Pattern (sequence and timing) of central nervous output to the peripheral musculature |
| 4 Central nervous system structure | Anatomy of brain and spinal cord nuclei and their interconnections |
| 5 Central nervous system circuits | Neural networks and the pattern of neuronal excitation |

**Evolution at different levels: behaviour, structure and function**

One way to analyse the extent to which any complex morphological system forms an integrated whole is to break it down into multiple levels and ask if any one of these levels can be modified independently of the others. In the context of animal biomechanics and evolution there are several natural levels into which one can divide a complex system. Table 1 illustrates several of these levels.

Any complex morphological system is used by animals to perform behaviours, and the behavioural level (table 1: level 1) is perhaps the most general level at which one might study form and function. If one is interested in studying the form and function of bird wings, then this level of analysis might involve analysing the kinematic patterns of wing strokes during flight, mating behaviour, feeding and nesting. Behavioural acts such as these have a morphological basis, systems of muscles and bones that are directly responsible for the kinematic patterns seen, and this forms the next level (table 1: level 2) at which one could study a complex morphological system. Continuing with the example of bird wings, one would then dissect the bones, muscles and ligaments of the wing and determine their form, interrelations and mechanical properties. In order to execute a behavioural act, the muscles must be activated by the nervous system in a particular pattern and with particular levels of force. The pattern of central nervous activation of peripheral musculature, called the motor pattern, can thus be analysed (table 1: level 3) to provide information at the next 'lower' level of organization. In the study of bird wings, the investigator would measure the sequence and amplitude of muscle electrical activity with electromyographic recordings of wing musculature during the behaviours of interest.

The motor pattern is communicated to the peripheral musculature by the

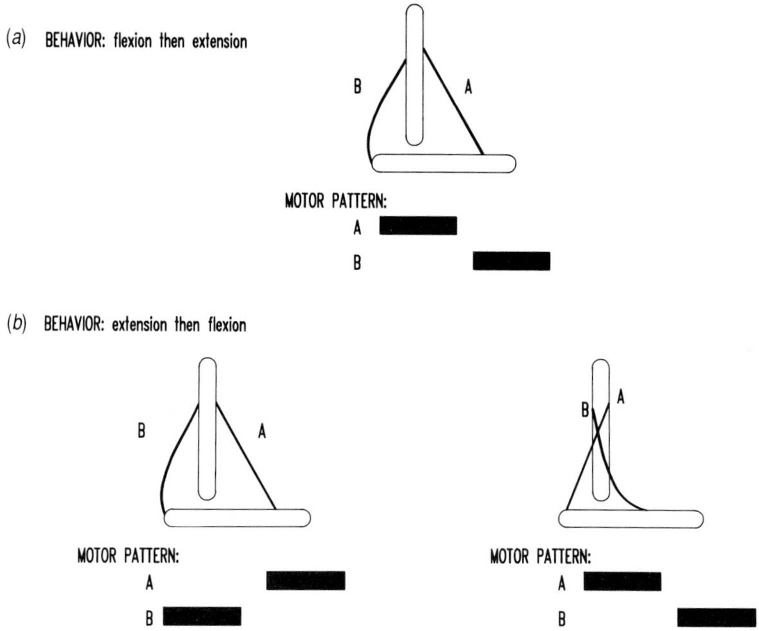


Figure 2. Diagram of two bones in several hypothetical species, and two muscles (black lines), A and B, moving those bones. (*a*) One species in which a behaviour, flexion followed by extension, is produced with the motor pattern shown: muscle A contracts for a time (indicated by the black bar) and then muscle B contracts. (*b*) Two alternative ways for a novel behaviour, extension followed by flexion, to be produced. This behaviour could be produced by altering the motor pattern such that muscle B contracts before muscle A (left), or by altering the peripheral morphology and retaining the same motor pattern (right). Muscle B now causes flexion because of alterations in its insertion.

central nervous system, and one could carry the analysis of any complex morphological system into the nervous system (table 1: level 4) to study the interconnections of nerve nuclei that are responsible for the motor pattern studied previously. Thus, one might make horseradish peroxidase injections into the peripheral muscles of the bird wing and study the central projections of the motor neurons to these muscles. Finally, one might want to use electrophysiological techniques to carry the analysis down to the cellular level in the nervous system and determine the precise neuronal circuit that produces the motor pattern (table 1: level 5). Of course, neuronal circuits have their origin in the biophysical properties of nerve cells and cell membranes (not shown in table 1), and this is yet another potential level of analysis.

If complex morphological systems are truly integrated, and if the

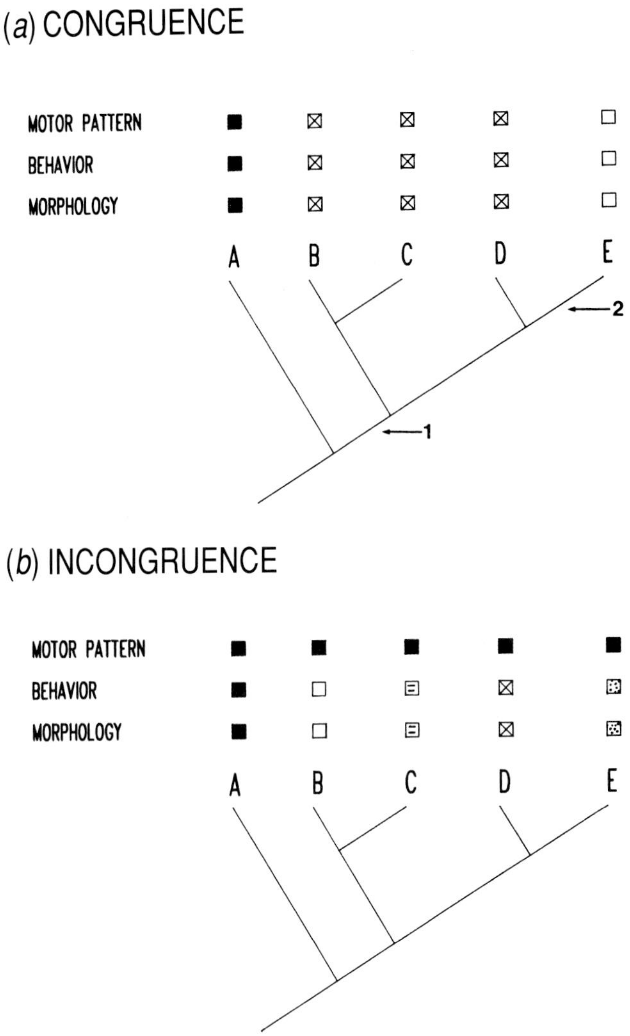


Figure 3. Diagram of historical congruence (*a*) and non-congruence (*b*) in the evolution of levels of morphological design. For the sake of simplicity, only three of the levels outlined in table 1 are shown. The arrows in (*a*) indicate the phylogenetic level at which congruent change occurred. In (*b*) the motor pattern has remained phylogenetically conservative, while behaviour and morphology have changed. This is a non-congruent pattern of historical transformation in levels of organization.

correlation among parts in such systems is high, then one would expect that a change at any one of these levels would necessitate a change in other levels. On the other hand, if a particular complex system is loosely integrated, then one would predict that changes could occur in any one of these levels without

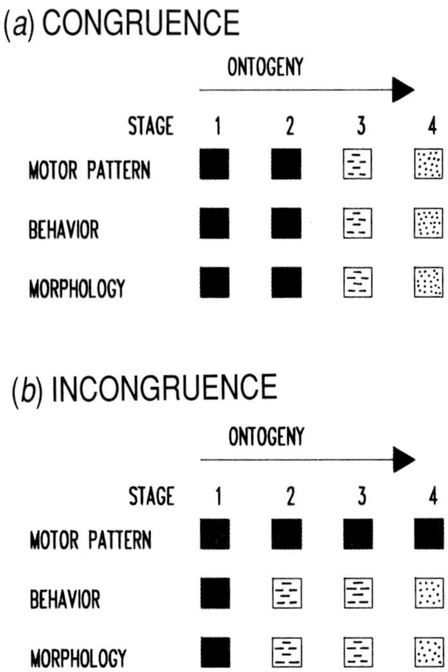


Figure 4. Diagram of (*a*) ontogenetic congruence and (*b*) non-congruence in the transformation of levels of morphological design. In (*a*) modification of three levels of design occurs congruently across ontogenetic stages. Changes in all three levels occur between ontogenetic stages 2 and 3, and between stages 3 and 4. In (*b*) one of many possible non-congruent patterns is shown, in this case with the motor pattern remaining conservative throughout ontogeny.

substantially affecting the others. Levels of design may thus be decoupled from one another.

We may now state an important problem in biomechanics and evolution: *to what extent are changes at different levels of biological design independent of one another during ontogeny and phylogeny*?

A simple example will serve to illustrate how changes could occur at one level independently of another level (figure 2). One species (figure 2*a*) has a morphological system composed of two bones connected on each side by two muscles, and exhibits a behaviour of flexion followed by extension. This results from the motor pattern shown (figure 2*a*): muscle A contracts followed by muscle B. Another closely related species might exhibit a different behaviour: extension followed by flexion. This could result from changes either at the motor pattern level or at the peripheral morphology level (table 1; figure 2*b*). A simple reversal in the motor pattern such that muscle B is activated prior to muscle A would produce the reversal of the behaviour seen in species A (figure 2*b*). A change at the motor pattern level would thus affect behaviour without affecting other levels. Alternatively, the motor pattern could remain the same, with the new behaviour in species B being entirely the result of alterations in the design of the peripheral musculo-skeletal system. If muscle A moves to attach to the opposite side of the bone, and muscle B does the same, then the motor pattern of species A will have the effect opposite to that which it used to have (figure 2*b*).

Depending on which of these results was obtained in a comparative analysis of these two species, one would say that either the motor pattern or the peripheral musculo-skeletal system was conserved in the evolution of these species.

The analysis of these different levels of design (table 1) involves physiological and biomechanical techniques to determine the structure and function appropriate for each level. However, to discuss the modification of these levels, we need to introduce an historical or ontogenetic analysis. For an historical analysis, the question of which levels are modified independently of others (if any) reduces to the following question: *are changes at different levels of design phylogenetically congruent*? This question can be answered by mapping changes in these different levels onto a phylogeny of a group of related species. Figure 3 shows two simple examples. If changes at the motor pattern and musculo-skeletal levels are decoupled historically, then one would expect to see a phylogenetic pattern where changes in these two levels at successive hierarchical nodes are not concordant (figure 3*b*). Alternatively, historical concordance, indicating that complex morphological systems are highly integrated, would result in a congruent pattern (figure 3*a*) where changes at any given node involve many levels of design.

A similar analysis could be conducted ontogenetically, and examples are shown in figure 4. Again, if all levels of organismal design are highly integrated, then when behaviour changes during ontogeny we expect to see concomitant changes at lower levels (figure 4*a*). On the other hand, non-congruent changes at the different levels would produce the pattern shown in figure 4*b*.

To date, no analyses have been completed even for one case study that would enable us to say unequivocally whether changes at *all* these different levels are congruent or not during ontogeny and phylogeny. However, two sets of ongoing research projects, involving both ontogenetic and phylogenetic analyses, allow us to address these issues at least for some of the possible levels of analysis. These studies involve the functional morphology and biomechanics of prey capture in lower vertebrates.

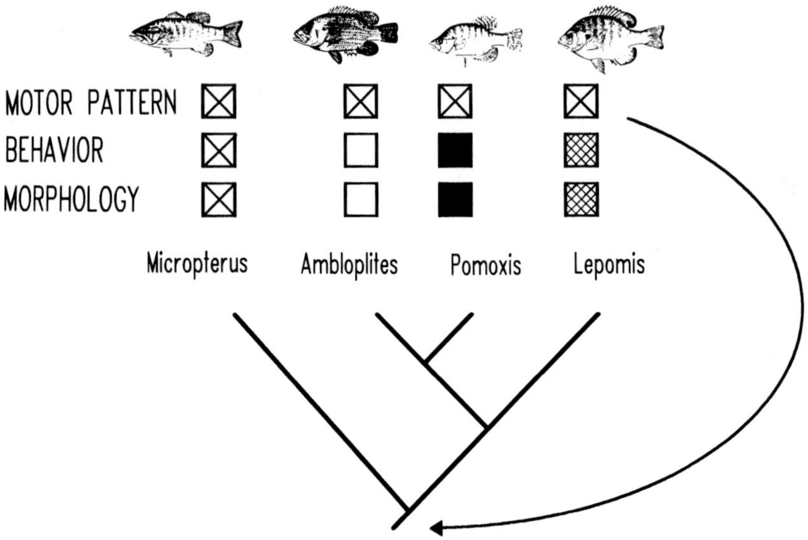


Figure 5. Diagram of the phylogenetic relationships of four genera of sunfishes (Centrarchidae) after Humphries & Lauder (unpublished results). Despite differences in mouth shape and size and differences between genera in feeding performance on different prey, the motor pattern used by these genera is extremely similar (Wainwright & Lauder 1986). Only one of eleven variables measured from the feeding motor pattern in four cranial muscles differed among the genera. Thus, the motor pattern is a primitive feature of this clade, as indicated by the arrow.

First, feeding behaviour in fishes has been used as a case study in the phylogenetic transformation of a complex morphological system, the feeding mechanism (Lauder 1980, 1982*b*, 1983, 1985; Wainwright & Lauder 1986). Second, we have used the ontogeny of feeding behaviour in tiger salamanders as a system to investigate ontogenetic congruence among behavioural, musculo-skeletal and motor pattern levels (Lauder & Shaffer 1986; G. V. Lauder & H. B. Shaffer, unpublished results; Shaffer & Lauder 1985*a*,*b*).

The results from a biomechanical and phylogenetic analysis of prey capture in sunfishes, an endemic North American family containing about 32 species, are shown schematically in figure 5. The phylogeny of the four genera shown derives from work by Humphries & Lauder (unpublished results). Each of the four genera differs considerably from the others in the gross morphology of the mouth (figure 5) and in feeding performance on different prey. For example, the bass (*Micropterus*) feeds on minnows to a greater extent than the other species and has a large gape and an enlarged mouth cavity, especially relative to genera such as *Lepomis*. Thus, the genera shown differ considerably in feeding behaviour and morphology. Wainwright & Lauder (1986) conducted an analysis of the muscle activity pattern (motor pattern) used by these species to determine if they differed. Small electrodes were implanted in four head muscles while the fishes were anaesthetized, and the fishes allowed to feed when they had recovered. The relative timing and duration of electrical activity in four head muscles involved in opening and closing the mouth was measured in all four genera. Statistical analysis showed that the motor pattern used by the four genera during prey capture was extremely similar, differing in only one of eleven variables (Wainwright & Lauder 1986). This case study argues strongly that the differences seen among the species in feeding performance and behaviour are a consequence of the morphological differences among the species, not of any differences in the neural output to the musculature. Behavioural variation in this clade is thus a consequence of morphological variation, not of changes in the output of the nervous system.

The results of our work on the ontogeny of feeding behaviour in salamanders is summarized in table 2. In these studies, the feeding behaviour of tiger salamanders (*Ambystoma tigrinum*) was studied during ontogeny. Larval tiger salamanders feed in the water by using suction feeding (Lauder 1985; Lauder & Shaffer 1985): prey are captured by rapidly expanding the volume of the mouth cavity (creating a negative pressure) and drawing water and the prey into the mouth. After metamorphosis, tiger salamanders will feed either on land or in the water.

Among numerous morphological changes in the feeding mechanism at metamorphosis, the most significant is the change from a unidirectional to a bidirectional feeding system (Lauder & Shaffer 1986). Larval tiger salamanders feed by drawing water into the mouth anteriorly and expelling it over the gill slits posteriorly. Water flow through the mouth is thus unidirectional, from anterior to posterior. During metamorphosis the posterior gill openings close off, eliminating the exit for water. Thus, when metamorphosed animals feed in the water, they must expel the water captured with the prey anteriorly through the mouth, and flow is bidirectional.

The transformation to a bidirectional feeding design results in greatly reduced feeding performance in metamorphosed tiger salamanders feeding in the water (table 2): larvae are much more efficient at capturing elusive prey (Lauder & Shaffer 1986). The results of G. V. Lauder & H. B. Shaffer (unpublished) show that there has been no change in the pattern of electrical activity in the jaw muscles with metamorphosis. Thus, the motor pattern does not change with metamorphosis, despite major changes in the morphology of the feeding mechanism (table 2).

Table 2. *Changes in the feeding system of* Ambystoma tigrinum, *comparing larvae feeding in the water with metamorphosed animals feeding in the water*

| Level of analysis | Change from larva (aquatic) to metamorphosed (aquatic) | Comments |
|---|---|---|
| 1 Behavioural/performance | Yes | Larvae have higher performance |
| 2 Peripheral morphology | Yes | Major changes in the hydrodynamic design of the feeding mechanism |
| 3 Motor pattern | No | Electro-myography of six jaw muscles shows no changes between these two stages |

Although neither of these case studies has yet been developed to the point where we can unequivocally say that the only change that produces an alteration in behaviour or performance is at the peripheral morphological level, both case studies strongly indicate that the motor pattern used to produce behaviour may be very conservative in ontogeny and phylogeny. It is thus not necessary to alter this level of biological design to effect a novel behavioural pattern. To this extent, then, we may say that complex morphological systems such as the feeding apparatus in lower vertebrates may change in behaviour (performance) without any necessary concomitant alteration in motor pattern.

## Biomechanical analyses as causal links

It is common for biologists interested in particular morphological features or behaviours to propose explanations for why these structures or behaviours have arisen, and to propose that the acquisition of a morphological feature conferred a particular selective advantage to the individuals or species that possessed this feature. Such hypotheses can be very difficult to test. However, it is possible to define two criteria (involving both historical and biomechanical analyses) which can serve as a set of boundary conditions on hypotheses that certain structures confer advantages or allow new behaviours.

If several species share the presence of a certain feature, such as a shortened humerus, one might wish to propose that this shortening confers an increase in digging efficiency in those species with the shorter bone. If one further wished to propose that the shortening of the bone is an adaptation for digging that arose and has been maintained in evolution because of the functional advantage obtained by shortening the humerus, then at least two criteria need to be met.

First, the feature proposed to confer the functional advantage must have arisen phylogenetically in a congruent fashion with the improved behavioural efficiency (figure 6*a*). This contrasts with the situation in which the morphological feature has an incongruent origin with the behavioural advantage (figure 6*b*). Historical discordance in origin of the morphology and the behaviour refutes any hypothesized advantage attributable to the morphological feature. As discussed by Greene (1986), it is minimally necessary for hypotheses of adaptation that historical concordance be true for the characters under consideration.

However, the criterion of historical concordance alone is insufficient to demonstrate a causal connection between morphological features and behavioural advantage. As shown in figure 6*c*, there are *many* derived morphological features at each phylogenetic level. On any corroborated phylogenetic hypothesis, there are numerous shared derived attributes that characterize lineages. Which of these many features is related to the behavioural shift shown by species A, B and C (figure 6*c*)? Without an external covering law or model (Hull 1974) there is no way to distinguish which of the many *correlated* characters that arose at a particular phylogenetic level is in fact *causally* related to the change in behaviour.

Thus, the second criterion that needs to be met is that a biomechanical model be available that predicts precisely which morphological feature should cause the alteration in behaviour and why they should do so. This biomechanical model plays the role of a covering law: it provides an external standard by which to separate out correlated from causal relationships (figure 6*d*).

Returning to our example of digging behaviour and the shortening of the humerus, we may find that indeed both an increase in digging efficiency and shortening of the humerus have a correlated historical origin. (This would be determined by examining bone length and digging speed in each of the six species shown in figure 6*a*; species A, B and C, and only these species, should share both decreased humerus length and increased digging speed.) Even though the criterion of historical congruence has been met, we still have no evidence that in fact the decreased length of the humerus has *caused* the increase in digging efficiency unless we can produce a biomechanical model that predicts *a priori* that shortening the humerus should produce an increase

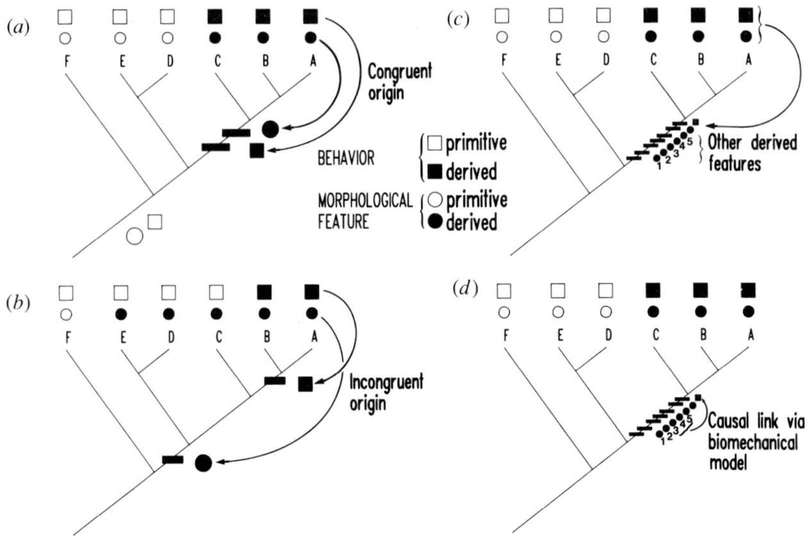


Figure 6. Four phylogenies illustrating the idea that the analysis of the historical relation between morphological and behavioural features is incomplete without a biomechanical model describing the causal linkage between specific morphological features and the behaviour. In all parts of the figure behaviours are indicated by boxes and morphological features by circles; derived states are represented by filled symbols. The transition from open to filled symbols indicates the origin of a novel feature. Open symbols characterize the ancestral condition in all panels. (*a*) Phylogenetic relationships of six species (A to F). Three species, A, B and C, share both derived behavioural and morphological conditions relative to ancestral traits. Both these traits are inferred to have evolved together prior to the origin of species C, as indicated by the arrows and black bars on the phylogeny. (*b*) Theoretical case where the distribution of a particular morphological feature is wider than that of a behaviour possessed by species A and B. The historical origin of the morphological and behavioural features (arrows) is incongruent, in contrast to the situation in (*a*). (*c*) At any level on a phylogeny, there will be many characters (out of the infinite number that organisms possess) that have evolved concordantly (indicated by filled circles 1 to 5), and thus many characters that are correlated with the historical origin of a particular behavioural feature. (*d*) Only some of those features sharing an historical origin with a behavioural feature (in this case filled circles 3, 4 and 5) will be predicted by any given biomechanical model to be causally related to the behavioural feature. Only then can we ascribe causal relations to the congruent origin of morphological and behavioural features.

in digging efficiency. This second criterion of biomechanical causation must be met in order for us to have any confidence in arguments for historical adaptation. In the digging example, such a biomechanical model might take the form of a diagram resolving the impact forces on the forelimb when the hand contacts the ground. This model could be used to predict the mathematical relation between length of the humerus and the force applied to the ground. If the model were sufficiently general, then it could also be used to evaluate the effect of changing other morphological features (such as the length of each of the other limb bones) on the force produced on the ground.

## Conclusions and prospectus

In this chapter I have focused on two areas in which biomechanical and historical analyses can be mutually illuminating. These are areas that have not received much attention to date, but which have considerable potential for guiding future research.

Both biomechanical and historical analyses are needed in our attempts to clarify the problem of complexity raised by Darwin, and to discover how ontogenetic and phylogenetic change occurs in complex morphological systems. If evolutionary change at different levels of organization is independent, then the 'problem of complexity' raised by intricate organs such as the vertebrate eye (Darwin, 1859) is greatly reduced.

Similarly, both biomechanical and historical analysis are needed to constrain and test hypotheses of adaptation. By integrating biophysical and historical approaches, we will be better able to make progress in understanding the actual causal factors underlying the diversity of organismal form and function.

## Acknowledgments

I thank Peter Wainwright and Steve Reilly for the helpful comments that they provided on the manuscript. This chapter could not have been written without the joint research conducted with Brad Shaffer, Peter Wainwright and Steve Reilly. The research described here was supported by NSF grants BSR 84–20711 and DCB 86–02606.

## References

Alexander, R.McN. 1968 *Animal Mechanics*. Seattle: University of Washington Press.

Alexander, R.McN. 1971 *Size and Shape*. London: Edward Arnold.

Currey, J.D. 1970 *Animal Skeletons*. London: Edward Arnold.

Cuvier, G. 1826 *Histoire des Progrés des Sciences naturelles depuis* 1789. Paris.

Darwin, C. 1859 *On the Origin of Species*. London: John Murray.

Gans, C. 1960 Studies on amphisbaenids (Amphibia, Reptilia). I. A taxonomic revision of the Trogonophinae and a functional interpretation of the amphisbaenid adaptive pattern. *Bulletin of the American Museum of Natural History* **119**, 129–204.

Gans, C. 1968 Relative success of divergent pathways in amphisbaenian specialization. *American Naturalist* **102**, 345–362.

Greene, H. 1986 Diet and arboreality in the emerald monitor, *Varanus prasinus*, with comments on the study of adaptation. *Fieldiana Zoology* (N.S.) **31**, 1–12.

Hull, D. 1974 *Philosophy of Biological Science*. New Jersey: Prentice Hall.

Kingsolver, J.G. & Koehl, M.A.R. 1985 Aerodynamics, thermoregulation, and the evolution of insect wings: differential scaling and evolutionary change. *Evolution* **39**, 488–504.

Lauder, G.V. 1980 Evolution of the feeding mechanism in primitive actinopterygian fishes: a functional anatomical analysis of *Polypterus*, *Lepisosteus*, and *Amia*. *Journal of Morphology* **163**, 283–317.

Lauder, G.V. 1981 Form and function: structural analysis in evolutionary morphology. *Paleobiology* **7**, 430–442.

Lauder, G.V. 1982*a* Historical biology and the problem of design. *Journal of Theoretical Biology* **97**, 57–67.

Lauder, G.V. 1982*b* Patterns of evolution in the feeding mechanism of actinopterygian fishes. *American Zoologist* **22**, 275–285.

Lauder, G.V. 1983 Functional design and evolution of the pharyngeal jaw apparatus in euteleostean fishes. *Zoological Journal of the Linnean Society of London* **77**, 1–38.

Lauder, G.V. 1985 Functional morphology of the feeding mechanism in lower vertebrates. In *Functional Morphology in Vertebrates*, ed. H.-R. Duncker & G. Fleischer, pp. 179–188. Stuttgart and New York: Gustav Fischer Verlag.

Lauder, G.V. 1986 Homology, analogy, and the evolution of behavior. In *The Evolution of Behavior*, ed. M. Nitecki & J. Kitchell, pp. 9–40. Oxford University Press.

Lauder, G.V. & Shaffer, H.B. 1985 Functional morphology of the feeding mechanism in aquatic ambystomatid salamanders. *Journal of Morphology* **185**, 297–326.

Lauder, G.V. & Shaffer, H.B. 1986 Functional design of the feeding mechanism in lower vertebrates: unidirectional and bidirectional flow systems in the tiger salamander. *Zoological Journal of the Linnean Society of London* **88**, 277–290.

Liem, K.F. 1973 Evolutionary strategies and morphological innovations: cichlid pharyngeal jaws. *Systematic Zoology* **22**, 425–441.

Lombard, R.E. & Wake, D.B. 1986 Tongue evolution in the lungless salamanders, Family Plethodontidae. IV. Phylogeny of plethodontid salamanders and the evolution of feeding dynamics. *Systematic Zoology* **35**, 532–551.

Rayner, J.M.V. 1981 Flight adaptations in vertebrates. *Symposia of the Zoological Society of London* **48**, 137–172.

Rayner, J.M.V. 1985 Mechanical and ecological constraints on flight evolution. In *The Beginnings of Birds*, ed. M.K. Hecht, J.H. Ostrom, G. Viohl & P. Wellnhofer, pp. 279–292. Eichstätt: JuraMuseum.

Russell, E.S. 1916 *Form and Function, a Contribution to the History of Animal Morphology*. London: John Murray. [Reprinted, University of Chicago Press, 1982.]

Schaefer, S.A. & Lauder, G.V. 1986 Historical transformation of functional design: evolutionary morphology of feeding mechanisms in loricarioid catfishes. *Systematic Zoology* **35**, 489–508.

Shaffer, H.B. & Lauder, G.V. 1985*a* Patterns of variation in aquatic ambystomatid salamanders: kinematics of the feeding mechanism. *Evolution* **39**, 83–92.

Shaffer, H.B. & Lauder, G.V. 1985*b* Aquatic prey capture in ambystomatid salamanders: patterns of variation in muscle activity. *Journal of Morphology* **183**, 273–284.

Taylor, C.R. & Weibel, E. 1981 Design of the mammalian respiratory system. I. Problem and strategy. *Respiration Physiology* **44**, 1–10.

Thompson, D'A.W. 1917 *On Growth and Form*. Cambridge University Press.

Wainwright, P. & Lauder, G.V. 1986 Feeding biology of sunfishes: patterns of variation in prey capture. *Zoological Journal of the Linnean Society of London* **88**, 217–228.

Wainwright, S.A., Biggs, W.D., Currey, J.D. & Gosline, J.M. 1976 *Mechanical Design in Organisms*. New York: John Wiley.

Weibel, E. 1984 *The Pathway for Oxygen*. Cambridge, Massachusetts: Harvard University Press.

JULIAN F. V. VINCENT & GEORGE JERONIMIDIS

# The mechanical design of fossil plants

## Introduction

In the study of fossils we are often limited in our interpretations by our knowledge and perception of current organisms. Nowhere can this be more true than in the application of biomechanics to fossil plants, since our knowledge of the biomechanics of modern plants is so scanty. We know many of the properties of wood, and something of the principles of the design of trees (Jeronimidis 1980). We also know a lot about the mechanical properties of cellulose and a little about its properties in the walls of a single cell (Haughton & Sellen 1969; Hettiaratchi & O'Callaghan 1978; Preston 1974). We know a little about the design principles of a few plants (Niklas & Kerchner 1984; Vincent 1982; Wainwright 1970; Wainwright *et al.* 1976), but much of the work in this area has been ill-informed and superficial or was done so long ago that the ideas, although by no means wrong, need to be brought up to date and incorporated into the present body of knowledge of biological materials and structures. In particular we know little of the engineering design principles of herbaceous plants. The first (and still the major!) worker in this area was Schwendener (1874), whose work is most readily available – especially to those who do not read German – in translation in texts such as that by Haberlandt (1914; first German edition in 1884). More recently, some work has been done in the general area of the fluid mechanics of plants (Rand 1983). In this chapter we have used a model for the mechanical design of turgor systems to compare the design of some modern and fossil plants. The problem of analysing stems which combine turgor and lignification with pre-stressed fibres has yet to be solved.

To begin at the beginning, a plant needs light and most light is gained by the tallest and most shading plant. Therefore the plant needs to have a stem which is as high as possible per unit of metabolic cost. The stem has to support a variety of mechanical loads. The most important is the side loads due to wind; stresses arising from the weight of the plant itself are trivial by

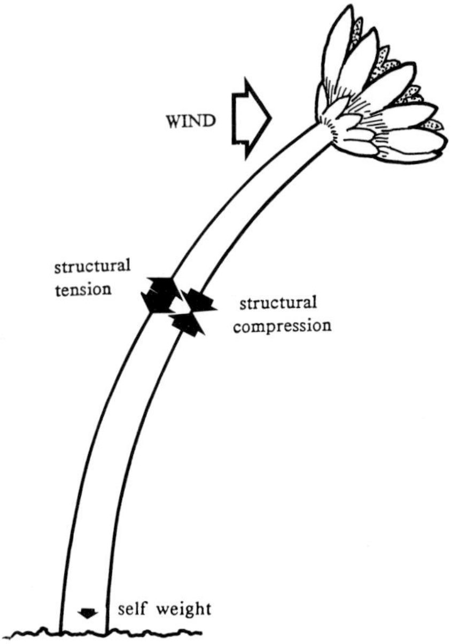


Figure 1. Diagram showing the distribution of forces due to the weight of the stem and to bending caused by side loads such as wind. The size of the arrows suggests the relative magnitude of the forces.

comparison with the bending stresses induced by drag forces. The plant stem must therefore be stiff to stand erect and it is limited by its ability to withstand the compressive forces experienced on the concave or compression face when it bends (figure 1). It is highly unlikely that any plant will fail on the tension side since cellulose has a very high tensile strength (Wainwright *et al.* 1976). The most likely mode of failure (although it has not been explicitly observed in non-woody plants) is local buckling of the fibres in the cell walls, leading to a compression crease at the macroscopic level. The fibres buckle because they have little or no lateral support. This illustrates the basic impossibility of the 'Indian rope trick'.

The problem of the instability of cell wall fibres under compression can be overcome in a number of ways. First, the fibres can be stabilized laterally by being glued together. Plants achieve this with lignin. Second, the stress can be reduced by increasing the load-bearing area, that is by having thicker cell walls. Third, the fibres can be rotated so that they are not being loaded on end, a solution rooted in composite theory and a phenomenon observed in the compression wood of trees. In fact all three of these methods are found in wood and together they remove any mechanical size constraint from the stem. Even the largest tree does not fall through compressive failure of the stem. However, the fourth solution to the problem – the introduction of prestress – is found not only in trees and other plants which are secondarily thickened (where it is due to differential shrinkage related to rates of growth; Jeronimidis 1980), but also in turgor systems. In these systems the compressive forces in the liquid within the cells emerge as tensile forces in the cell wall. These tensile forces must be paid off on the compression face of the bent stem before the cell wall fails in compression. The disadvantage of turgor is that it seems to be available only to smaller plants; the limit on the height of the stem is about 20 to 30 cm. Our (unpublished) experiments with various species of hogweed (*Heracleum*) show that one cannot tell whether or not the stem is lignified simply by measuring its stiffness. The younger parts of the stem have little or no lignin and rely solely on turgor for their stiffness and strength. This is common among young, small plants. As the stem ages it lignifies and increases in diameter, thereby increasing its flexural rigidity, or $EI$ (where $E$ is stiffness and $I$ is a shape factor, the second moment of area), by changing only $I$. In other words the decision whether or not to lignify seems to depend on what the compressive loads are going to be: lignification does not affect the stiffness.

## Developing a model

The effects of turgor in single cells and even in parenchymatous tissues are probably moderately well understood (Hettiaratchi & O'Callaghan 1978). However, the mechanical design of turgor systems as found in stems has not been investigated. We have experimented on the flowering stem of the dandelion (*Taraxacum officinale* agg.) to develop a mathematical model of its mechanical behaviour. That this stem uses a turgor system is shown by the fact that strips cut from it curl by differing amounts when immersed in solutions of sucrose varying from 0 to 15% (an experiment used to measure the osmotic pressure of the cell sap; see, for example, McLean & Ivimey-Cook 1962), and by the observation that in a totally wilted stem the epidermis is crinkled, whereas in the fully turgid stem it is smooth. We have used these observations as the basis for our model, making the following assumptions: (1) all cells are at the same turgor and have the same osmotic potential; (2) all cell walls have the same Young modulus, proportional to the amount of cellulose in them; (3) all cells are cylindrical along the length of the stem; (4) the cells are firmly stuck together. In addition we have chosen to neglect the vascular bundles. In *Taraxacum* flowering stem these are relatively little developed and have few or no fibres and very little evidence of lignification.

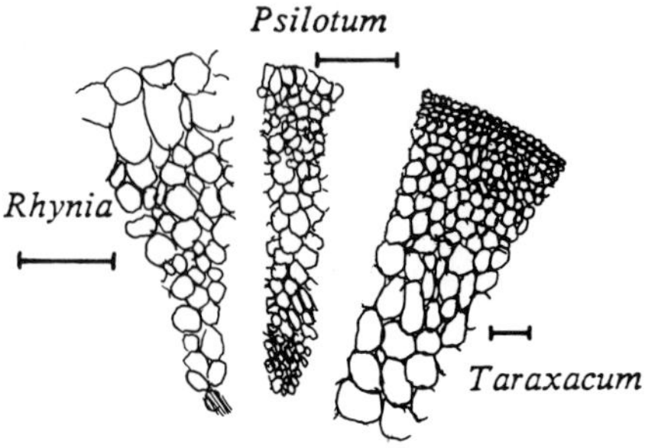


Figure 2. Stem sections of *Taraxacum*, *Rhynia* and *Psilotum*, drawn to different scales. Relative thickness of the cell walls is not included. Scale bars represent 0.1 mm.

The distribution of cellulose across the stem is very heterogeneous. On the outside of the stem the cells are small (6 μm diameter) with thick cell walls (about 3 μm); on the inside of the stem they are larger (about 60 μm diameter), with cell walls less than 1 μm thick. This is illustrated diagrammatically in figure 2 and graphically in figure 3, where the ratio of cell radius $r$ to the thickness of the cell walls $t$ is plotted against position $z$ across the stem. Measurements of $z$ are referred to the mid-plane of the section ($z = 0$), with the positive direction being radially outwards. The volume fraction of cell wall material across the stem is then given by

$$V_f^{cw}(z) = \text{(area of cell wall) / (area of stem)}, \\ = \tfrac{1}{4}\pi\ [(2\Phi(z)-1)/(\Phi(z)^2)], \quad (1)$$

where $\Phi(z) = r(z)/t(z)$, that is the ratio of the external radius of the cell to the thickness of the cell wall. $V_f^{cw}$ varies from about 0.5 on the outside of the stem to about 0.013 at the inside, with an average of 0.0844. This corresponds to a mean mass fraction of cellulose of 8.8%, assuming that the cell walls contain 30% of water in the cellulose, which itself has a dry density of 1500 kg/m$^3$. This compares with a measured dry matter of $8.73 \pm 0.533$% (mean $\pm$ standard deviation). The longitudinal Young modulus of the stem $E_x(z)$ will be a function of position across the thickness and is given by the product of the Young modulus of cell wall material and the local value of the volume fraction. For convenience, the stem is notionally divided into a large number of thin strips so that within each strip $V_f^{cw}$ can be considered constant. Each strip is identified by the distance of its middle from the origin ($z = 0$), and its Young modulus is

$$E_x(z) = V_f^{cw}(z)E_{cw}, \quad (2)$$

where $E_{cw}$ is the Young modulus of the cell wall material in the axial

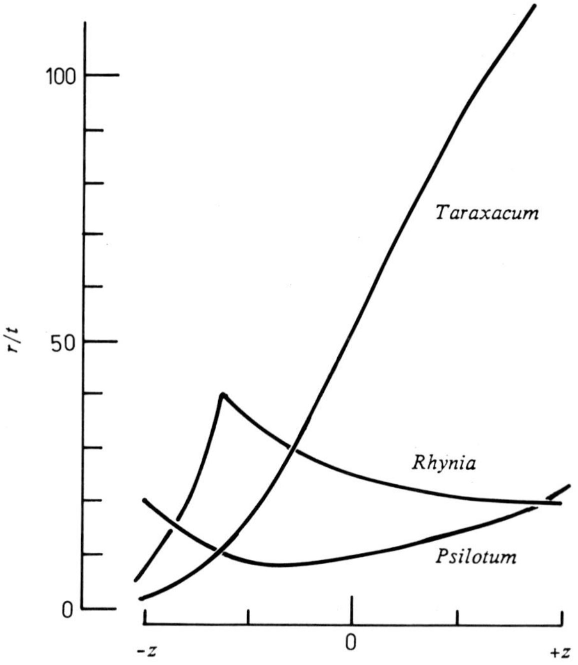


Figure 3. The variation in the ratio between cell radius and wall thickness ($r/t$) through the thickness of the stem, shown as negative (to outside) and positive (to inside) distance (values of $z$) from the centre of the section. The maximum distance of $+z$ to $-z$ is 0.8 mm for *Taraxacum* and *Rhynia*, and 0.4 mm for *Psilotum*.

direction. This can be estimated from measurements of the longitudinal modulus of the stem, giving a value of about 500 MPa, which compares well with the value of 700 MPa obtained for *Nitella* cell wall by Kamiya, Tazawa & Takata (1963). Note, in passing, that neither $E_{cw}$ nor the modulus $E_{stem}$ of the stem depend on turgor pressure, a result that has been obtained both theoretically and experimentally.

The response of the multi-layered stem to external forces and moments depends on three stiffness parameters which are derived from considerations of force and moment by using the theory for laminate composite beams. These are computed by the following three equations:

$$A = \int_{-\frac{1}{2}h}^{+\frac{1}{2}h} E_x(z)\, dz, \tag{3a}$$

$$B = \int_{-\frac{1}{2}h}^{+\frac{1}{2}h} E_x(z)\, z\, dz, \tag{3b}$$

Table 1. *Calculated parameters used in the mathematical model of the stem. See text for details. In computing* $N_x^P$ *and* $M_x^P$, *the restraint* β *has the value* 0.5

| | *Taraxacum* | *Psilotum* | *Rhynia* |
|---|---|---|---|
| $A$ (N/mm) | $67.5 \times 10^{-3} E_{cw}$ | $53.2 \times 10^{-3} E_{cw}$ | $44.8 \times 10^{-3} E_{cw}$ |
| $B$ (N) | $16.5 \times 10^{-3} E_{cw}$ | $0.9 \times 10^{-3} E_{cw}$ | $0.9 \times 10^{-3} E_{cw}$ |
| $D$ (N mm) | $6.0 \times 10^{-3} E_{cw}$ | $3.1 \times 10^{-3} E_{cw}$ | $2.8 \times 10^{-3} E_{cw}$ |
| $N_x^P$ (N/mm) | $594.5 \times 10^{-3} \Delta P$ | $294.5 \times 10^{-3} \Delta P$ | $470.0 \times 10^{-3} \Delta P$ |
| $M_x^P$ (N) | $1.4 \times 10^{-3} \Delta P$ | $1.3 \times 10^{-3} \Delta P$ | $4.55 \times 10^{-3} \Delta P$ |

and

$$D = \int_{-\frac{1}{2}h}^{+\frac{1}{2}h} E_x(z)\, z^2\, dz, \tag{3c}$$

The parameter $A$ in equation (3*a*) is used to relate the axial force $N_x$ per unit width to the mid-plane ($z = 0$) strains; $B$ (equation 3*b*) takes into account the fact that in a non-symmetrical beam the neutral axis of bending is not located at the centroid of the section (the mid-plane); $D$ relates the bending moment per unit width $M_x$ to the curvature of the mid-plane. The axial force and bending moment are related to mid-plane strains and curvatures by

$$N_x = A\varepsilon_x^0 + BK_x, \tag{4a}$$

and

$$M_x = B\varepsilon_x^0 + DK_x, \tag{4b}$$

where $\varepsilon_x^0$ is the mid-plane normal strain and $K_x$ is the mid-plane curvature. By substitution from equation (2) into (3*a–c*) and by numerical integration (which required the thickness to be divided into at least 600 strips), values for $A$, $B$ and $D$ were obtained (table 1).

In the absence of *external* forces and moments, a longitudinal strip of tissue cut from the stem acquires a non-zero curvature $K_x$ when the cells are under internal turgor pressure. Hence the effect of turgor can be expressed as a set of equivalent forces and moments in much the same way that thermal forces and moments are used in the analysis of the mechanics of a bimetallic strip. An additional expression is required to introduce the internal turgor pressure $P$ of the cells into the relations. The axial strain $\varepsilon_x^{cw}$ induced in the cylindrical cells by *free* axial expansion due to turgor, assuming that radial expansion is limited both by neighbouring cells and by the presence of an epidermis, is given by

$$\varepsilon_x^{cw} = \beta Pr/E_{cw}t. \tag{5}$$

The coefficient $\beta$ is a measure of the degree of restraint in the radial direction. For full restraint,

$$\beta = \tfrac{1}{2}(1-\nu_{yx}\nu_{xy}). \tag{5a}$$

Assuming that the Poisson ratio $\nu_{xy}$ in the wall is 0.4, then $\nu_{yx}$ will be at least ten times smaller and hence $\nu_{yx}\nu_{xy}$ will be small compared to 1, and $\beta$ can be taken as $\frac{1}{2}$. This equation can be used to define a coefficient of pressure expansion $\alpha_x^P(z)$, analogous to a coefficient of thermal expansion, in the form

$$\begin{aligned}\alpha_x^P(z) &= \varepsilon_x^{cw}/P = \beta r/E_{cw}t,\\ &= \beta\Phi(z)/E_{cw},\\ &= \tfrac{1}{2}\Phi(z)/E_{cw}.\end{aligned} \tag{6}$$

This relation can be used to calculate the equivalent forces and moments which produce the curvature in the strip under turgor pressure $\Delta P$ (with reference value $P = 0$ in the wilted state); thus

$$N_x^P = \int_{-\frac{1}{2}h}^{+\frac{1}{2}h} E_x(z)\alpha_x^P(z)\Delta P \,\mathrm{d}z, \tag{7}$$

and

$$M_x^P = \int_{-\frac{1}{2}h}^{+\frac{1}{2}h} E_x(z)\alpha_x^P(z)\, z\, \Delta P \,\mathrm{d}z. \tag{8}$$

Numerical values for $N_x^P$ and $M_x^P$ for the various plants considered are obtained by introducing equations (1), (2) and (6) into (7) and (8), and then integrating (table 1). These data can be used to calculate the mid-plane strains $\varepsilon_x^{0P}$ and curvatures $K_x^P$ due to turgor pressure, by using equations (3*a*–*c*) and (4*a*, *b*), as

$$K_x^P = [(AM_x^P - BN_x^P)/(AD-B^2)]\,\Delta P/E_{cw}, \tag{9}$$

and

$$\varepsilon_x^{0P} = [(DN_x^P - BM_x^P)/(AD-B^2)]\Delta P/E_{cw}. \tag{10}$$

Since curvature $K_x^P$ of the excised strip is the inverse of the radius of curvature of the strip, the effect of putting a strip cut from the stem into mannitol can now be calculated if the value of $P$ is known. The concentration of mannitol required to cause an excised strip of *Taraxacum* flowering stem to straighten has an osmotic potential of about 20 atmospheres, or 2 MPa (figure 4). Thus this will be the turgor pressure exerted against the cell walls

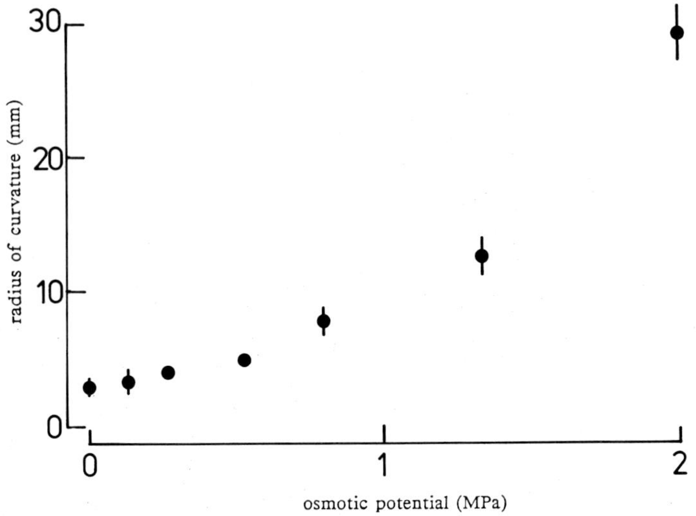


Figure 4. Variation in radius of curvature of excised strips of stem of *Taraxacum* with changing concentrations of mannitol, expressed as osmotic potential.

when the same strip is immersed in distilled water. The calculation for *Taraxacum*, based on the figures presented so far, gives a radius of curvature of 3.4 mm for a strip immersed in distilled water, compared with an experimental value of 3 mm. Thus we have modelled, quantitatively, the effect of turgor pressure on the mechanics of the stem in one major characteristic, and are confident that we can interpret the morphology in terms of the distribution of stress in the intact stem. It is appropriate here to point out that our model dismisses two of the interpretations of the structure and mechanics of the herbaceous stem which are commonly found in textbooks of botany. First, there is no requirement for differences in turgor pressure across the stem; second, tissues in the centre of the stem are not held in compression in the intact stem. The reason why the excised strip curls is that the cell walls of the inner cells are much thinner than those of the outer cells. They will thus stretch further under the same turgor pressure, since the stress will be greater on the thinner wall. Thus the excised strip will curl with the thinner walled cells on the outside of the cell.

Omitting some algebra, the resultant strain at any point throughout the thickness of the intact stem is the sum of the strains due to turgor pressure and to the bending moments reflecting the constraints in the stem which hold each excised strip straight (figure 5). Thus

$$\{\varepsilon_x(z)\}_{\text{total}} = (BK_x^{\text{P}}/A + \varepsilon_x^{0})\Delta P/E_{\text{cw}}. \qquad (11)$$

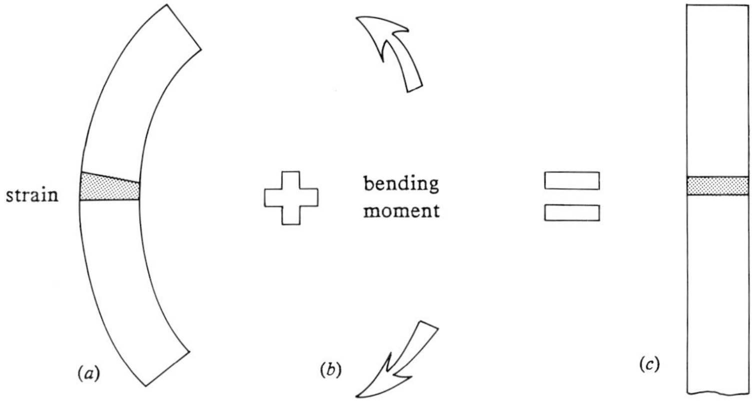


Figure 5. Diagram illustrating the summation of strains in an isolated portion of pre-stressed stem, showing how the pre-strain due to turgor pressure causes the stem to bend, giving a gradient in strain across the stem (*a*). The constraint provided within the intact stem can be represented as bending moment strain (*b*), leading to uniform strain across the intact stem (*c*).

This is independent of $z$ (strain does not depend on position across the stem) which is to be expected since the stem is straight. The net stress in the stem can now be obtained by assuming that, for small strains, the stress is proportional to the strain, a fact confirmed experimentally in tensile tests of whole stems. Thus we apply Hooke's law in each layer, and multiply the total strain by the local Young modulus of the stem. Since the stiffness of the stem is a product of $E_{cw}$ and $V_f^{cw}$, it will vary across the stem in the same manner as the volume fraction: thicker cell walls will allow a greater contribution to total stem stress. This is shown for *Taraxacum* in figure 6.

## Applying the model

The outcome of this model, in terms of the original formulation of the importance of turgor in the plant stem, is that any side load will have to pay off this high stress on the compression side. Thus if the plant stem can arrange to concentrate its cell wall material at the periphery of the stem then the tensile stress will be at a maximum just where the compressive stress due to bending will be highest. This can be illustrated by summing the stresses (figure 7). Admittedly this mechanism puts much greater tensile loads on the tension side of the stem, but the cell wall cellulose can take loads of the order of 1 GPa before it breaks (Wainwright *et al.* 1976). An additional advantage of this design is that the small, thick-walled cells which take the highest

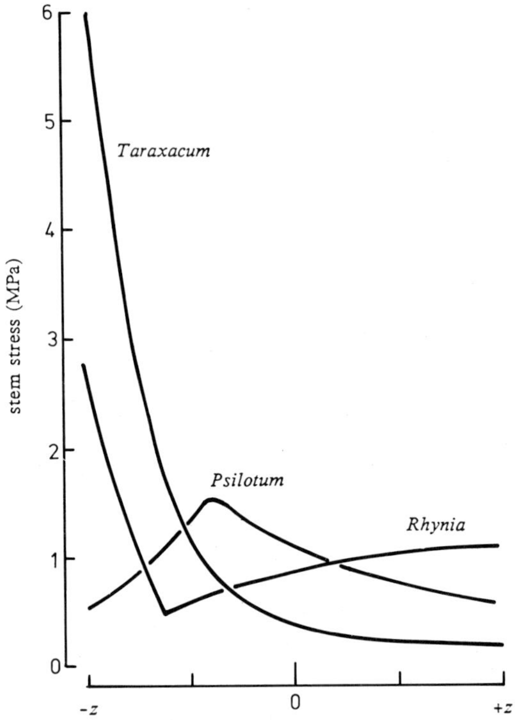


Figure 6. Calculated stress in the stem due to turgor pressure related to distance *z* from the centre of the stem. The outside of the stem is to the left.

compressive stress are also much less likely to fail by buckling than the larger, thinner-walled, cells found nearer the centre of the stem. Thus the gradient in $r/t$ has a two-fold effect in resisting buckling of the stem, acting both at the level of the cell and at the level of the whole structure.

We have made similar measurements on the distribution of cell size and cell wall thickness in stems of *Rhynia gwynne-vaughani* (a Devonian plant of the Rhynie chert in which the cell walls are exceptionally well preserved; Kidston & Lang 1917, 1920, 1921) and the extant primitive plant *Psilotum nudum*. The specimens of *Rhynia* were in rock sections provided by Dr R. Goldring of the Department of Geology and Sedimentology, University of Reading. We have assumed for calculation that the other required parameters – namely $E_{cw}$ and $\Delta P$ – are as measured on *Taraxacum*. Hence we calculate $A$, $B$, $D$, $M_x^P$ and $N_x^P$ in exactly the same way (table 1). Our main prediction for experimental confirmation is the radius of curvature of an excised strip of stem. Our calculated value for *Psilotum* is 210 mm (table 2), which is so large that in experimental terms it is indistinguishable from

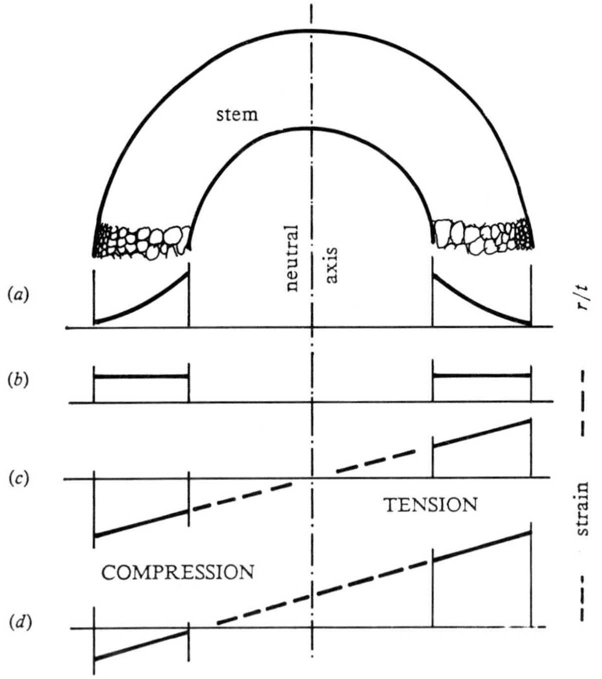


Figure 7. The interaction (*d*) between turgor pre-strain (*b*) and strain induced by bending due to a side load (*c*). The cells which will receive the greatest compressive load are the most resistant to buckling, being of small diameter and thick walled (low $r/t$ ratio (*a*)).

*Table 2. Calculated results for stems of the three plants investigated. See text for details. Position of the neutral axis is relative to* $z = 0$*, and is calculated as* $B/A$ *(mm). Stem strains computed for intact stem*

| | *Taraxacum* | *Psilotum* | *Rhynia* |
|---|---|---|---|
| Neutral axis | 0.244 | 0.017 | 0.02 |
| $K_x^P$ (mm$^{-1}$) | $-73.8\Delta P/E_{cw}$ | $-1.19\Delta P/E_{cw}$ | $-1.76\Delta P/E_{cw}$ |
| $\varepsilon_x^0$ | $26.7\Delta P/E_{cw}$ | $5.6\Delta P/E_{cw}$ | $10.5\Delta P/E_{cw}$ |
| $R_x$ (mm) | 3.42 | 210.0 | 142.0 |
| Stem strains (%) | 3.536 | 2.232 | 4.184 |

infinity. Strips of the stem of *Psilotum* show no detectable curvature changes in solutions of different osmolarity, so presumably the prediction is confirmed.

As may be seen in figure 2, the morphology of these two stems is very different from that of *Taraxacum*. Most obviously the gradient of cell size and cell wall thickness is much less pronounced and much more variable. *Rhynia* has cell walls which are slightly thicker in the outer layers of cells (3 μm) than the inner (1 μm), but suffers, our model predicts, by having a distinct zone (the inner cortex) made of one or more layers of large cells just beneath the outer layer (Edwards 1986). In this critical zone $V_f^{cw}$ is reduced, and this forms a major contribution to the apparently bad distribution of stress across the stem. The other bad feature of its mechanical design is the presence of much smaller cells towards the centre of the stem, leading to a higher $V_f^{cw}$ along the neutral axis in bending, where it will see very little of the bending loads. *Psilotum* also has thin-walled cells towards the outside of the stem, but internal to this there is an extensive zone of vertically elongated cells with thicker cell walls which are more or less lignified. Continuing towards the centre, the cells are somewhat larger and thinner walled. Figure 3 summarizes these observations as the ratio $r/t$, normalized to the same axes as *Taraxacum*. The calculated stem strains are presented in table 2, and appear to be little different between the three plants. The calculated distribution of stem stress is presented in figure 6, which shows that the maximum stem stresses ascribable to turgor in the two primitive plants are less than in *Taraxacum*. In *Psilotum* this probably does not matter quite so much, since the cells which contribute the most to the stem stress are also the lignified ones. In *Rhynia* the pre-stress in the centre of the stem is much higher than in *Taraxacum*: this seems rather a bad design since pre-stress in this position can contribute so little to bending stiffness. Rough calculation shows that the pre-stress in the dandelion stem is just about sufficient to resist side loads due to wind. With its lower pre-stress, *Rhynia* could not support anything at the tip of its axis which was likely to generate drag forces. Indeed, it is tempting to suggest that *Rhynia* was actually partly submerged, like many plants found today growing at the edge of lakes and ponds. In a short (unpublished) study we have found that the leaves of *Iris pseudacorus*, which grows on the banks of lakes and rivers, and of *Acorus calanus*, which grows partly submerged, have very different mechanical properties and failure rates. While *Iris* leaves fail comparatively rarely and have a safety factor of the order of 4, the leaves of *Acorus* are much more prone to failure (more than 30% of leaves were found to have buckled and folded irreversibly) and have a safety factor of about 0.7, suggesting that if they were to grow under the same conditions as *Iris* they would simply not survive. In fact they gain extra support partly because they are submerged for the lower third of their length, and partly because they grow very closely together, thus gaining mutual support.

Applying similar ideas to *Rhynia*, it is unlikely that they grew totally or continually partially submerged, since they have stomata along the entire length of the stem (Kidston & Lang 1917). However, they are always found densely packed in the chert with their stems laid parallel as if they were growing in clumps. So they might have needed to grow in dense stands simply to make up for the shortcomings in the design of their stem. Since there are so many specimens of *Rhynia* in such good preservation, it would be very interesting to look for evidence of modes of failure of the axis.

**General aspects of design**

At the beginning of this chapter we said that our model could not cope with the various types of thickened and lignified cells found in most herbaceous stems. We can still treat them in a qualitative manner. Such fibres are presumably pre-stressed by the same mechanism that pre-stresses other parts of the stem (i.e. turgor or growth), and so will contribute to the tension which must be paid off before the cell walls fall in compression. In this respect they could be considered much like the pre-stressed rods found in reinforced concrete beams and columns, except, of course, that the source of the pre-stress is within the column, not the rods. They will have their greatest effect when they are well away from the neutral axis, thus contributing most to the second moment of area. This generalization needs some qualification: in herbaceous stems the tendency is to have the tissue with the longest fibre, the sclerenchyma, arranged in close association with the conducting tissue, which is therefore not on the largest possible radius and not, therefore, presenting as large a second moment of area as it might. The conclusion must be that the stem must have more design requirements than the need to be as tall and stable as possible. It may be significant that the neutral axis of an excised strip of *Taraxacum* stem coincides with the conducting tissue, so that even if the stem splits and curls the conducting tissue will not only not be stretched or compressed but will not have to carry any extra load. Thus the flower (and seeds) can survive quite extensive damage to the stem. There may be other mechanical consequences of this observation, and the apparently anomalous position of the most fibrous (and therefore, presumably, most resistant to tension) cells in the stem may be a clue. We have been unable to solve the problem further. In addition, we have the impression that in the flowering stem of *Taraxacum* the conducting tissue appears to be mechanically isolated. Pressures within the xylem and phloem are going to be, if anything, negative, and thus will not contribute to the turgor system. However, the parenchyma cells around the conducting tissue are smaller, thus giving increased $V_f^{cw}$ in this area. One can then consider that the conducting tissue makes no contribution to the mechanical stability of the stem and depict it as a 'hole' in the mean distribution of cellulose across the stem, the reduced $r/t$ ratio around the conducting tissue compensating for this lack, and making the conducting tissue mechanically invisible. In this context the stelar morphology of *Selaginella* is particularly relevant, since the conducting tissue is almost completely mechanically isolated from the rest of the stem, being supported within tubes by relatively weak trabecular cells.

Rather less enigmatically, collenchymatous cells occur in ridges on the surface of many stems, thus contributing maximally to the second moment of area. The fact that they are shorter than sclerenchyma cells suggests that they may be more important in resistance to compression. This is supported by the experiments of Ambronn (1881), reported by Haberlandt (1914), who found that collenchyma fails in tension at about half the stress at which sclerenchyma fails, but deforms plastically at one tenth of the stress. As a corollary, any conducting tissue which occurs in the centre of the stem, the common arrangement in primitive plants, will be relatively free of mechanical influences, a state which is necessary for the effectiveness of such tissue in conducting fluids. This may well have been adopted as the simplest design solution in early land plants.

Just as it is possible to make numerical deductions about the mechanical properties of stems of fossil plants, given sufficiently good preservation, so it would be possible to derive more information about the mechanics (and therefore the habitat) of other organs of fossil plants if we had more information about the mechanics of present-day plants. Once again we find that Schwendener was there first and is considered to be correct simply because no-one has followed him; more particularly, no-one has been prepared to test Schwendener's morphological insights by experiment. It is therefore dangerous to extrapolate from our current understanding of plant mechanics – which is largely theoretical – to the mechanics of fossil plants. Schwendener had, it seems, a penchant for explaining all plant structures in terms of I-beams, even if the web (the central upright portion of the I) could not be found. But the whole problem of fibres in plant structures can be approached from the viewpoint of composite theory, an approach of which Schwendener seemed aware but for which he lacked the necessary theory. This approach has proved to be useful in understanding the mechanical design of grass leaves (Vincent 1982) and will doubtless be useful in other aspects of the design of plants.

Clearly there are many more aspects of the design of fossil plants which can be analysed from the mechanical point of view. The usefulness of this approach can be judged only if it yields new insights, and for that to be possible we need to know more about the mechanical design of present-day plants.

## References

Ambronn, H. 1881 Ueber die Entwicklungsgeschichte und die mechanischen Eigenschaften des Collenchyms. Ein Beitrag zur Kenntnis des mechanischen Gewebensystems. *Jahrbuch für Wissenschaftliche Botanik* **12**, 473–541.

Edwards, D.S. 1986 *Aglaophyton major*, a non-vascular land-plant from the Devonian Rhynie chert. *Botanical Journal of the Linnean Society* **93**, 173–204.

Haberlandt, G. 1914 *Physiological Plant Anatomy* (trans. from 4th German edn by M. Drummond). London: Macmillan.

Haughton, P.M. & Sellen, D.B. 1969 Dynamic mechanical properties of the cell walls of some green algae. *Journal of Experimental Botany* **20**, 516–535.

Hettiaratchi, D.R.P. & O'Callaghan, J.R. 1978 Structural mechanics of plant cells. *Journal of Theoretical Biology* **74**, 235–257.

Jeronimidis, G. 1980 Wood, one of nature's challenging composites. In *The Mechanical Properties of Biological Materials*, ed. J.F.V. Vincent & J.D. Currey. *Symposia of the Society of Experimental Biology* **34**, 169–182.

Kamiya, N., Tazawa, M. & Takata, T. 1963 Relation of turgor pressure to cell volume in *Nitella* with special reference to the mechanical properties of the cell wall. *Protoplasma* **57**, 501–521.

Kidston, R. & Lang, W.H. 1917 On Old Red Sandstone plants showing structure, from the Rhynie chert bed, Aberdeenshire. I. *Rhynia gwynne-vaughani* Kidston and Lang. *Transactions of the Royal Society of Edinburgh* **51**, 761–784.

Kidston, R. & Lang, W.H. 1920 On Old Red Sandstone plants showing structure, from the Rhynie chert bed, Aberdeenshire. II. Additional notes on *Rhynia gwynne-vaughani*, Kidston and Lang; with description of *Rhynia major* n. sp., and *Hornea lignieri* n. g., n. sp. *Transactions of the Royal Society of Edinburgh* **52**, 603–627.

Kidston, R. & Lang, W.H. 1921 On Old Red Sandstone plants showing structure, from the Rhynie chert bed, Aberdeenshire. IV. Restorations of the vascular cryptograms and discussion of their bearing on the general morphology of the Pteridophyta and the origin of the organisation of land-plants. *Transactions of the Royal Society of Edinburgh* **52**, 603–627.

McLean, R. & Ivimey-Cook, W.R. 1962 *Textbook of Theoretical Botany*. London: Longmans, Green.

Niklas, K.J. & Kerchner, V. 1984 Mechanical and photosynthetic constraints on the evolution of plant shape. *Palaeobiology* **10**, 79–101.

Preston, R.D. 1974 *The Physical Biology of Plant Cell Walls*. London: Chapman & Hall.

Rand, R.H. 1983 Fluid mechanics of green plants. *Annual Review of Fluid Mechanics* **15**, 29–45.

Schwendener, S. 1874 *Das Mechanische Prinzip im Anatomischen Bau der Monocotylen.* Leipzig: Engelmann.

Vincent, J.F.V. 1982 The mechanical design of grass. *Journal of Materials Science* **17**, 856–860.

Wainwright, S.A. 1970 Design in hydraulic organisms. *Naturwissenschaften* **57**, 321–326.

Wainwright, S.A., Biggs, W.D., Currey, J.D. & Gosline, J.M. 1976 *Mechanical Design in Organisms.* London: Edward Arnold.

DEREK E. G. BRIGGS, JOHN E. DALINGWATER & PAUL A. SELDEN

# Biomechanics of locomotion in fossil arthropods

## Introduction

Only a limited number of studies of fossil arthropods have employed a biomechanical approach. When direct comparison with living relatives is possible, as for example in many Crustacea, theoretical or experimental biomechanical studies are not usually necessary to interpret functional morphology. The approaches have therefore been most widely applied to extinct arthropod groups or to taxa whose morphologies differ significantly from that of their living relatives. In such cases biomechanical studies may allow the mode of life of the arthropod to be elucidated more precisely than is possible from comparisons with living relatives or analogues, from deductions based on morphology, or from the evidence of external factors (such as the associated sediments and organisms, or distribution as an indication of benthic or nektonic habit). However, strictly biomechanical and other approaches should be complementary, the one providing a method of testing the other. Some biomechanical studies are of value to our general understanding of arthropod evolution.

The arthropods owe their outstanding evolutionary success in large part to their firm, jointed exoskeleton. Arthropod cuticle is a complex tissue with a variety of functions, but its use in conjunction with internal muscles in a simple, elegant and highly modifiable lever system is the most obvious characteristic of the arthropod grade of organization. A biomechanical approach to the study of arthropods implies consideration of the structures and movements of the exoskeletal system in a quantitative or semi-quantitative way.

Detailed biomechanical studies of the cuticle and limbs of fossil arthropods are relatively few, although there is a comparatively extensive literature on Recent arthropod material (see, for example, Herreid & Fourtner 1981). The approaches involved cover a wide area, impinging, for example, on the physics of lever systems, fluid mechanics, and material sciences. In this chapter we present a review of biomechanical studies on fossil arthropods and the methodologies involved, with a view to identifying directions for future investigation in this nascent research area.

Because of limitations of space, and because the biomechanical factors influencing the different major arthropod groups are similar, our treatment is functionally rather than taxonomically based. We consider the biomechanics of fossil arthropods under two major headings: walking and swimming.

## Walking

Walking has been extensively treated in living arthropods (see, for example, Manton (1977 and references therein) on uniramians and Hessler (1981, 1982) on eumalacostracan crustaceans), based essentially on observation and on the morphology of muscles and articulations. Manton also used trackways in her studies of uniramian gaits. The articulations of arthropod limbs are well preserved in fossils only occasionally, and direct evidence of the configuration of muscles, such as the tendons preserved in the eurypterid *Baltoeurypterus* (figure 1; Selden 1981), is rarely available. Restoration of the musculature must usually rely on the configuration of podomeres and the articulations between them. In some cases trace fossils can be used to deduce likely gait patterns.

Manton (1977 and references therein) demonstrated the two simple movements employed by arthropods in walking. These are promotor–remotor movements at the coxa–body junction (or coxa–trochanter in Chelicerata) and levator–depressor movements which are effected by the articulations between the more distal podomeres. Most arthropod limbs conform to a plane (the 'limb-plane') in which the levator–depressor movements take place. This plane may be orientated in different attitudes to the body during the promotor–remotor movement at the coxa–body junction. Movement which is not confined to the limb plane usually occurs in appendages which are used for functions other than walking (e.g. grooming, feeding, burrowing). Walking as a result of movement in a limb plane is more effective than that achieved by bending the limb plane (whether the limb functions as an extensible strut, or as a lever (Hessler (1982, p. 291) gives a brief discussion)).

Manton (1977 and references therein) has elucidated the controls on the gaits of living myriapods. The stepping movements of long series of limbs have to be regulated in metachronal waves to avoid interference between successive appendages. The phase difference between appendages and the

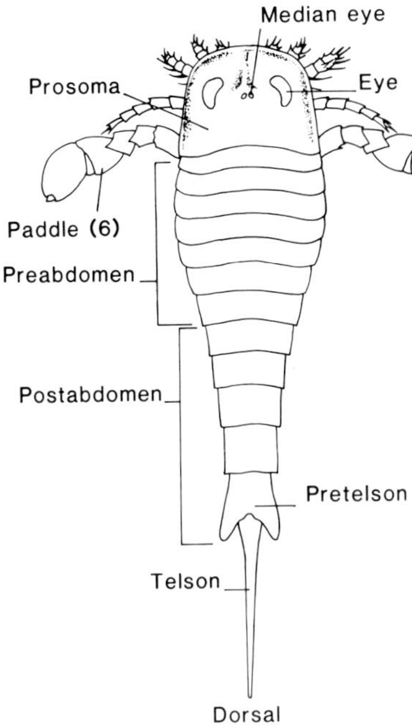


Figure 1. The Silurian eurypterid *Baltoeurypterus tetragonophthalmus*, seen in dorsal view.

consequent length of the metachronal wave can be varied to some extent in response to functional requirements. The ratio of recovery forward stroke to propulsive backstroke will also influence the force exerted by the arthropod on the substrate; a millipede will use a much lower geared gait (large proportion of limbs in contact with the ground at any time) when it is pushing through vegetation than when it is running on open ground.

The gaits employed by fossil arthropods have been studied in two ways: the first is based on functional analysis including aspects of biomechanics, the second on trace fossils.

### *Functional analysis*

Analysis of gaits relies on evidence for the nature of the limbs. Such an approach can therefore be applied only to arthropod groups with well sclerotized and therefore readily preserved appendages such as eurypterids and arthropleurid myriapods. Rarely, exceptional circumstances lead to the preservation of the lightly sclerotized limbs of other arthropods such as trilobites.

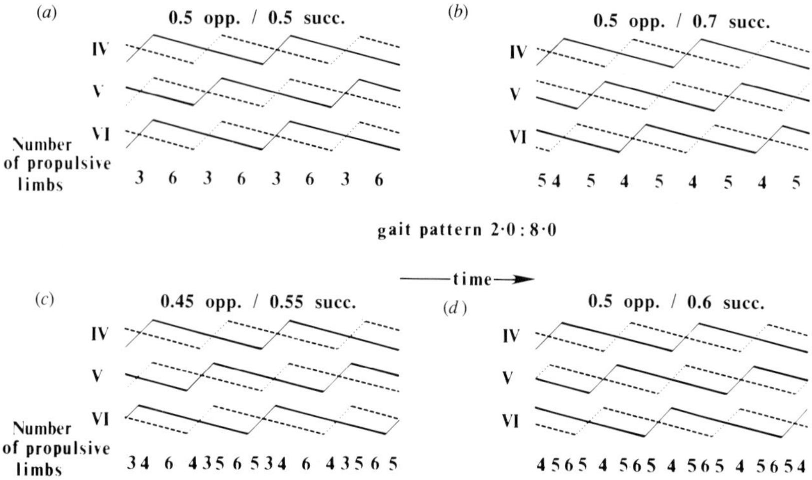


Figure 2. (*a–d*) Diagrams of four possible gaits in *Baltoeurypterus* in the style of Manton (e.g. fig. 7.3 of Manton 1977). Right limbs are denoted by solid lines: thin in the forward (recovery) stroke and thick in the backward (propulsive) stroke. Left limbs are denoted by broken lines: dotted in the forward stroke and dashed in the backstroke. Two cycles are shown in the diagram for limbs 4, 5 and 6. (After fig. 15 of Selden 1981, phase difference notation modified to conform to Manton's (1977) conventions.)

The most detailed analysis of eurypterid gaits based on functional analysis is that of Selden (1981) on *Baltoeurypterus* (figure 1) from the Silurian of Estonia and Gotland. The evidence both from morphology (length of limbs, position of centre of gravity) and from trace fossils indicates that eurypterids employed hexapodous gaits. Assuming hexapody, the range of possible gaits can be discussed on the basis of considerations of stability. Walking is more efficient with the limbs on opposite sides of the body out of phase. The gait patterns are likely to have been low geared: this would have ensured a relatively long overlap in the time during which successive limbs are in contact with the substrate, thus conferring greater stability. Speed could have been increased by reducing the pace duration as in arachnids, rather than by altering the gait pattern as in uniramians.

It is likely that eurypterids would usually have walked with a ratio of forward to backstroke of about 2.0:8.0 (that is 80% of the limbs would have been in contact with the substrate at any one time). Assuming this pattern, Selden generated a series of gait diagrams (figure 2) in the style of Manton (1977 and references therein) varying the phase differences between opposite and between successive limbs. The number of limbs in contact with the substrate at different points in the diagram gives an indication of the stability of the gait. Selden (1981) showed how a three-dimensional graph (figure 3) could be plotted to show the number of limbs in contact with the substrate over the possible range of phase differences in both opposite and successive limbs and of ratios of forward to backstroke duration. The diagram clearly shows that phase differences around 0.5 opposite and 0.5 successive produce the most stable gaits. High or low phase differences result in gaits which include intervals when no limbs are in contact with the substrate or when all three limbs in contact are on the same side of the eurypterid!

In analysing walking in *Baltoeurypterus*, Selden (1981) had the advantage of well preserved isolated cuticle, which permitted accurate restoration of the articulations between the podomeres. His analysis suggested minor flaws (Selden 1981, p. 29) in the gaits previously deduced from morphology for *Parastylonurus* (Waterston 1979) and for *Mixopterus* (Hanken & Størmer 1975) based on a fossil trackway. Both these studies, however, also indicated a hexapodous gait with the limbs on opposite sides of the body out of phase.

*Olenoides*, from the Middle Cambrian Burgess Shale of British Columbia, is perhaps the most completely known of fewer than ten trilobites so far discovered with well preserved appendages. Trilobites, unlike myriapods, were incapable of lateral flexure. Whittington (1975) argued that walking in *Olenoides* involved all the biramous appendages apart from the last two in the pygidium. The gait which he deduced was based largely on a comparison with the gaits of living myriapods (and the work of Manton (1977 and references therein)). The opposing limbs of each pair probably moved in phase. The nature of the limb articulations in *Olenoides* is unknown; Whittington argued that the coxa–body articulation may have been followed by two pivot joints and then hinge joints. This deduction is based largely on the requirement for a hanging stance to promote stability. Considerations of leg length and the likelihood of slow movements close to the substrate led Whittington to postulate a relative duration of forward and backward strokes of 3 to 5 (a gait pattern of 3.75:6.25). An angle of swing of 16° was selected (4° in front of a transverse plane, 12° behind) and a phase difference of 0.125. By using this postulated gait, Whittington was able to generate a theoretical trail, but a closely similar trackway has yet to be discovered in the fossil record.

### *Trace fossils*

Seilacher (1955) made an early contribution to the study of fossil arthropod trackways in his interpretations of trilobite trails from the Cambrian of the Salt Range in Pakistan. Osgood (1970) attempted to assign

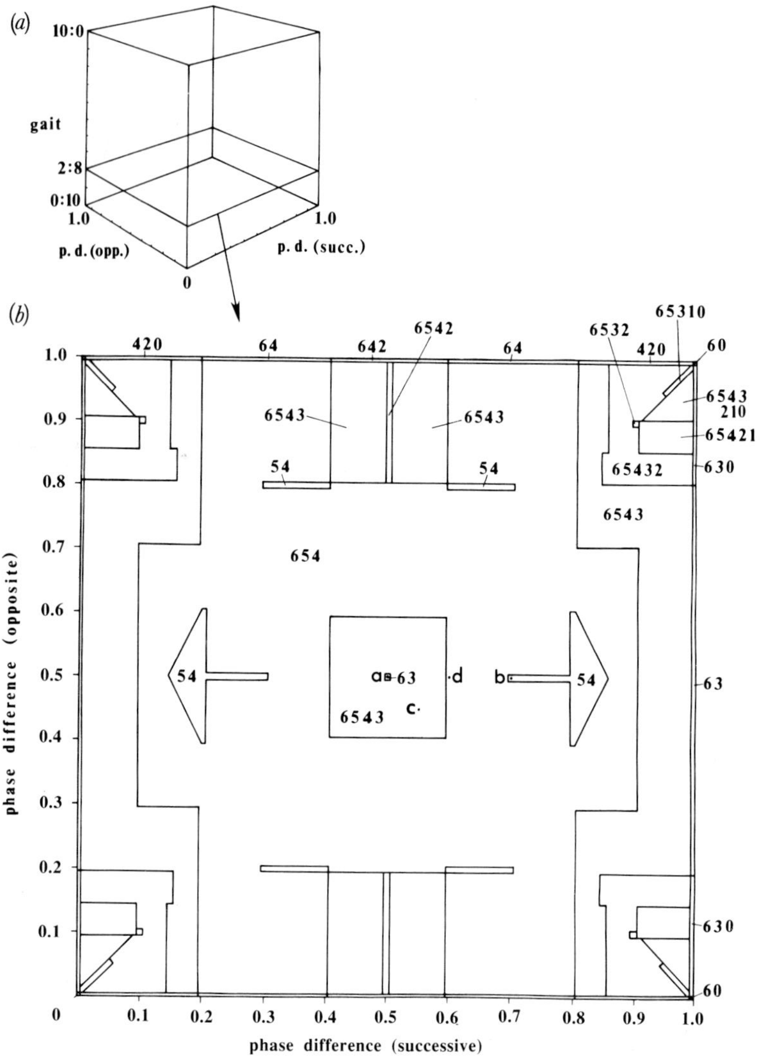


Figure 3. Gaits in *Baltoeurypterus*. (*a*) Three-dimensional graph of phase difference between opposite limbs ($x$ axis), phase difference between successive limbs ($y$ axis), and gait pattern ($z$ axis). The position of the two dimensional slice at gait 2.0:8.0 shown in (*b*) is indicated. (*b*) Matrix of the number of propulsive limbs during hexapodous walking in the gait pattern 2.0:8.0 related to the phase differences between opposite and between successive limbs. The numbers (654, 63, etc.) refer to the number of propulsive limbs at all points in the cycle of the gait; for example, in figure 2*a* the number of propulsive limbs covering all points in the cycle is referred to as 63. The positions a, b, c, and d refer to the gaits in figure 2. Minimum stability is afforded by three limbs; no gait with a number less than this is possible. Equally, gaits (such as 6543 on the left and right sides of the

trail makers to arthropod trace fossils from the Ordovician of Ohio based essentially on number of imprints in a set, size, and associated (and therefore 'available') body fossils. Briggs & Rushton (1980) identified the influence of bottom currents on the formation of a trilobite trail from the Upper Cambrian of north Wales. A major shortcoming of all such studies is the absence of independent evidence on the nature of the limbs. Indeed, owing to the different conditions required for their preservation, body fossils of organisms are rarely found in association with the traces which they produce. Thus the two most useful lines of evidence for the gaits of fossil arthropods, preserved appendages and trackways, can rarely be used in association.

On rare occasions it is possible to identify the arthropod responsible for a particular trail. This necessitates first a knowledge of the limbs, and second a characteristic morphology which results in a trail with diagnostic attributes. Such a combination is usually confined to arthropods with well sclerotized appendages. The trackway of the Silurian eurypterid *Mixopterus* (Hanken & Størmer 1975) has been briefly mentioned above. The most detailed analyses of walking based on trace fossils are those of the giant Carboniferous myriapod *Arthropleura* (figure 4).

*Arthropleura* is one of the largest arthropods known, reaching lengths of up to 1.8 m. Trackways are known from the Upper Carboniferous of Arran (Scotland), Montceau-les-Mines (France) and New Brunswick (Canada). The best preserved of these is the example from Arran (Briggs, Rolfe & Brannan 1979; Briggs 1986*b*). The Arran trackway allowed the identification of repeated imprints of the shorter limbs which plotted inside the rest and were not obscured by subsequent footfalls. This indicated a stride length of 22 cm. Identifying the stride length made the following possible:

*(1) Estimation of the number of limbs used in walking.* All of the limbs must plot within the stride. Briggs *et al.* (1979) identified 23 imprints within the stride in the Arran trace. Where appendages are evident near the posterior of some of the small arthropleurids from Montceau-les-Mines, two pairs appear to correspond to each tergite (Almond 1986) as opposed to the one previously reconstructed

---

matrix) in which three limbs of one side only are in contact with the ground at one point in the cycle are ruled out. Note that the matrix is symmetrical, but that the shapes of the enclosed areas would be different at different gait patterns ($z$ values). The matrix for 8.0:2.0 gait pattern would look the same, but have different numbers in the enclosed areas. (After fig. 14 of Selden 1981, phase difference notation modified to conform to Manton's (1977) conventions.)

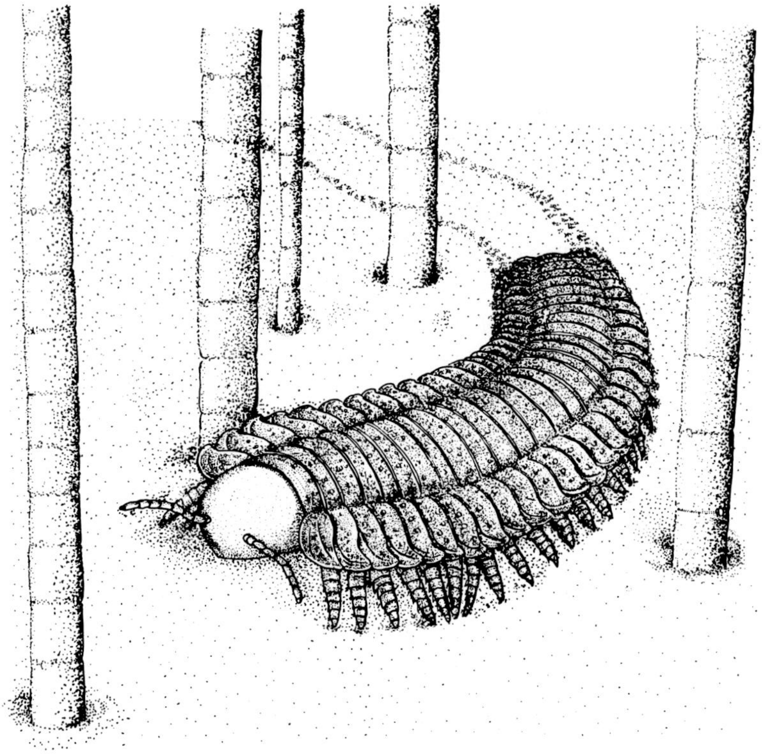

Figure 4. The giant late Carboniferous myriapod *Arthropleura armata* making the New Brunswick trackway as it walked through a stand of calamites. The detailed morphology of the head is unknown (and reconstructed differently by Hahn *et al.* (1986)). This reconstruction shows one pair of walking limbs per tergite (see discussion in text). (After fig.5 of Briggs *et al.* 1984.)

(Rolfe 1969). In addition, W. D. I. Rolfe (personal communication) has drawn our attention to evidence of an excess of paired limbs over tergites in the largest known body fossil of *Arthropleura* illustrated by Hahn, Hahn & Brauckmann (1986). Their figure (Pl. 2 of Hahn *et al.* 1986) of a portion of the ventral surface shows eight limbs corresponding in position to not more than five, possibly only four, of the dorsal tergites (see their Pl. 1). None the less, Hahn *et al.* (see their fig. 2) restored *Arthropleura* with only one pair of limbs per tergite. If the number of paired walking limbs in *Arthropleura* did exceed that of tergites, the interpretation of 23 pairs of walking limbs in the Arran trace will require revision (diplotergites would imply an individual with perhaps as few as 12 tergites, but a width between the limb extremities of about 36 cm!). It is possible that Briggs *et al.* (1979) underestimated the number of limbs plotting within the stride, and that some of the depressions represent multiple footfalls (Briggs 1986*b*).

(2) *Demonstration that opposing limbs moved in phase.*

(3) *Analysis of the gait.* Assuming a maximum angle of swing and length of limb (based on body fossils) it was possible to calculate that the backstroke is unlikely to have exceeded 10 cm. Given a stride length of 22 cm this indicates a gait of 12:10 (that is 5.5:4.5). Such a gait is much faster (higher geared) than is suggested by the usual habit of the arthropod, pushing through coal-swamp vegetation. It is likely, however, that *Arthropleura* employed a different range of gaits when walking on more open ground.

(4) *Plotting of a theoretical trail.* This trail compared closely with the trace-fossil and confirmed the major conclusions of the analysis.

The New Brunswick trail (Briggs, Plint & Pickerill 1984) confirmed that the trunk of *Arthropleura* was laterally flexible and showed that the anterior appendages were used to probe laterally as the arthropod changed direction (figure 4). Thus, where the trail maker can be identified, trackways can yield important insights into the functional morphology and biomechanics of the arthropod.

### *Walking on land: limitations on body size*

Dalingwater (1985) attempted to assess whether large Carboniferous eurypterids could have walked on land. He used extant *Limulus* as a model, as it is primarily an aquatic animal but capable of excursions onto land. Although many of the mechanical properties of *Limulus* cuticle are known, calculation of the maximum stresses imposed on limb podomeres while the animal is walking would require a knowledge of limb postures and the effects of muscular action. Experiments with some type of force platform would probably produce more accurate results, but that approach is clearly not applicable to fossil arthropods! Calculation of the stresses on the limbs of extinct eurypterids in an indirect biomechanical approach requires many assumptions and approximations (about the mechanical properties of the cuticle, its original thickness, limb postures, gaits etc.) but an alternative approach, still with *Limulus* as a model, involves a consideration of scaling exponents. This provides more useful, and considerably more general, conclusions.

Measurements were made from a size series (32–147 mm carapace width) of *Limulus* to document the changes in relative dimensions (length, diameter and cuticle thickness) of a particular limb podomere with increasing mass. The exponent for increase in cuticle thickness against carapace width was 1.15. This lies within the range reported by Prange (1977) for wolf-spiders (1.0) and cockroach (1.38), but is well below that expected from theoretical considerations if the increase in cuticle mass is achieved solely by an increase in cuticle thickness (Dalingwater 1985; Anderson, Rahn & Prange 1979). Thus one of the factors traditionally considered to limit the size of land arthropods – that in order to maintain mechanical similarity with increasing size, cuticles would become too thick to leave sufficient space for muscles within the limbs – may not be such a crucial factor after all. Clearly larger arthropods offset potentially unacceptable stresses on limbs in other ways, for example by changes in shape or in skeletal materials, or by a reduction in locomotory performance.

The soft cuticle of a newly moulted arthropod of even moderate size would be unlikely to support its body mass on land, let alone resist the stresses imposed in crawling out of the exuvium. (Dalingwater (1985) calculated that the newly formed cuticle of tibia 1, walking leg 3, of a *Limulus* with 75 mm carapace width could support the static axial load resulting from the weight of the animal, but that with increased size even the static axial load would become near-critical. He concluded that it is unlikely that even a relatively small *Limulus* could moult on land.) Thus truly terrestrial arthropods could not even approach the cow-sized mass conjured up by Neville (1975) to illustrate the constraints on moulting in larger arthropods. *Arthropleura* (figure 4) had to return to water in order to moult. Célérier (1986) showed that exuvia production represents 5.2–16.2% of growth production for the whole period of post-embryonic development in six spider species. It reaches 50% of growth production in large adult females of the theraphosid spider *Scodra griseipes*, which, unusually for spiders, moults throughout life (few spider species moult after reaching adulthood). Thus the energy costs of exuviation may have limited the size of arthropods. To this must be added Dalingwater's (1985) suggested size constraints on aquatic arthropods, some of which are also related to moulting. These include: (1) the time taken to produce the cuticle, since the surface area of the epidermal cells produces a volume of material which becomes disproportionately greater with increasing size; (2) the cells must also produce increasing volumes of moulting fluid; and (3) the effects of impact and of wear and tear on the cuticle become more critical with increasing size and concomitantly longer periods between moults. It must be conceded, however, that none of these factors, all of which are related to the properties of the exoskeleton, may be critical in limiting size. Others, such as respiration, might be more important. Size may well be limited by a cumulative combination of the disadvantages accruing from various factors.

It seems unlikely on theoretical grounds that a spider could have achieved a body length of about 35 cm (Hünicken 1980), or that eurypterids with body lengths around 2 m could have thrived (Plotnick & Baumiller 1988), albeit in water where there were fewer limitations on large size than on land. That such sizes were achieved, in spite of the number of possible constraints, indicates that there were considerable ecological and evolutionary advantages in being large during certain periods of geological time. The advantages may have differed at different times: Rolfe (1980) has suggested that large size in some Devonian eurypterid stocks may have served to reduce desiccation in periods of drought, or to facilitate regular forays onto land. Large size was presumably an adaptive response to optimal environmental conditions which were stable for significant intervals of time, as giant arthropods would have been particularly susceptible to resource fluctuations (Briggs 1985).

Biomechanical considerations thus indicate that larger eurypterids are likely to have been at best amphibious rather than fully terrestrial. In addition Selden (1985) has argued that the true gills of eurypterids are not known, and that the so-called gill tract preserved in some eurypterids may be an accessory respiratory organ for air-breathing. The area of the gill tract relative to body mass is an order of magnitude lower than that of the gills in living Crustacea. If the eurypterid gill tract is an air-breathing organ it is closer in morphology to the pseudo-tracheae which occur in the limited number of terrestrial isopods and crabs rather than the book-lungs of scorpions, the development of which led to a much more extensive terrestrial radiation.

There is also limited (and equivocal) evidence from palaeoecology and trace fossils that eurypterids were not fully terrestrial. A population of eurypterid trackways from the late Silurian of the Murchison district of Western Australia which were apparently made sub-aerially indicate that the animals used a 'lurching' hexapodous gait with the opposing appendages in phase. Although there is no median groove produced by dragging the abdomen, readjustment of the gait at intervals indicates that the arthropods were not well adapted for walking on land (personal communication by K. J. McNamara, who is studying these trace fossils). A large, probably terrestrial, eurypterid trackway 79 cm wide from the Mississippian of Pennsylvania (Briggs & Rolfe 1983) suggests that this giant eurypterid was not primarily adapted for locomotion on land. The trace fossil preserves a median groove, and the right and left tracks are in phase. This indicates that the arthropod was unable to keep the trunk clear of the substrate, and adopted a powerful low-geared gait keeping most of the appendages in contact with the ground at any time. This allowed it to walk slowly out of water, dragging the abdomen along. Thus large eurypterids with thick cuticles may have invaded coal-swamp forests during the early Carboniferous, when such a relatively new habitat would have provided little opposition or competition. Small eurypterids occur inside hollow Carboniferous (Westphalian) tree-stumps at Joggins, Nova Scotia (Rolfe 1980), which they may have inhabited to avoid desiccation. Trackways of some of the later Carboniferous forms would be more significant in the search for evidence of terrestrial eurypterids.

## Swimming

The assumption that an extinct arthropod could swim may be tested by biomechanical analysis but this invariably involves comparison with swimming models derived from studies on living animals. If virtually the whole animal is preserved, as in many eurypterids, then the entire natatory system can be investigated; if, as in most trilobites, the appendages are not preserved then only certain aspects of the model can be tested. Three approaches are reviewed here: (1) measurements of body drag on arthropods without preserved appendages; (2) studies of fossil arthropods with tail flukes; and (3) investigations of the swimming system as a whole in well preserved arthropods, illustrated by the eurypterids.

### *Body drag*

Fortey (1985) used pelagic trilobites to illustrate the methods used in deducing the modes of life of extinct arthropods. He recognized three important criteria: (1) analogy with living forms, (2) mechanical design, and (3) geological evidence. Only the second of these involves biomechanics, but the other two can sometimes provide an independent test of an hypothesis. Fortey (1985, p. 223) identified two categories of pelagic trilobite: poorly streamlined, generally spinose forms, and well streamlined forms. Body drag was measured by using resin casts suspended in a water current and the flow around the body was studied with potassium permanganate. Drag was calculated on the basis of the deflexion of a thread supporting the trilobite in the current (figure 5). The absolute values of drag are largely irrelevant to this study, but the relative values supported the hypothesis of active

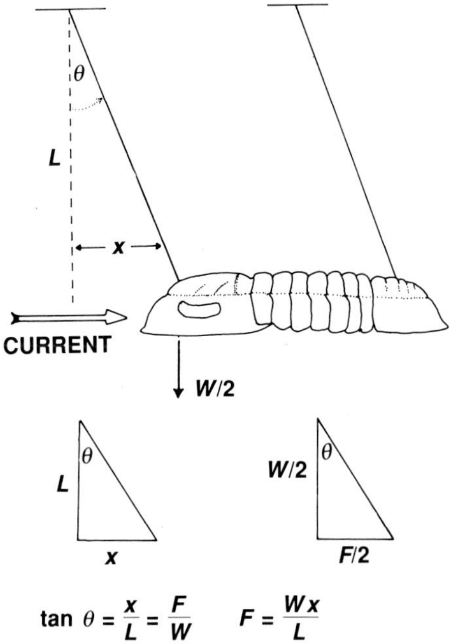


Figure 5. Experimental measurement of total drag $F$ on a trilobite model suspended in a current. $W$ is the immersed weight of the model, $x$ the deflexion of the thread measured through a travelling microscope at given length $L$. (After fig. 8 of Fortey 1985.)

swimming in the well streamlined group. Additionally, one of the 'control' genera, *Ogygiocarella*, bucked and planed even in a gentle current, but proved to be very resistant to lifting even by strong currents when settled on the bottom of the flume, thus supporting a benthic mode of life.

Fisher (1975) investigated swimming in Recent *Limulus* and the related Jurassic *Mesolimulus*, a more flattened form. These arthropods swim inverted, presenting the carapace to the water in the form of the hull of a boat. Fisher, like Fortey (1985) after him, made models of the animals and studied the flow patterns (by using oxygen bubbles) in a tank, at various angles of attack. By observing at which angle of attack and flow speed the wake vortex offered least resistance to propulsion, Fisher deduced the optimum values for normal swimming. Observations of live *Limulus* swimming at the same speed (10–15 cm/s) and angle of attack (20–30°) provided an independent test of the experimental results. The flatter carapace of *Mesolimulus* would have enabled it to swim at 15–20 cm/s with an angle of attack of 0–10°. The stroking rate of the propulsive appendages ($f_s$) should equal the vortex shedding rate ($f_v$), and calculations and observations on *Limulus* showed both rates to be about 2 Hz. Equally the measured $f_v$ of *Mesolimulus* of 1.7 Hz should approximate the $f_s$ of its appendages in life. Thus it is possible to investigate the action of the appendages of an extinct arthropod on the basis of fossils which lack them!

Fisher (1977) also used resin models to investigate the function of the prosomal spines in the Carboniferous limuline *Euproops danae*. He prepared models of enrolled *Euproops* at two sizes and with various lengths of spine. During passive settling through a water column, those closest in morphology to the real animal at both juvenile and adult sizes equilibrated most rapidly and descended with the least amount of lateral oscillation. Vertebrate visual and lateral line systems are tuned to detect the lateral movements of potential prey rather than passively falling objects. Fisher therefore concluded that the *Euproops* prosomal spines were adapted to produce stable passive settling in the enrolled animal as a device to avoid detection by vertebrate predators.

### *Tail flukes*

Undulations of the body (axial swimming) are rare in arthropods, which more commonly use paired appendages (paraxial locomotion) in swimming. The reverse is true in vertebrates (Herreid 1981; Riess & Frey, this volume). Decapod crustaceans such as lobsters and crayfish use a tail fan (telson and uropods) in rapid backward escape-responses, but their abdominal morphology does not allow undulatory swimming. Some of the eurypterids, the pterygotids, had broad, dorso-ventrally flattened telsons which might have been used in undulatory swimming, in a tail-flip escape-response, or to aid stability. These eurypterids had flipper-like sixth prosomal appendages which clearly functioned in swimming. Some pterygotid telsons had a dorsal, vertical, median keel in addition to the lateral flukes (figure 6); in these species the telson probably functioned in steering and stabilizing (Størmer 1936; Briggs 1981) and is unlikely to have been used in axial propulsion (Plotnick & Baumiller 1988). A triradiate tail also occurs in the non-swimming eurypterid *Parastylonurus ornatus* and was used to provide stability and lift for the opisthosoma during walking (Waterston 1979). A unique triradiate tail occurs in the large Middle Cambrian nektonic arthropod *Odaraia* (Briggs 1981). The vertical fluke and very high aspect ratio indicate that the tail was used in stabilizing (particularly critical with an essentially tubular carapace) and steering. Further investigations of fluked tails in eurypterids should consider three paradigms: (1) a steering/stability organ, (2) an undulatory swimming device, and (3) a tail-flip escape mechanism. Analyses to date suggest that the first was the normal function.

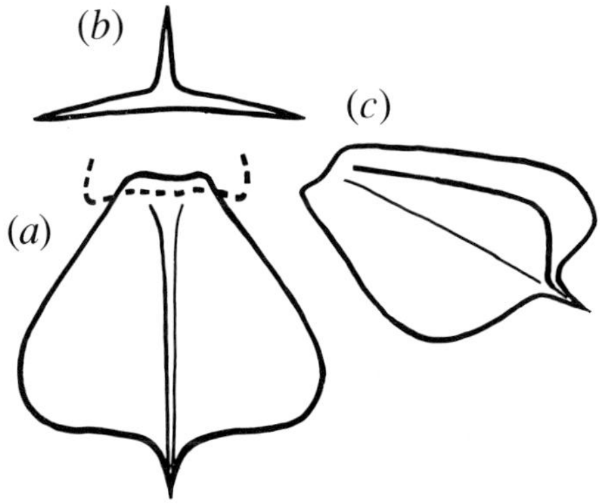


Figure 6. Telson of the pterygotid eurypterid *Jaekelopterus rhenaniae*. (*a*) Dorsal view, with dashed outline of pretelson. (*b*) Cross-section. (*c*) Oblique view. (After fig. 6 of Størmer 1936.)

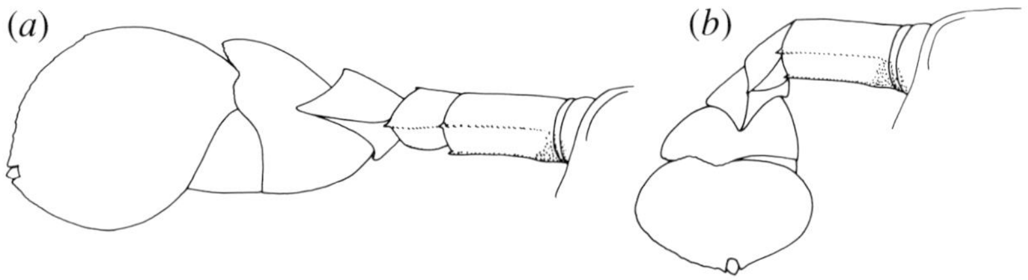


Figure 7. Left swimming appendage (6) of *Baltoeurypterus tetragonophthalmus* viewed from above and behind. (*a*) Outstretched halfway through the propulsive stroke. (*b*) Collapsed halfway through the recovery stroke. (After fig. 7 of Selden 1981.)

*Studies of swimming considering the arthropod as a whole: eurypterids*

The eurypterids included the largest known arthropods, ranging up to 2 m in length; thus at least the larger ones must have swum at Reynolds numbers (*Re*) which are more suited to a lift-based rather than a drag-based swimming system (fig. 2 of Nachtigall 1978). Eurypterids are often sufficiently well preserved to allow their swimming appendages (modified sixth prosomal limbs) to be reconstructed in detail, and none more so than the Silurian eurypterid *Baltoeurypterus* (figures 1, 7). This eurypterid reached lengths of 45 cm and thus had a lower *Re*, in the range in which either lift– or drag-based swimming was appropriate (Selden 1984).

A comparison of the morphology of the swimming appendage in *Baltoeurypterus* (figure 7) with drag-based (paddle) and lift-based (flipper) mechanisms in living animals indicates that the *Baltoeurypterus* limb functioned as a paddle (but was also used in walking; Selden 1981). The evidence for this includes: (1) the proximal joint morphology, which allowed anterior–posterior but not dorso-ventral movement; and (2) the long proximal podomeres and broad 'oarblade' termination, which had neither a flipper-like shape nor a hydrofoil section. The 'blade' is not broadest distally as in an ideal oar, owing to constructional constraints and the need to use it in walking. Complex joints enabled the blade to be folded in the recovery stroke to reduce drag; the drag on the cylindrical femur cannot be reduced in this stroke in any model. Tilting of the paddle would have allowed steering up, down and sideways.

The morphology of the *Baltoeurypterus* paddle and that of portunid crabs is remarkably convergent. Plotnick (1985) studied swimming in these crabs by using high-speed filming and measurements of lift and drag in a flow tank. He also made a model of a *Baltoeurypterus* – like eurypterid to determine body drag and mass. Plotnick's model of eurypterid swimming differed from that of Selden (1981) only in the introduction of a small amount of lift in the recovery stroke to counteract the presumed negative buoyancy (see Briggs (1986*a*) for discussion). This combination of drag in the propulsive phase and lift in the recovery phase is also a feature of Godfrey's (1984) model of plesiosaur locomotion, which is based on a comparison with otariid pinnipeds (see also Riess & Frey, this volume). Clearly drag- and lift-based modes of swimming need not be mutually exclusive. The largely drag-based mechanism deduced for *Baltoeurypterus*, which swam with a relatively high *Re*, remains anomalous. Investigations of swimming in the much larger pterygotid eurypterids will prove interesting in this context. The functional and constructional requirements for using the paddle in walking may have constrained any potential for evolving lift-based underwater flying.

Selden (1984) investigated the possible swimming speed of *Baltoeurypterus*. His estimates of *Re* ($1.65 \times 10^4$) and body drag ($3.0 \times 10^{-3}$N) compare well with those obtained experimentally by Plotnick (1985) with a wax model of a eurypterid at *Re* between $1.0 \times 10^4$ and $4.1 \times 10^4$, which gave $2.5 \times 10^{-3}$N. Selden's (1984) analysis was based on a simple rowing model (Alexander 1968), and compared the possible mean power-output of the coxal muscles with the mean power requirements of the oarblade system to obtain an estimate of maximum sustainable speed. A mean body velocity of about 38 cm/s and a mean paddle velocity of 109 cm/s were obtained. The efficiency was 0.29. For an animal about 16.5 cm long, a maximum speed of 2.5 times its body length per second does not seem to be unreasonable, and also conforms to that predicted by Hughes (in Selden 1984) based on an extrapolation of Nachtigall's (1977) studies of speeds in water beetles.

While such an estimate of swimming speed is intrinsically interesting, it also provides some corroboration of Selden's (1981) original model of drag-based rowing in *Baltoeurypterus*. However, a number of simplifications have been introduced into the calculation. A steady-state rowing mechanism (Alexander 1968) is assumed, although the animal proceeds in a sequence of faster propulsive phases and slower recovery phases under the effect of momentum. Total musculature was used in the calculation, although that for propulsion is more massive (as evidenced by the strong antero-inferior coxa–trochanter articulation; Selden 1981) than that for recovery. In addition, neither the power output of the muscles nor how it changed through the stroke cycle is known. Some power must have been used to provide lift, as the animal was certainly slightly denser than water. However, specific gravity and the lift requirement were probably minimized in an actively swimming eurypterid such as *Baltoeurypterus*, as in actively swimming Crustacea (Spaargaren 1980).

## Conclusions

Even the small number of investigations which have so far used a biomechanical approach in research on fossil arthropods illustrate the important insights which such studies can yield. The paucity of applications to date may stem partly from a lack of appreciation of the potential of biomechanics, but may also reflect a suspicion of results far more precise than is apparently warranted by the nature of the data. However, as this review shows, the absolute values obtained in a biomechanical calculation are usually irrelevant to the broader questions addressed. Was *Arthropleura* relatively agile when moving across open ground compared to its presumed low-geared gait for pushing through coal-swamp vegetation? Could very large arthropods walk on land? Were certain widely distributed trilobites streamlined? How did eurypterids swim? Although we ultimately retreat from the exactitude of results obtained and draw conclusions on safer ground, the biomechanical approach is in no sense devalued. It introduces a precision into comparisons that is unobtainable with more qualitative methods of interpreting functional morphology. However, all biomechanical analyses, whether on living or fossil organisms, involve assumptions and simplifications. We have therefore emphasized the necessity of using a range of independent approaches to test conclusions (the shotgun approach of Brower & Veinus 1974). Thus not only should mathematical models or inferences drawn from analogy be tested by experiments, where appropriate, but geological evidence (trace fossils, distribution patterns etc.) should also be considered. Biomechanics has considerable potential for refining our interpretations of the palaeobiology of arthropods.

### Acknowledgements

We are grateful to the Royal Society of Edinburgh and the Palaeontological Association for permission to reproduce figures.

### References

Alexander, R. McN. 1968 *Animal Mechanics*. London: Sidgwick and Jackson.

Almond, J.E. 1986 Les arthropleurides du Stephanien de Montceau-les-Mines, France. *Bulletin Trimestriel de la Société d'Histoire Naturelle et des Amis du Muséum d'Autun* (*France*) **115**, 59–60.

Anderson, J.F., Rahn, H. & Prange, H.D. 1979 Scaling of supportive tissue mass. *Quarterly Review of Biology* **54**, 139–148.

Briggs, D.E.G. 1981 The arthropod *Odaraia alata* Walcott, Middle Cambrian, Burgess Shale, British Columbia. *Philosophical Transactions of the Royal Society of London* B **291**, 541–584.

Briggs, D.E.G. 1985 Gigantism in Palaeozoic arthropods. *Special Papers in Palaeontology* **33**, 157.

Briggs, D.E.G. 1986*a* How did eurypterids swim? *Nature*, *London* **320**, 400.

Briggs, D.E.G. 1986*b* Walking trails of the giant arthropod *Arthropleura* (Traces d'*Arthropleura* géants). *Bulletin Trimestriel de la Société d'Histoire Naturelle et des Amis du Muséum d'Autun* (*France*) **117**, 141–148.

Briggs, D.E.G., Plint, A.G. & Pickerill, R.K. 1984 *Arthropleura* trails from the Westphalian of eastern Canada. *Palaeontology* **27**, 843–855.

Briggs, D.E.G. & Rolfe, W.D.I. 1983 A giant arthropod trackway from the Lower Mississippian of Pennsylvania. *Journal of Paleontology* **57**, 377–390.

Briggs, D.E.G., Rolfe, W.D.I. & Brannan, J. 1979 A giant myriapod trail from the Namurian of Arran, Scotland. *Palaeontology* **22**, 273–291.

Briggs, D.E.G. & Rushton, A.W.A. 1980 An arthropod trace fossil from the Upper Cambrian Festiniog Beds of North Wales and its bearing on trilobite locomotion. *Geologica et Palaeontologica* **14**, 1–8.

Brower, J.C. & Veinus, J. 1974 The statistical zap versus shotgun approach. *Mathematical Geology* **6**, 311–332.

Célérier, M.-L. 1986 Energy cost of exuviae production of spiders. *Bulletin of the British Arachnological Society* **7**, 77–82.

Dalingwater, J.E. 1985 Biomechanical approaches to eurypterid cuticles and chelicerate exoskeletons. *Transactions of the Royal Society of Edinburgh Earth Sciences* **76**, 359–364.

Fisher, D.C. 1975 Swimming and burrowing in *Limulus* and *Mesolimulus*. *Fossils and Strata* **4**, 281–290.

Fisher, D.C. 1977 Functional significance of spines in the Pennsylvanian horseshoe crab *Euproops danae*. *Paleobiology* **3**, 175–195.

Fortey, R.A. 1985 Pelagic trilobites as an example of deducing the life habits of extinct arthropods. *Transactions of the Royal Society of Edinburgh Earth Sciences* **76**, 219–230.

Godfrey, S.J. 1984 Plesiosaur subaqueous locomotion, a reappraisal. *Neues Jahrbuch für Geologie und Paläontologie Monatshefte* **11**, 661–672.

Hahn, G., Hahn, R. & Brauckmann, C. 1986 Zur Kenntnis von *Arthropleura* (Myriapoda: Ober-Karbon). *Geologica et Palaeontologica* **20**, 125–137.

Hanken, N.M. & Størmer, L. 1975 The trail of a large Silurian eurypterid. *Fossils and Strata* **4**, 255–270.

Herreid, C.F. 1981 Locomotion in arthropods: an evolutionary phantasmagoria. In *Locomotion and Energetics in Arthropods*, ed. C.F. Herreid & C.R. Fourtner, pp. 1–7. New York and London: Plenum Press.

Herreid, C.F. & Fourtner, C.R. (eds.) 1981 *Locomotion and Energetics in Arthropods*. New York and London: Plenum Press.

Hessler, R.R. 1981 Evolution of arthropod locomotion: a crustacean model. In *Locomotion and Energetics in Arthropods*, ed. C.F. Herreid & C.R. Fourtner, pp. 9–30. New York and London: Plenum Press.

Hessler, R.R. 1982 The structural morphology of walking mechanisms in eumalacostracan crustaceans. *Philosophical Transactions of the Royal Society of London* B **296**, 245–298.

Hünicken, M.A. 1980 A giant fossil spider (*Megarachne servinei*) from Bajo De Veliz, Upper Carboniferous, Argentina. *Boletin do la Academia Nacional de Ciencias, Cordoba, Argentina* **53**, 317–340.

Manton, S.M. 1977 *The Arthropoda: Habits, Functional Morphology and Evolution*. Oxford University Press.

Nachtigall, W. 1977 Swimming mechanics and energetics of locomotion of variously sized water beetles – Dytiscidae, body length 2 to 35 mm. In *Scale Effects in Animal Locomotion*, ed. T.J. Pedley, pp. 269–283. London: Academic Press.

Nachtigall, W. 1978 Evolutionsaspekte und physikalische Grenzwerte. *Zoologische Jahrbücher Abteilung für Anatomie* **99**, 109–114.

Neville, A.C. 1975 *Biology of Arthropod Cuticle*. Berlin: Springer Verlag.

Osgood, R.G. 1970 Trace fossils of the Cincinnati area. *Palaeontographica Americana* **6** (41), 279–444.

Plotnick, R.E. 1985 Lift based mechanisms for swimming in eurypterids and portunid crabs. *Transactions of the Royal Society of Edinburgh Earth Sciences* **76**, 325–338.

Plotnick, R.E. & Baumiller, T. 1988 The pterygotid telson as a biological rudder. *Lethaia* **21**, 13–27.

Prange, H.D. 1977 The scaling and mechanics of arthropod exoskeletons. In *Scale Effects in Animal Locomotion*, ed. T.J. Pedley, pp. 169–181. London: Academic Press.

Rolfe, W.D.I. 1969 Arthropleurida. In *Treatise on Invertebrate Paleontology*, part R, ed. R.C. Moore, pp. 607–620. Boulder, Colorado & Lawrence, Kansas: Geological Society of America and University of Kansas.

Rolfe, W.D.I. 1980 Early invertebrate terrestrial faunas. In *The Terrestrial*

*Environment and the Origin of Land Vertebrates*, ed. A.L. Panchen, pp. 117–157. *Systematics Association Special Volume* **15**. London: Academic Press.

Seilacher, A. 1955 Spuren und Lebensweise der Trilobiten. In *Beiträge zur Kenntnis des Kambriums in der Salt Range (Pakistan)*, by O. Schindewolf & A. Seilacher. *Abhandlungen der Mathematisch-naturwissenschaftlichen Klasse, Akademie der Wissenschaften. Mainz* (1955) pp. 342–372.

Selden, P.A. 1981 Functional morphology of the prosoma of *Baltoeurypterus tetragonophthalmus* (Fischer) (Chelicerata: Eurypterida). *Transactions of the Royal Society of Edinburgh Earth Sciences* **72**, 9–48.

Selden, P.A. 1984 Autecology of Silurian eurypterids. *Special Papers in Palaeontology* **32**, 39–54.

Selden, P.A. 1985 Eurypterid respiration. *Philosophical Transactions of the Royal Society of London* B **309**, 219–226.

Spaargaren, D.H. 1980 Hydrodynamic properties of benthic marine Crustacea. II. Energy requirements for maintaining vertical position. *Marine Ecology Progress Series* **2**, 153–156.

Størmer, L. 1936 Eurypteriden aus dem Rheinischen Unterdevon. *Abhandlungen der Preussischen Geologischen Landesanstalt* **175**, 1–74.

Waterston, C.D. 1979 Problems of functional morphology and classification in stylonuroid eurypterids (Chelicerata, Merostomata), with observations on the Scottish Silurian Stylonuroidea. *Transactions of the Royal Society of Edinburgh* **70**, 251–322.

Whittington, H.B. 1975 Trilobites with appendages from the Middle Cambrian, Burgess Shale, British Columbia. *Fossils and Strata* **4**, 97–136.

JOHN A. CHAMBERLAIN JR

# Cephalopod locomotor design and evolution: the constraints of jet propulsion

The locomotor systems of undulatory fish and jet-propelled cephalopods are radically different in anatomy and performance characteristics. Shelled cephalopods typically have large buoyancy devices, small mantle cavities and little propulsive muscle. Equilibrium is based on hydrostatic stability. Drag is high, and maximum swimming velocities are generally low. Lacking a heavy external shell, soft-bodied cephalopods require only small buoyancy devices, or even none at all; the mantle cavities are large, and the propulsive muscles massive. Equilibrium depends on body and fin motions. Fish also have light skeletons and little need for large buoyancy systems. Since they accelerate ambient water, they require no internal water reservoirs, they can devote much internal space to propulsive muscle, and are generally more powerful propulsors than cephalopods. Locomotor design appears to be a key factor directing the course of cephalopod evolutionary history. Successive replacement of dominant shell morphologies among Palaeozoic nautiloids and ammonoids can be attributed to improvement in locomotor adaptive complexes. Interactions with fish appear to be important evolutionary factors for cephalopods during the Devonian–Carboniferous diversification of coiled shell morphology and during the Cretaceous when coiled forms largely disappear.

## Introduction

Two main locomotor mechanisms occur among vagile, aquatic organisms: jet propulsion and body/fin undulation. Jet propulsion appears to be the more ancient of these two systems. Jet-propelled cnidarian medusoids have been recovered from Ediacara rocks about 650 million years in age. The earliest undulators appear to be presumptive primitive chordates preserved in the 540 million year old Burgess Shale. Cephalopods and fish, groups which eventually carry jet and undulatory design to their most sophisticated levels, both appear nearly simultaneously in the late Cambrian.

Table 1. *Buoyancy systems and lift generation mechanisms commonly observed in aquatic animals*

| Buoyancy system | Lift generation | Examples |
|---|---|---|
| Hydroplaning | Pectoral fins or other cambered body parts | Sharks, scombroids, probably pelagic squid |
| Body fluids | Internal reservoirs of low density fluid | Ammonical squid |
| Lipids | Deposits of low density body fat | Sharks, some myctophid fish |
| Gas bladders | Internal gas inclusions | Many teleost fish |
| Camerate shell | Gas-filled chambers of shell | *Nautilus*, fossil ectocochleates, some endocochleates, e.g. *Spirula* |

The subsequent history of these two groups is marked by complex patterns of profound morphologic change, and is punctuated by radiations and extinctions of major proportion. This rich and variegated panoply of events is thought to reflect the interplay of adaptive opportunity and adaptive constraint in locomotor design.

My main purpose here is to explore locomotor design and evolutionary history in swimmers. I focus in this discussion on cephalopods, and in particular on shelled, or ectocochleate, cephalopods. *Nautilus* figures prominently because it is the only modern ectocochleate genus, and thus serves as a valuable analogue for its extinct relatives.

The approach I take is three-fold. I first review information relevant to five aspects of ectocochleate locomotion: buoyancy, equilibrium control, drag, propulsion and swimming speed. Second, I compare their locomotor design to that of soft-bodied or endocochleate cephalopods, and to that of fish. Third, I use this information, together with cephalopod diversity data, to argue that some events in cephalopod history may indeed be attributed to aspects of locomotor design.

## Buoyancy

Buoyancy is important to swimmers because it relieves them of the need to exert muscular effort to maintain position in the water column. An advantageous balance of buoyancy and deadweight forces can also reduce cost of transport, as in glide-and-swim locomotion (Weihs 1973; Weihs & Webb 1983). Modern swimmers exhibit a wide variety of buoyancy

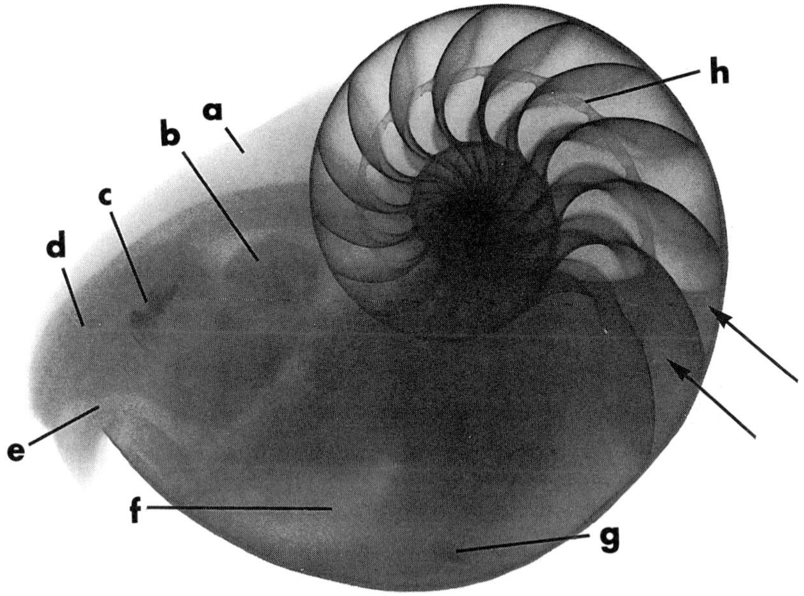


Figure 1. X-radiograph of a live *Nautilus* showing organization of shell and body. The shell is divided into a large body chamber, which houses the body, and a phragmocone consisting of a series of smaller, gas and fluid-filled chambers. a, hood; b, eye; c, jaws; d, tentacles; e, funnel passage; f, mantle cavity; g, renal appendage; h, siphuncle; arrows indicate chambers containing cameral water. Note meniscus on free fluid surface. Actual shell diameter at aperture 14 cm.

mechanisms (table 1), and many rely on a combination of agents to provide necessary buoyancy forces. Fossil forms were similarly diverse.

Figure 1 shows the ectocochleate cephalopod solution to the problem of buoyancy: the camerate shell. The gas-filled chambers of the shell (the camerae) provide the buoyancy force necessary to offset the animal's deadweight. The shell and siphuncular tube physically isolate the camerae from the external environment so that *Nautilus* need not equilibrate internal cameral gas pressure and ambient hydrostatic pressure. In fact, measurements of cameral gas pressure by Bidder (1962) and Denton & Gilpin-Brown (1966) show that internal pressure never exceeds about 1 atm ($\approx 10^5$Pa). This low value derives from the fact that, unlike many deep-water fish (Gee (1983)), *Nautilus* cannot concentrate gas; instead gas enters the camerae by simple diffusion from seawater via the animal's blood (Denton & Gilpin-Brown 1966). The fundamental unity in shell form among all ectocochleates strongly implies that the buoyancy mechanism of fossil forms operated similarly.

Ectocochleate buoyancy generation, based as it is on pressure-isolated gas reservoirs, has several interesting properties. First, the vertical mobility of *Nautilus*, and presumably of ectocochleates generally, is not inhibited by pressure/volume constraints inherent in pliant swimbladders (Gee 1983 gives a comprehensive discussion of swimbladder design). The ectocochleate buoyancy system thus facilitates lifestyles involving diurnal vertical migration. Recent remote telemetry tracking of several free-ranging *Nautilus* populations indicates that daily vertical movements of several hundred metres are common (Carlson, McKibben & DeGruy 1984; Ward *et al.* 1984), and many fossil cephalopods are believed to have shown similar behaviour (Kennedy & Cobban 1976). Second, the ultimate limit on vertical descent is determined by the strength of the shell and siphuncle, as it is these structures that bear the full load of the internal/external pressure head. Measurement of shell and siphuncle strength give maximum depths of about 800 m for *Nautilus* (Denton & Gilpin-Brown 1966; Saunders & Wehman 1977; Kanie *et al.* 1980; Ward, Greenwald & Rougerie 1980; Chamberlain & Moore 1982; Kanie & Hattori 1983; Chamberlain & Chamberlain 1985). Other ectocochleates appear to have had different depth limits (Westermann 1973, 1982; Hewitt 1984; Hewitt & Westermann 1986), although a quantitative method for determining these limits is presently unavailable (Chamberlain & Chamberlain 1985, 1986).

Figure 1 also shows small volumes of cameral water as shadows in the last-formed chambers. These fluid reservoirs are used to provide support for new septa and to trim buoyancy during growth (Denton & Gilpin-Brown 1966; Ward *et al.* 1977; Ward & Martin 1978; Collins, Ward & Westermann 1980). Water is transported into or out of the camerae by activating an osmotic pump located in the siphuncular epithelium (Denton & Gilpin-Brown 1966; Greenwald, Ward & Greenwald 1980; Greenwald, Cook & Ward 1982). However, measurements of the rate of cameral water transport (Ward, Greenwald & Magnier 1981; Chamberlain & Moore 1982; Ward & Chamberlain 1983; Ward 1985, 1986) indicate that it is far too slow to assist diurnal migration by modifying buoyancy. Nevertheless, there is some indication that this may have been possible in some fossil forms, as for example in the Devonian nautiloid *Archiacoceras* in which prominent invaginations of the siphuncular wall increase the surface area for osmotic exchange (Crick & Teichert 1979), perhaps raising fluid flux sufficiently to assist vertical movement.

The camerate shell is an effective buoyancy provider, particularly in the context of vertical migration. Yet the fact that the camerae are not pressurized – the feature that underwrites the advantages of this system for vertical movement – produces consequences of significant negative outlook for active horizontal movement. The main problem here is that the shell must be robustly constructed, and hence quite heavy, to resist the load resulting from the unpressurized camerae. In *Nautilus*, for example, the shell comprises about 30% of total air weight and, because of the high density of shell carbonate, fully 90% of total weight in water (Chamberlain 1987). Shell weight data (Trueman 1941; Raup & Chamberlain 1967; Saunders & Shapiro 1986) show that this situation holds for shelled cephalopods generally, although specific shell/total weight ratios vary somewhat from that of *Nautilus*. Thus, not only must the buoyancy device be inordinately large, consuming space that could be used more propitiously for other locomotory purposes (see below) but moreover the bulk of the buoyancy force buoys the buoyancy apparatus, not the animal. Ectocochleates stand alone in subscribing to this unique condition. No other group of swimmers, not even oil-filled odontocetes or fluid-filled mid-water fish and squid, maintain buoyancy systems so heavy and wasteful of space and buoyancy force as ectocochleates.

### Equilibrium

Because orderly, directed motion is impossible without it, equilibrium control is essential to any form of aquatic locomotion. Figure 2 compares the equilibrium control systems of shelled cephalopods (*a–c*) and of most soft-bodied swimmers, represented in *d–f* by fusiform fish. Although both systems impart the requisite degree of control, they are distinguished by the role of hydrostatic stability. In ectocochleates stability is generally high and forms the operational basis of the control system, whereas in soft-bodied swimmers stability is usually low and is of relatively little importance.

In shelled cephalopods, buoyancy and weight, acting vertically through their respective centres, elicit a rotational couple when they are not precisely aligned (figure 2*a*, *b*). The effect of this couple is to restore the animal to equilibrium (figure 2*c*), in which the lines of action of the two forces coincide. During locomotion, thrust developed by the jet, acting at the position of the funnel, and the drag force acting through the centre of dynamic pressure elicit a movement that rotates the animal away from its equilibrium (figure 2*c*). The animal swings in this manner until the magnitude of the stabilizing moment balances the jet-drag moment. As thrust declines with the termination of the jet pulse, the animal returns to equilibrium under the action of the stabilizing moment. Chamberlain (1980*a*, 1981) observed that the angle through which *Nautilus* rotates amounts to about 15°, and

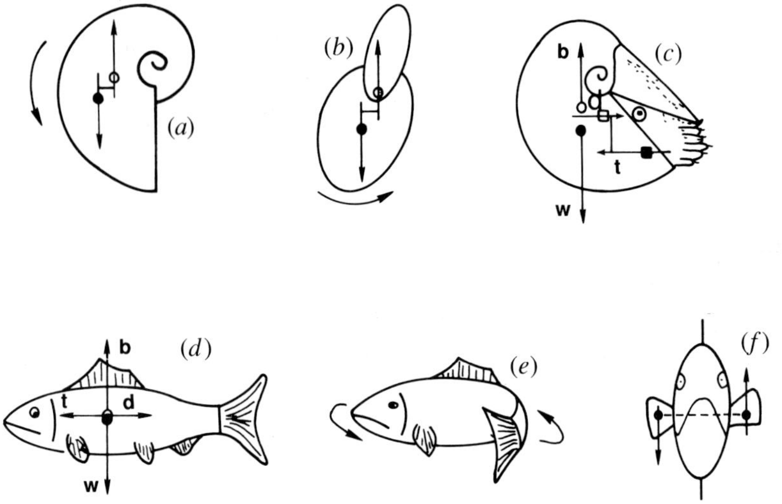


Figure 2. Equilibrium control mechanism of ectocochleate cephalopods (*a–c*) and fusiform fish (*d–f*). **b**, buoyancy force; **w**, weight; **t**, thrust; **d**, drag; ●, centre of mass; ○, centre of buoyancy; ■, point of application of thrust; □, centre of dynamic pressure. Thin arrows give lines of action of forces. and stabilizing moment elicited when centres of mass and buoyancy are not aligned in (*a*) pitch and (*b*) roll. (*c*) Equilibrium position of *Nautilus*. Buoyancy and weight are colinear and have opposite directions. Effect of jet-drag couple is to rotate shell to produce orientation shown in (*a*); rotation stops when stabilizing moment balances nett jet thrust and drag moment. (*d*) Equilibrium of fish. Forces are approximately colinear owing to axisymmetric body and medial location of swim bladder. Centres of buoyancy, mass, thrust and drag nearly coincide, producing low hydrostatic stability. Equilibrium control achieved through movements of body (*e*) and fins (*f*). (From fig. 6 of Chamberlain 1987)

suggested that fossil forms may have behaved similarly. The interaction of these two movements produces the prominent rocking motion of swimming *Nautilus* described by Bidder (1962), Chamberlain (1976, 1981, 1987), Packard, Bone & Hignette (1980) and many other students of this animal.

Most soft-bodied swimmers emphasize low, rather than high, stability. Modern fish, for example, lack the heavy skeleton typical of ectocochleates. Since tissue density differs only slightly from sea water density, soft swimmer deadweight approximates buoyancy force, rather than being 20–30% greater as in ectocochleates. Flotation devices employed to balance these small differences between weight and buoyancy are perforce only a small fraction of total body volume. Mass distribution is virtually uniform in this situation, and hence the centres of buoyancy and mass nearly coincide (figure 2*d*). In contrast to ectocochleates (figure 2*c*), hydrostatic stability and its contribution to equilibrium is minimal. Instead, instantaneous instabilities during locomotion are suppressed by compensatory movements of the flexible body and fins (figure 2*e*, *f*). Two factors are especially significant. First, the axisymmetric fish body ensures that drag and thrust act near the midline of the body so that their disequilibrating moments are minimized, and hence more readily dampened. Second, fins are particularly valuable as stabilizers because they are located at the periphery of the trunk, and consequently have considerable moment arms (figure 2*f*). Fin deployment can therefore create large stabilizing moments that quickly dampen potentially disequilibrating moments as soon as they appear.

Ectocochleate soft tissues play no active role in equilibrium maintenance during jet propelled swimming. The shell encloses far too great a portion of the body to allow body flexibility to come into play or to provide advantageous placement of stabilizing fins or tentacles. The scanty fossil evidence concerning tentacle placement and the possible possession of fins (see, for example, Flower 1955; Sturmer 1970) supports this view by indicating that fossil ectocochleates lacked fins and were basically similar to *Nautilus* in tentacle configuration. With a prominent external shell, no appreciable flexibility and no fins, ectocochleates are forced to rely solely on hydrostatic stability for equilibrium control. High stability and the uneven distribution of mass that produces it (buoyant phragmocone; tissue-filled body chamber) constitute an important adaptive asset promoting locomotion in these animals. However, as Trueman (1941), Raup & Chamberlain (1967) and Raup (1967) have shown, not all fossil forms were well endowed with regard to stability. Many shell geometries, notably serpenticonic and planulate forms, were quite deficient in this property. Since jet locomotion in ectocochleates appears to require hydrostatic stability, it would seem that such forms had lifestyles in which swimming was of little importance. It is interesting to note that the rapidly expanding shell geometry characteristic of *Nautilus* ranks among the most stable shell configurations.

Palaeontologists have also examined the relation between shell geometry and equilibrium orientation. Trueman (1941), Raup & Chamberlain (1967), Raup (1967) and Saunders & Shapiro (1986) have demonstrated that equilibrium orientation in coiled forms varies widely but systematically with shell geometry. Most geometries had orientations with down-turned apertures (as in figure 2*c*), but as shells become more serpenticonic the aperture becomes more upturned. In straight-shelled orthoconic forms weight distribution is especially critical because it forces the animal into a vertical orientation, with downturned aperture, not easily reconciled with horizontal swimming (figure 3*a*). This problem was solved early in

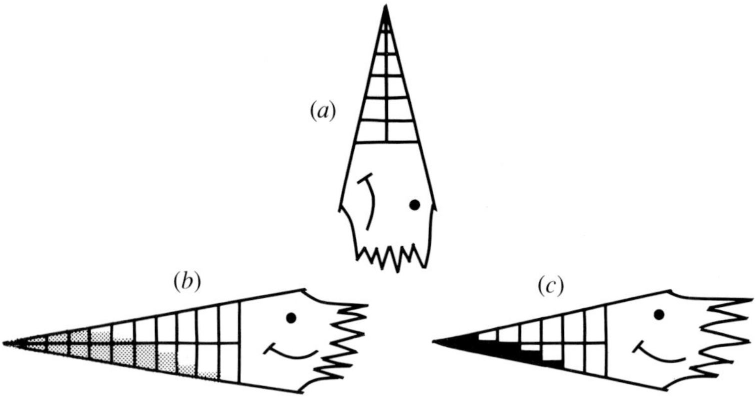


Figure 3. Effect of ballast on orthocone equilibrium orientation. (*a*) No ballast: tissue filled body chamber lies beneath buoyant phragmocone producing vertical orientation, detrimental to effective horizontal swimming. (*b*) Water ballast: presence of cameral water counterweight (dots) in apical end of shell balances tissue weight in body chamber and produces horizontal orientation. (*c*) Mineral ballast: carbonate deposits (solid colour) near shell apex balance tissue weight and produce horizontal orientation. Ballasted shells have small body chambers relative to phragmocone size owing to the space needed for the ballasting material and for the larger cameral gas volume necessary to offset the added weight of the ballast.

ectocochleate history when some orthocone lineages developed counterweights in the form of cameral water or carbonate deposits localized in the apical end of the shell, frequently positioned off-centre (as in figure 3*c*) to provide additionally some hydrostatic stability. This redistribution of weight brought the animals into the horizontal attitude necessary for swimming (Teichert 1933; Flower 1957; Teichert & Moore 1964).

Although ballasting is necessary for orthocone locomotion, it is not achieved without cost. Filling chambers with ballast necessitates phragmocones of increased volume relative to body size to ensure that the shell retains sufficient gas volume to offset the weight of the ballast (Westermann 1977), especially if the ballast used is cameral water. Although absolutely necessary for orthocone locomotion, ballasting seriously compromises orthocone locomotion because a relatively smaller body means smaller, less effective propulsion and steering mechanisms. In addition, the metabolic expense of secreting ballast, particularly carbonate, and of providing the additional carbonate needed for a more expansive shell, adds considerably to the general energy budget of such cephalopods.

**Drag**

Drag force opposes the motion of swimmers. When rapid efficient motion is important drag force must generally be minimized. Experimental analysis of ectocochleate drag began nearly 60 years ago with the work of Schmidt (1930) and Kummel & Lloyd (1955), but the bulk of current knowledge in this area stems from more recent research (Chamberlain 1969, 1976, 1980*b*, 1981, 1987; Chamberlain & Weaver 1978; Chamberlain & Westermann 1976; Weaver & Chamberlain 1976). Ectocochleate drag can be thought of as consisting of several components: (1) drag produced by the shell, (2) drag effects of exposed body parts and (3) inertial drag generated by velocity changes and water entrainment during movement. The total drag on an animal is determined by a complex and time-varying interaction of these factors.

*Shell drag*

Shell drag derives from two main sources: (1) shell shape and (2) shell surface characteristics. The first depends on shell geometry, the latter on the configuration of ribs, tubercles and spines with which the shell surface may be ornamented.

Drag due to shell shape can be evaluated with smooth shells. Figure 4 shows the pattern of fluid flow around a typical smooth shell. Water flows across the venter and flanks of the shell. The flow separates from the shell surface when it reaches the umbilical shoulder and peristome, producing a pronounced turbulent wake, visible in figure 4 as swirls of dye behind the shell. The evident intensity of wake turbulence implies that the shell creates significant pressure drag. This flow structure characterizes all ectocochleate shells so far studied (Chamberlain 1976).

Tow tank measurements of drag and velocity for a wide variety of coiled shell forms (Chamberlain 1976) suggest that this high drag assessment is correct. Table 2 summarizes shell drag coefficients for a variety of ectocochleates as well as other animals. Ectocochleate drag coefficients are generally similar to those of blunt geometrical bodies like spheres. They are not in the range of drag coefficients for elongate streamlined objects. Among ectocochleates, drag coefficient varies by a factor of about 6, with disk-like compressed oxycones and straight-shelled orthocones having the lowest values. Actually, the ectocochleate minimum may be somewhat lower than has been reported because highly compressed, narrowly umbilicate oxycones (which have not been tested) probably have drag coefficients as low as 0.1. Some orthocones may also have had equally low values, but Hoerner's drag data on cones (see fig. 21 of Hoerner 1965, pp. 3–12) indicate the inaptness

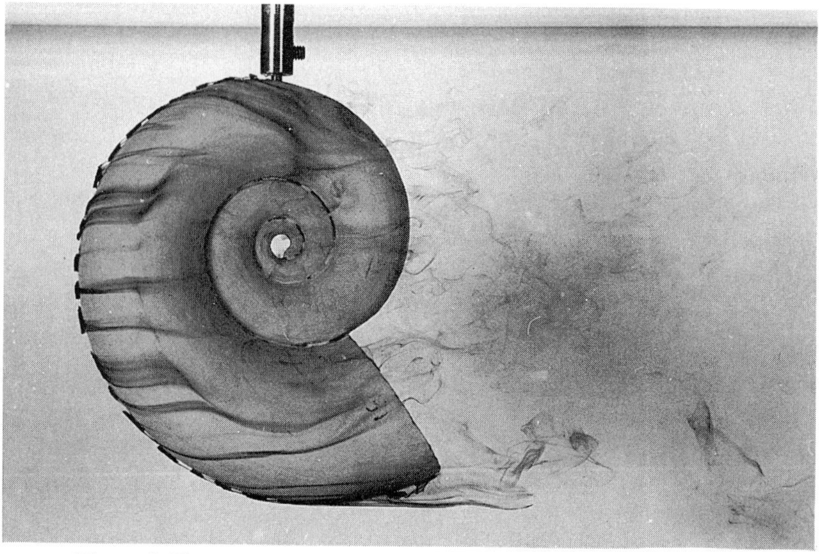

Figure 4. Flow pattern around a plastic shell model. Flow lines made visible by dye bleeding from crystals of potassium permanganate cemented to venter. Flow velocity 10 cm/s. Deflexion of dye streams along umbilical shoulder indicates boundary layer separation. (From plate 84, fig. 1 of Chamberlain 1976)

of the widespread view (see, for example, Furnish & Glenister 1964; Holland 1987) that orthocone drag coefficients are significantly less than those of coiled forms.

Roughening the surface of an object can significantly alter drag. For streamlined bodies a rough surface increases drag because it forces the boundary layer to become turbulent. This increases the frictional drag component and, since pressure drag is inconsequential, elevates total drag. This is why fish are smooth. For blunt objects such as cephalopod shells the situation is more complex. Surface roughness can increase drag when individual roughness elements protrude through the boundary layer and interact with ambient water. If roughness elements remain immersed in the shell boundary layer, however, they can lower drag by prematurely inducing transition to turbulent boundary layer flow, which usually entails a large reduction in pressure drag. Chamberlain & Westermann (1976) concluded that many fossil cephalopods with sculptured shells could have profited from this phenomenon.

Tow tank tests involving plastic shells ornamented with ribs made of thread indicate that the predicted benefits of shell sculpture do indeed occur (Chamberlain 1981, and table 3). The results for the compressed shell model

Table 2. *Drag coefficients of geometrical bodies, cephalopods and fish*
$C_D$, drag coefficient. The area column gives the representative area upon which calculation of $C_D$ was based: FA, frontal area; V, $(\text{volume})^{2/3}$; WA, wetted (=surface) area. Sources: 1, Hoerner (1965); 2, Chamberlain (1976); 3, Chamberlain (1980*b*); 4, Webb (1975); 5, Webb (1986); 6, this paper; 7, Vogel (1981).

| | $C_D$ | Area | Sources |
|---|---|---|---|
| **Geometrical bodies** | | | |
| Sphere | 0.47 | FA | 1, 7 |
| Circular cylinder | 1.2 | FA | 1, 7 |
| Streamlined body | 0.04 | WA | 1, 7 |
| **Ectocochleates** | | | |
| Gyrocone | 1.1 | V | 2 |
| *Nautilus pompilius* | 0.48 | V | 2 |
| Cadicone | 0.71 | V | 3 |
| Oxycone | 0.25 | V | 2 |
| Orthocone | 0.25 | FA | 6 |
| **Endocochleates** | | | |
| Fusiform squid | 0.2 | FA | 6 |
| | 0.004 | WA | |
| **Fish** | | | |
| bowfin (*Amia calva*) | 0.005 | WA | 4 |
| dogfish (*Squalus* sp.) | 0.014 | WA | 4 |
| sturgeon (*Acipenser fulvescens*) | 0.053 | WA | 5 |
| rainbow trout (*Salmo giardneri*) | 0.015 | WA | 4 |
| mackerel (*Scomber scombrus*) | 0.004 | WA | 4 |
| pike (*Esox* sp.) | 0.008 | WA | 4 |

suggest several general points. First, coarse ribs increase drag coefficient, probably because they distort fluid movement in the boundary layer (see fig. 1C, D of Chamberlain & Westermann 1976). Finer grades of ribbing produce the hypothesized decline in drag coefficient, although there is some hint that rib spacing influences the magnitude of this decrease. Data for the depressed shell model show a similar relation between rib morphology and drag coefficient, although coarse ribs merely elicit a smaller decrease in drag coefficient than do fine ribs, rather than increasing drag coefficient as in the compressed shell model. This inconsistency requires more study, but may perhaps derive from the strong disparity between the shell models in basic geometry.

Table 3. *Effect of simulated ribs on drag coefficient of two plastic shell models*

Ribs made of thread of constant diameter glued in rectiradiate pattern to shell surface. Rib spacing is distance between adjacent ribs relative to rib height (thread diameter). Rib size is height of ribs relative to shell size. *K*, rib height; λ, distance between adjacent ribs at venter; *D*, shell diameter; $C_D$, total drag coefficient of test configuration, $\%\Delta C_D$, percentage change in $C_D$ relative to $C_D$ of base test configuration, i.e. to shell model without ribs. Values for compressed and depressed shell models are based on tow tank measurements reported by Chamberlain (1980*b*). Rib data are partly from Chamberlain (1981) and partly new. CSM, compressed shell model; DSM, depressed shell model.

| Test configuration | Rib spacing $K/\lambda$ | Rib size $K/D$ | $C_D$ | $\%\Delta C_D$ |
|---|---|---|---|---|
| CSM | | | 0.26 | |
| CSM + coarse ribs | 0.10 | 0.0063 | 0.33 | +27 |
| CSM + fine, wide-set ribs | 0.08 | 0.0012 | 0.22 | −15 |
| CSM + fine, close-set ribs | 0.02 | 0.0012 | 0.16 | −38 |
| DSM | | | 0.71 | |
| DSM + coarse ribs | 0.10 | 0.0116 | 0.64 | −10 |
| DSM + fine, wide-set ribs | 0.08 | 0.0024 | 0.58 | −18 |
| DSM + fine, close-set ribs | 0.02 | 0.0024 | 0.53 | −25 |

### *Body drag*

Chamberlain (1980*b*) examined the drag contribution of the exposed parts of the body and tentacles by simulating these soft tissues with moulded prostheses. Tow tank tests of shell models with and without these simulated body parts produced the results in table 4. These data show that extension of the body outward from the aperture has a beneficial effect in reducing drag, particularly when a streamlined body fills much of the dead water region behind the shell aperture. Unstreamlined bodies, and body extension associated with depressed shells, appear to elicit smaller drag reductions. According to the arguments of Chamberlain (1980*b*), the 26% drag reduction reported for compressed forms with streamlined bodies probably represents a maximum value for ectocochleates as a whole.

Table 4 also provides data on tentacle extension in *Nautilus*, referring to its habit of extending two or more digital tentacles into the surrounding flow while moving. Similar behaviour occurs among endocochleates, and there is some circumstantial evidence in the form of tentacle impressions (Flower

Table 4. *Effect on drag coefficient of soft tissue extension*
Same shell models as in table 3. Body prostheses consist of carved wooden models. $C_D$, total drag coefficient of test configuration; $\%\Delta C_D$, percentage change in $C_D$ relative to $C_D$ of base test configuration, i.e. of the shell model without extended soft parts. Values for *Nautilus* tentacle extension based on calculation of the drag of individual tentacles by using the cross-flow principle discussed by Taylor (1952) and Hoerner (1965). Basic drag of *Nautilus* is taken from the $C_D$ of its shell (as in table 2) modified by the estimated effect of body extension as discussed by Chamberlain (1980*b*). CSM, compressed shell model; DSM, depressed shell model.

| Test configuration | $C_D$ | $\%\Delta C_D$ |
|---|---|---|
| CSM | 0.26 | |
| CSM + unstreamlined body prosthesis | 0.24 | −8 |
| CSM + streamlined Body prosthesis | 0.20 | −26 |
| DSM | 0.71 | |
| DSM + body prosthesis | 0.64 | −10 |
| *Nautilus* | 0.43 | |
| *Nautilus* with 2 extended tentacles | 0.45 | +5 |
| *Nautilus* with 10 extended tentacles | 0.54 | +26 |
| *Nautilus* with 20 extended tentacles | 0.65 | +51 |

1955) that fossil forms also practised this behaviour. Tentacles prolonged in this manner generate their own drag, and the hydrodynamic effect of such tentacles is pronounced. Extending two tentacles, as *Nautilus* often does to maintain tactile contact with the substrate, increases total drag by about 5%, while forming an array of 10–20 radiating tentacles, commonly done to enhance chemoreception during food searches, augments drag by as much as 50%. Such energy losses would severely compromise performance. Thus it is not surprising that *Nautilus*, like most other swimmers, exhibits a variety of locomotory behaviours. Search webs characterize slow movements used in locating food items, but when maximum speed is required the tentacles withdraw into their protective sheaths to minimize drag. Presumably fossil forms acted in a similar fashion.

### *Inertial drag*

Research on cephalopod locomotion has traditionally treated drag as a steady force, and its time variation has for the most part been neglected. Yet cephalopod locomotion is characterized by prominent periodic accelerations and decelerations, and such a view is at best oversimplified. In analysing *Nautilus* propulsion (Chamberlain 1987) I took a step towards remedying this situation by considering inertial drag produced as a consequence of periodic velocity change. My approach was to apply to *Nautilus* some generalized observations about inertial drag made by Daniel (1984). Several points merit mention in the broader context of ectocochleate locomotion. First, added mass coefficients for shelled cephalopods, estimated from Daniel's data on spheres and ellipsoids, probably fall in the range 0.8 to 0.2, with depressed shells and compressed shells forming the upper and lower limits, respectively. Second, in accelerating from rest, the inertial force developed by an ectocochleate probably comprises 60–80% of its total drag. This is somewhat more than the 48% that Daniel obtained for juvenile squid.

Inertial drag is clearly a parameter of prime significance. Yet it is important to note that for an animal swimming at constant average velocity, although accelerating and decelerating with each pulse of the propulsive organs, the inertial forces cancel, so that average performance is governed by steady-state drag. Since *Nautilus*, probably like many of its fossil antecedents, relies on extended periods of locomotion rather than on the rapid acceleration characteristic of squid, the practice of focusing on steady-state aspects of drag may not be as great an analytic shortcoming for ectocochleates as it might appear.

### *Drag of cephalopods and fish*

Drag coefficient is an index of the efficacy with which propulsive force is converted to velocity. Comparison of ectocochleate drag coefficients to those of other swimmers is thus an instructive means of surveying this aspect of locomotor design (table 2). However, the drag coefficient values are based on different reference areas and are comparable only in a very general sense. Other difficulties emerge as well. In his treatment of the fish data, reproduced here in table 2, Webb (1975) emphasized strongly the extreme difficulty in obtaining suitably accurate drag coefficients for pliant fish bodies. In his opinion the figures for fish drag coefficient represent maximum values only, with the exception of the data for *Scomber*, which appear to be reasonably accurate. The drag coefficients for squid are estimates based on analogy to streamlined fish bodies, modified by Hoerner's (1965) data on drag consequences of truncating streamlined objects. The smaller squid value assumes flow remains attached over most of the body, as in fish. The higher figure gives the estimated drag coefficient if flow is forced to separate at the collar formed by the mantle edge just behind the head.

The general impression is that both coiled and straight-shelled ectocochleates have drag coefficients considerably higher than those of either fish or squid. This can be quantified somewhat by following Vogel's recommendation for converting disparate drag coefficients to a common basis of evaluation (Vogel 1981, pp. 112–114). Choosing $(\text{volume})^{2/3}$ as the common reference area, and following Vogel, the drag coefficient for a sphere, for example, becomes 0.56. Converting drag coefficient for fish bodies is unfortunately not as straightforward, but assuming that elongate bodies of revolution are reasonable facsimiles for fish, the $(\text{volume})^{2/3}$ equivalent drag coefficients for fish, and probably squid also, are in the region of 0.02. Thus, the existence of a significant dichotomy in drag coefficient between ectocochleates and slender soft-bodied swimmers becomes evident. Most ectocochleates have drag coefficients about 25–35 times higher than those of fish, and even the best streamlined shelled cephalopods exceed fish by a factor of 10. Allowing for drag gains due to body extension and shell sculpture reduces this disparity to about 5. The biological implication of these estimates is clear. Shelled cephalopods are very much less efficient than fish or squid in converting propulsion to velocity. The extent of this difference is illustrated by a simple calculation: for a given propulsive force, ectocochleates have velocities only about one fifth to one half of those of fish of equal volume.

The higher drag coefficients of ectocochleates can be attributed to three main factors, all of which ultimately derive from that fundamental feature of 'ectocochleateness', the shell. These factors are (1) shell geometry, (2) the external character of the shell and (3) the large phragmocone.

Coiling in cephalopod shells produces objects that are roughly equidimensional, rather than elongate. As a result coiled shells are generally too short in the direction of motion to allow the flow to close posteriorly. Instead a prominent wake, with attendant pressure drag, develops behind the shell (figure 4). Moreover, the inwardly curving umbilicus, an unavoidable consequence of shell coiling, forces premature flow separation over much of the anterior portion of the shell (figure 4) and further augments pressure drag. Orthoconic geometries are also subject to the problem of excess pressure drag because these shells, like their coiled counterparts, have a sharp trailing edge at the aperture. Unless body extension masks this feature, early separation results.

The external character of the ectocochleate shell precludes use of the drag-reducing mechanisms upon which many soft-bodied swimmers may rely. Webb (1975) examined such phenomena, and while sceptical about some, considered both mucous skin secretion and boundary-layer vortex damping

Table 5. *Effect of size of buoyancy device on total drag*
In this table, the drag due to a buoyancy device is expressed as the buoyancy drag ratio, which is defined as the ratio of the drag due to the buoyancy device to the total drag. The numbers in each column of the table give buoyancy drag ratio at the swimming speed indicated. Drag due to hydrodynamic lift is the induced drag of pectoral fins. Swimbladder drag is the added drag component resulting from body size increase due to the presence of the swimbladder in the body. Phragmocone drag is defined as the drag component resulting from increase in total size (shell + body) due to presence of the phragmocone. (bc/p) is the ratio of body chamber volume to phragmocone volume. Values for hydrodynamic lift and swimbladder buoyancy drag ratio are based on data given by Alexander (1972, 1977) and Gee (1983). Values for cephalopod buoyancy drag ratios are based on data given by Chamberlain (1976, 1981) and Trueman (1941). (bc/p) values of 2 and 4 are volume ratio limits within which Trueman found most of his ammonite shells to lie.

| | Swimming speed (cm/s) | | | |
|---|---|---|---|---|
| Buoyancy device | 10 | 34 | 70 | 100 |
| Hydrodynamic lift | 0.59 | 0.13 | 0.048 | 0.026 |
| Swimbladder | 0.04 | 0.04 | 0.04 | 0.04 |
| Phragmocone (bc/p) = 2 | 0.16 | 0.16 | 0.16 | 0.16 |
| Phragmocone (bc/p) = 4 | 0.08 | 0.08 | 0.08 | 0.08 |

to be effective drag reducers among fish and cetaceans. But these mechanisms – as well as those judged by Webb to be of unproven value – require a pliant, porous skin. With their prominent, rigid shell covering most of their surface, ectocochleates do not have the capacity to invoke such drag reduction techniques. However, drag minimization is not totally beyond the reach of shelled cephalopods, since some can reduce drag by employing shell sculpture to generate turbulence.

To provide buoyancy the phragmocone makes the animal, and thus the drag it generates, larger than would otherwise be the case. Following Alexander (1977) and Gee (1983), this effect can be quantified by calculating the drag of an animal when size is reduced by an amount equivalent to phragmocone volume. Table 5 reports the results for several different fossil ectocochleates as well as for modes of buoyancy control used by fish. Up to 16% of total ectocochleate shell drag derives from this 'enlargement' component. At all swimming speeds tabulated the volumetrically small swimbladders of fish are more economical, in that they contribute a far smaller proportion of total drag than does a cephalopod shell. Since hydrodynamic lift is used almost exclusively by fast swimmers where this form of buoyancy generation is highly efficient (Vogel 1981), the data indicate that the cephalopod phragmocone is, in practice, the most uneconomical of these three common buoyancy systems.

## Propulsion

Propulsion is by far the must elusive aspect of ectocochleate locomotion. This is because measurement of propulsion performance parameters, necessary for quantitative analysis of locomotor function, requires live specimens. This is of course not possible in animals known only as lumps of stone. Only in *Nautilus* can ectocochleate propulsion be examined with the proper regard for objectivity, and for this reason *Nautilus* assumes prime importance in any treatment of this matter. In the present discussion I focus almost exclusively on *Nautilus*, trusting to the view that this modern ectocochleate gives a good representation of the general level of performance of the larger group to which it belongs, but admitting that in specific instances this notion is undoubtedly oversimplified and misleading. I would add, however, that observations on muscle arrangement in fossil cephalopods (Mutvei 1957, 1964, 1967; Mutvei & Reyment 1973) and on other aspects of soft tissue organization (Jordan 1968; Lehmann 1971; Kennedy & Cobban 1976) indicate that *Nautilus* is not atypical in its anatomy and thus give some substance to this view. My major objective here is to describe the organization and performance of the ectocochleate locomotor system, as exemplified by *Nautilus*, in the light of squid and fish propulsion. The present discussion is based largely on data given by Chamberlain (1987); Packard *et al.* (1980) also dealt with this matter.

### *Propulsive mechanism*

*Nautilus* develops thrust in a manner similar to that of soft-bodied cephalopods: water is taken into the mantle cavity under low pressure and expelled under high pressure. In detail, however, the ectocochleate locomotor system differs markedly from that of endocochleates owing to extensive anatomical differences between the two groups. For example, *Nautilus* lacks a highly muscular mantle, and so unlike endocochleates cannot generate thrust by powerful mantle contractions. Instead, thrust in *Nautilus* is produced by coordinated movements of the funnel and cephalic retractor muscles, the activation of which elicits the pressures necessary first to draw external water into the mantle cavity, and then to propel it outward through the funnel passage at high velocity.

Ventilation of the gills, and probably slow movement also, are produced by rhythmic beating of the funnel wings, but to maximize output the cephalic retractors are brought into play. Wells (1987) has described ventilatory movements in considerable detail. These consist of a series of contractile and expansive wing motions that produce a continuous flow over the gill filaments. These motions are modified during active propulsion by the addition of prominent in-and-out movements of the cephalic region as described by Wells (1987) and Chamberlain (1987). Retracting the head is induced by firing the longitudinal fibres of the cephalic refractors and elevates pressure by compressing the mantle cavity. This generates the thrust necessary to propel the animal. However, it also interferes with oxygen uptake, as Wells (1987) has documented, perhaps by interrupting flow through the gills. Thus, *Nautilus* is subject to the same kind of oxygen shortfall constraining peak output as in endocochleates (O'Dor 1982; Baldwin 1987; Wells 1987).

Nautilus, like other cephalopods, develops thrust in a pulsatile fashion, rather than continuously as in undulatory propulsion. This is because new water must be drawn into the mantle cavity at the end of a pulse, and because the funnel muscles and retractors must be restored to readiness for the next contraction. The pulse rate of the propulsive system has been studied by several observers. Packard *et al.* (1980) and I (Chamberlain 1987) both noted that pulse rate increases from about 0.5 Hz to 1 Hz as active locomotor pulsing replaces respiratory ventilation. These results are similar to the pulse kinematics of soft bodied cephalopods studied by Trueman & Packard (1968). Packard *et al.* (1980) and I (Chamberlain 1987) monitored instantaneous pressure and thrust generated during pulsing and observed that the two parameters are to some extent out of phase. With a much larger sample available for study, I showed that thrust generally peaks slightly ahead of pressure (Chamberlain 1987). I attributed this to idiosyncrasies of the muscle activation sequence, particularly to instantaneous change in funnel cross-sectional area. I also noted that, although there was some indication that pulse rate varied inversely with specimen size, the relation was statistically weak. This presents the interesting possibility that the rate at which the endocochleate locomotor system operates is largely independent of size. If true, this would stand in marked contrast to vertebrate undulatory propulsion in which the rate of operation is strongly dependent on size (Bainbridge (1958) and Wardle (1975) discussed tail beat frequency in carangiform swimmers).

Table 6. *Propulsive parameters for cephalopods and fish*
Muscle mass ratio is the ratio of propulsive muscle mass to total mass; values are based on hyponomic and cephalic retractor muscles for *Nautilus*, on mantle muscles for endocochleates, on trunk muscles for fish. Mantle cavity ratio = (mantle cavity volume)/(total volume), where total volume = (total mass)/(sea water density). Exhaust ratio = (mass of water in a single propulsive pulse)/(total mass). Exhaust ratio data for endocochleates are based on the assumption that 2/3 of mantle cavity emptied during pulse. Values for trout are based on mass of water accelerated by caudal fin during interval equivalent to one pulse period of cephalopods (about 1 s) and apply to swimming speeds of 10–50 cm/s. Propellant velocity is the velocity of water accelerated by animal to produce thrust. For cephalopods, propellant velocity is the maximum velocity of water expelled in jet. For trout, propellant velocity is the velocity of water accelerated by caudal fin at swimming speeds of 10–50 cm/s. Mantle cavity pressure is maximum pressure in mantle cavity during propulsive pulse. Data for *Nautilus* are valid for total mass above about 40 g. For endocochleates data are specific to total masses as follows: *Octopus*, 500 g; *Eledone*, 600 g.; *Loligo*, 350 g; *Sepia*, 750 g. Fish data refer to adults of average size. Data from Chamberlain (1987), Trueman & Packard (1968), Bainbridge (1960), Webb (1971) and Alexander (1977).

| | *Muscle mass ratio* | *Mantle cavity ratio* | *Exhaust ratio* | *Propellant velocity (cm/s)* | *Mantle cavity pressure (kPa)* |
|---|---|---|---|---|---|
| **Ectocochleates** | | | | | |
| *Nautilus pompilius* | 0.09 | 0.15 | 0.06 | 485 | 11.3 |
| **Endocochleates** | | | | | |
| *Octopus vulgaris* | 0.08 | 0.14 | 0.22 | 126 | 16.6 |
| *Eledone moschata* | 0.20 | 0.50 | 0.33 | 940 | 38.1 |
| *Loligo vulgaris* | 0.35 | 0.50 | 0.38 | 730 | 29.6 |
| *Sepia officinalis* | 0.30 | 0.27 | 0.16 | 275 | 9.5 |
| **Fish** | | | | | |
| *Salmo giardneri* | 0.60 | | 0.63–6.91 | 15–25 | |
| *Leuciscus leuciscus* | 0.55 | | | | |
| *Carassius auratus* | 0.40 | | | | |
| *Perca flavescens* | 0.33 | | | | |

*Dimensions of propulsion components*

The physical dimensions of a motor's components constrain its performance and limit its output. Thus, one way to evaluate locomotor performance in ectocochleates is to examine the size and capacity of the locomotor organs in *Nautilus*. Table 6 reports relevant data for *Nautilus*, along with figures for cephalopods and fish, which allow us to characterize ectocochleate propulsion.

The first column of table 6 gives values for propulsive muscle size relative to body size (termed muscle mass ratio). Only 9% of *Nautilus*' body mass is propulsive muscle, compared to up to 35% in soft-bodied cephalopods. *Nautilus* clearly is not as heavily muscled as its endocochleate relatives, and its propulsion therefore can be expected to be correspondingly less powerful. The apparent equivalency in muscle mass ratio of *Nautilus* and *Octopus* is misleading because *Octopus* is not an obligatory swimmer as is *Nautilus*, and cannot on this ground be expected to contain massive swimming muscles. Moreover, when *Octopus* does swim, it often beats its arms to assist its motion, so that, in a sense, the figure for *Octopus* muscle mass ratio in table 6 is artificially low because it does not include the mass of the arm muscles. Table 6 also gives muscle mass figures for several carangiform fish. It is evident that cephalopods possess relatively less propulsive muscle than do these fish. Only in *Loligo*, and presumably related epipelagic squid, does muscle mass ratio approach that in the undulatory swimmers included here.

Similar comparisons can be made for the other parameters in table 6. For example, with respect to mantle cavity ratio and exhaust ratio, the data reported in table 6 indicate that with the exception of *Octopus*, endocochleates have much larger mantle cavities, and expel a larger volume of water in each propulsive pulse, than does *Nautilus*. This implies a more forceful propulsive system for endocochleates for two reasons. First, the force exerted by the jet depends on its momentum and thus varies directly with jet mass. Second, massive jets require large propellant reservoirs to supply the necessary water. The mantle cavity of *Nautilus* is simply too small to support a powerful jet. Notice, also, that the exhaust ratio of the trout is significantly higher than that of any cephalopod listed. This is because cephalopods accelerate internal water to produce thrust, while fish accelerate ambient, external water. Internal water is limited to the contents of the mantle cavity and, of necessity, is less than total body volume, whereas ambient water is in theory unlimited, although it can actually be defined in terms of the volume accelerated by the caudal fin per unit time in carangiform and thunniform species (Webb 1975; Alexander 1977; Yates 1983). In practice this is often several times the volume of the fish's body. Thus fish would appear greatly to exceed cephalopods, and especially ectocochleates, in this facet of propulsion.

Jet thrust is determined by jet velocity as well as by jet mass. The last two columns in table 6 report data relevant to the velocity component of propulsive force. *Nautilus* expels fluid with far less speed than do endocochleates, again with the exception of *Octopus*. The low value for *Sepia* undoubtedly reflects incomplete sampling. This velocity discrepancy may be attributed to the lower mantle cavity pressure achieved in *Nautilus* than endocochleates. Again the value for *Sepia* is suspect. Notice also that the velocity of water accelerated by trout during propulsion is much less than that of the cephalopods. This simply reflects the fact that the great volume of water which fish accelerate to develop the momentum necessary for locomotion implies that the actual velocity increase given to their propellant can be quite small.

### *Propulsive performance*

Table 7 reports performance data for the propulsion systems of the three groups of swimmers presently under discussion. These data suggest that *Nautilus* develops much less thrust and power relative to its weight than do endocochleates, particularly rapid swimmers such as *Loligo*. A main consequence of the different locomotor configurations in these two kinds of cephalopod, and especially of the smaller dimensions of the ectocochleate propulsive organs, is now evident. *Nautilus* is a much weaker, less powerful propulsor than its soft-bodied relatives. The output of the funnel-retractor system of ectocochleates does not measure up to the mantle muscle system of endocochleates. However, both types of cephalopod appear to be less powerful than fish, although *Loligo* lies within the lower range of fish data in table 7. This latter situation may be more apparent than real because Webb (1975) note that methodological defects in acquiring the fish data led to the values in table 7 greatly underestimating the actual power output of piscine body undulators, with the possible exception of *Sphyraena*.

The data also suggest that fish have greater Froude (propeller) efficiencies than do cephalopods. This is because fish accelerate a large mass of propellant to low velocities to produce their motion, while cephalopods, having much smaller propellant mass, must accelerate this small mass to high velocity. The very low value for *Nautilus* reflects the fact that its jet mass is volumetrically the smallest of the cephalopods tabulated.

### *Constraints on propulsive design*

The foregoing observations give a strong impression of important disparities in the propulsion mechanisms among fish and cephalopods, stemming from design constraints inherent in cephalopod jet propulsion and

Table 7. *Performance parameters of cephalopods and fish*
Thrust ratio = (maximum propulsive thrust)/(total weight). Power ratio = (maximum propulsive power)/(total weight). Power ratio of *Nautilus* based on average swimming velocity of 25 cm/s. Power ratio of *Loligo* based on swimming speed of 60 m/s as reported by O'Dor (1982). Power ratio values for fish taken from Webb (1975, table 6), and according to him represent minimal values. Froude efficiency = (power used to overcome drag)/(total power), where total power is defined as the sum of powers used to overcome drag and to accelerate propellant. Froude efficiency values for *Nautilus* are based on a swimming velocity of 25 cm/s. Data for endocochleates assume 2/3 of mantle water expelled during pulse. Value for trout is based on fish of mass 220 g swimming at speeds of 10–50 cm/s. Cephalopod data probably represent minimal values because maximum jet velocity during pulse rather than average jet velocity was used to calculate values.

| | Thrust ratio | Power ratio | Froude efficiency |
|---|---|---|---|
| **Ectocochleates** | | | |
| *Nautilus pompilius* | 0.2 | 5 | 0.15 |
| **Endocochleates** | | | |
| *Octopus vulgaris* | 0.4 | | 0.22 |
| *Eledone moschata* | 1.3 | | 0.27 |
| *Loligo vulgaris* | 1.3 | 78 | 0.30 |
| *Sepia officinalis* | 0.7 | | 0.16 |
| **Fish** | | | |
| *Salmo giardneri* | | | 0.61–0.81 |
| *Salmo irideus* | | 35 | |
| *Esox lucius* | | 38 | |
| *Perca flavescens* | | 327 | |
| *Carcharhinus leucas* | | 100 | |
| *Sphyraena barracuda* | | 673 | |

in the mode of buoyancy control adopted by ectocochleates. These constraints are mainly in the form of size differences in three key components of the propulsive system: (1) in propulsive muscle, which generates motive force; (2) in the propellant reservoir, which provides the fluid upon which the motive force operates; and (3) in the buoyancy apparatus, which when present relieves the motor of providing power to offset sinking.

Table 8 shows how these propulsive elements are configured in the three types of swimmer. The data in the last column indicate the total fraction of body volume consumed by the propulsion mechanism, and represent the sums of the figures for component volumes given in the preceding columns.

Table 8. *Relative volume of components of propulsion mechanism expressed as a fraction of total body volume*

Buoyancy volume is volume fraction of buoyancy apparatus; value for *Sepia* is calculated from data given by Denton & Gilpin-Brown (1961). Reservoir volume is volume fraction of internal propellant reservoir, equivalent to the mantle cavity ratio in table 6. Muscle volume is volume fraction of propulsive muscle, or muscle mass ratio from table 6. Total volume of propulsive mechanism is the volume of all propulsive components relative to total volume.

| | Buoyancy volume | Reservoir volume | Muscle volume | Total volume of propulsive mechanism |
|---|---|---|---|---|
| *Nautilus* | 0.30 | 0.15 | 0.09 | 0.54 |
| *Loligo* | 0 | 0.50 | 0.35 | 0.85 |
| *Sepia* | 0.22 | 0.27 | 0.30 | 0.79 |
| *Salmo* | 0.04 | 0 | 0.60 | 0.64 |
| *Carassius* | 0.04 | 0 | 0.40 | 0.44 |

The totals for endocochleates seem unacceptably high and may reflect the need for better sampling. Also, many scombroids have muscle volumes exceeding those of the carangiform species, so that the fish data reported here do not necessarily indicate peak piscine locomotor construction. In any event, the data point toward a rough equivalency among the groups in propulsion system size, which presumably represents in each case an acceptable balance between the volumetric needs of the locomotor and non-locomotor organ systems. However, the sizes of the locomotor components vary considerably, and the performance discrepancies lie in these differences.

The cause of these component size differences is anatomy. Because they need a heavy shell to resist hydrostatic pressure, ectocochleates also require extensive cameral space to provide buoyancy (table 8, column 1). Since their total locomotor volume is limited, they must consequently sacrifice the size of the other two propulsion components to achieve this goal. In contrast, endocochleates and fish, having light skeletons, need only volumetrically small buoyancy generators, or none at all. Both kinds of cephalopod require large mantle cavities to maximize their motive force by maximizing jet mass (table 8, column 2). Since they accelerate ambient water, fish need not provide internal space for this purpose, but instead can pack more of their bodies with propulsive muscle. Hence, fish are much more powerful swimmers than cephalopods because they are not confronted with competing locomotor demands for internal space. Ectocochleates must set aside significant space for flotation and propellant, endocochleates for propellant, but fish for neither, and thus ectocochleates have relatively little propulsive muscle and are weak swimmers; endocochleates are more powerful; fish are more powerful still. Anatomy dictates design, and in so doing limits performance.

### Swimming velocity and manoeuvrability

The swimming speeds of fossil cephalopods, like their propulsion, is difficult to approach directly for the reason that dead animals do not swim. Of necessity, I again focus on *Nautilus* as an analogue for its defunct antecedents, but I also estimate swimming speeds for fossil forms by considering hydrostatic stability as a limit on propulsive thrust.

#### *Speed of* Nautilus

Cinematographic analysis of captive adult specimens of *N. pompilius* indicates a speed of the order of 25 cm/s (Chamberlain & Westermann 1976). Ward *et al.* (1977) obtained similar results in tests on freshly captured adult *N. macromphalus*. These values represent maximum velocities associated with escape behaviour because the animals were considerably stressed during testing. As noted above, lower velocities are associated with food searches and other non-stress activities (Chamberlain 1987 and references therein). It appears that *Nautilus* exhibits a variety of swimming modes, perhaps equivalent to sustained, prolonged and burst swimming in fish (Webb 1975). This inference is also explicit in the conclusion of Baldwin (1987) and Wells (1987) that *Nautilus* is confined to short intervals of top speed owing to its limited resistance to fatigue. When travelling over horizontal distances of the order of kilometres, as described by Saunders & Spinosa (1979), or migrating vertically as described by Carlson *et al.* (1984) and Ward *et al.* (1984), Wells (1987) estimates that speeds only about half maximum are possible. Such interpretations conform to observations of Packard *et al.* (1980), Zann (1984), Chamberlain (1987) and Jordan, Chamberlain & Chamberlain (1988) concerning the short durations of peak activity sustained by the specimens that they studied. Arguments advanced by Chamberlain (1987) regarding size scaling of *Nautilus* locomotor organs suggest that top speed varies as the 1/6 power of body mass. Extrapolating from adult speed and mass gives maximum velocities of about 10–15 cm/s for newly hatched juveniles.

Swimming speed is considerably lower in *Nautilus* than in other cephalopods and fish of comparable body size (fig. 7 of Chamberlain &

Westermann 1976), owing to the propulsive deficiencies of ectocochleates noted previously. However, Wells (1987) has shown that *Nautilus*' cost of transport is quite low, and is well-suited to a slow-moving, scavenging way of life in environments where competition with fast-moving fish is not severe. This is, of course, precisely how *Nautilus* lives (Saunders 1984; Saunders & Ward 1987; Hayasaka *et al.* 1987). The similarity in shape and size between *Nautilus* and fossil nautiloids, at least since the Jurassic, may indicate similar lifestyles for them also.

### *Speed of fossil ectocochleates*

Estimates for swimming velocity of fossil cephalopods can be obtained by evaluating the moments and forces acting during locomotion. For an animal swimming at constant velocity under the action of a constant application of thrust, drag equals thrust, and the moments due to drag, thrust, specimen weight and buoyancy sum to zero. This is an oversimplified view of cephalopod propulsion, but I assume that no gross errors result from it. Swimming velocity can be calculated by following the scheme of Chamberlain (1981) from a knowledge of shell geometry and the rotational angle described in figure 2, under the following three assumptions: (1) the distance separating the centres of buoyancy and mass approximates the distance between the former and the centre of dynamic pressure, (2) the rotational angle of fossil forms approximates the 10° maximum of *Nautilus*, and (3) the propulsive system of fossil forms is powerful enough to produce a similar 10° rotation. Chamberlain (1981) presents arguments supporting this approach. The computations (Chamberlain 1981) suggest that the velocities of most fossil cephalopods lie between 30% and 150% of the velocity of equally sized *Nautilus*. Animals at the lower end of the range have high drag shells or little stability. Those at the top consist of flattened, disk-shaped shells. It is important to note that the fastest ectocochleate velocities estimated in this way are only about 33% of those of equally sized carangiform fish (fig. 8 of Chamberlain 1981).

### *Manoeuvrability*

The ability to manoeuvre – to turn, reverse direction and brake – is a key facet of any locomotor system. Fish manoeuvrability has been studied by several workers, including Gray (1968) and Weihs (1972), who examined turning in fast-moving carangiform fish, and Blake (1976, 1978, 1983), who produced a series of papers on median and paired fin propulsors which in part deal with manoeuvring. Although much remains to be learned, it is evident that most fish are well advanced in this capacity. Yet, as Webb (1984) points out, manoeuvrability and speed in fish are to a degree antithetical. Characters that enhance one function detract from the other. Thus, compact, deep-bodied chaetodontiform fish have finely tuned, precision manoeuvre capability, but generally are slow swimmers, while fast-moving thunniform fish are only moderately agile.

Cephalopod manoeuvrability has not been investigated quantitatively. But from numerous observations reported in the literature (see, for example, Bidder 1962; Chamberlain 1980*b*, 1981, 1987; Zuyev 1966), it is evident that they too show respectable skill in this regard. This is especially true of *Sepia*, *Loligo* and many other endocochleates which use their lateral mantle fin-folds very much as piscine median fin propulsors use their enlarged dorsal and anal fins (Zuyev 1966). Moreover, the flexible funnel can orient in all directions, thus providing an effective mechanism for turning and reversing.

Nautilus, however, is a more awkward swimmer. As with endocochleates, turning and reversing are accomplished by everting the funnel. However, *Nautilus* lacks the flexible body and fins upon which fish, and presumably soft-bodied cephalopods, rely to provide control. Moreover, the pulsatile nature of the jet means that active control of its movements by a manoeuvring *Nautilus* is discontinuous and confined to the thrust-producing phase of the propulsive cycle. Since manoeuvring involves dealing with rapidly fluctuating moments and forces, this is clearly not an efficient mechanism for precision attitude control. Endocochleates also confront this problem, but benefit from flexible body and fins. In contrast, *Nautilus* is so awkward that – in my experience with captive animals – crashes with tank walls and missed food targets are commonplace. Undoubtedly fossil ectocochleates suffered similar difficulties. They were by no means the highly manoeuvrable swimmers that their reversible jets have implied to many workers (e.g. Flower 1955, Furnish & Glenister 1964, Holland 1987). In this regard Dunbar (1924) recognized many years ago that orthocones, being so elongate, would have an even greater difficulty effecting precision control than more compact, coiled forms like *Nautilus*.

## Evolutionary implications

In the light of the foregoing, we can rate swimmers according to sophistication in locomotor design. Shelled cephalopods rank lowest in this hierarchy largely owing to limitations inherent in their heavy shell. The large phragmocone, necessary for buoyancy, increases drag and restricts equilibrium control to a slow-acting system based on hydrostatic stability. It also consumes internal space that could otherwise go to enlarging propulsive muscle and mantle cavity volume. As a result ectocochleate propulsion is weak and inefficient. In losing the external shell, soft-bodied cephalopods have lost many of these limitations. They have powerful propulsive muscles, large mantle cavities and effective equilibrium, but they also have lower propulsive efficiency and less propulsive muscle than do fish. Cephalopods are trapped by their anatomy in a paradox which illustrates well the essential problem of jet propulsion in organisms. The large internal water reservoirs necessary to jet propulsors rob such animals of space needed for propulsive muscle. Fish do not waste internal space on large buoyancy devices or water reservoirs. They pack their bodies with muscle and operate in ambient water. In addition, vertebrate locomotion benefits from several favourable physiological attributes, among which are the greater oxygen capacity of vertebrate haemoglobin compared to molluscan haemocyanin, and the greater fatigue resistance of vertebrates. For a variety of cogent reasons, therefore, piscine undulatory propulsion ranks at the apex of the locomotory hierarchy, with endocochleate cephalopods occupying the middle position.

These considerations suggest the possibility that locomotor design may have been an important determinant of evolutionary success or failure among marine nektonic organisms. The underlying operational principle is that the outcome of inter group interactions, generally viewed in terms of interspecific competition, may have hinged on factors related to locomotor capability. Such interactions have been considered from several viewpoints. For example, Thomson (1971) and Webb (1982, 1984) focussed on interaction between fish taxa, while Packard (1972), in his now-classic paper on convergence, emphasized interactions between fish and cephalopods (chiefly endocochleates). My approach here is to look for patterns of possible evolutionary interaction based on locomotor design among shelled cephalopods. My analysis is based on the stratigraphic range data compiled for marine families by Sepkoski (Sepkoski 1982; updated to July 1986).

Figure 5 is a plot of cephalopod standing familial diversity through time. The small vertical arrows identify the five greatest Phanaerozoic mass extinctions, now much discussed in connection with extraterrestrial impact scenarios (see, for examples, Sepkoski & Raup 1986; Jablonski 1986 and references therein). The pattern observed is a familiar one (see, for examples, Arkell 1957; Teichert & Moore 1964; Teichert 1967); nautiloids appear first, diversify, and then decline as ammonoids diversify. Soft-bodied cephalopods appear to diversify last, but since their record is so fragmentary, especially for the pre-Jurassic, further discussion of their history is not warranted. Two

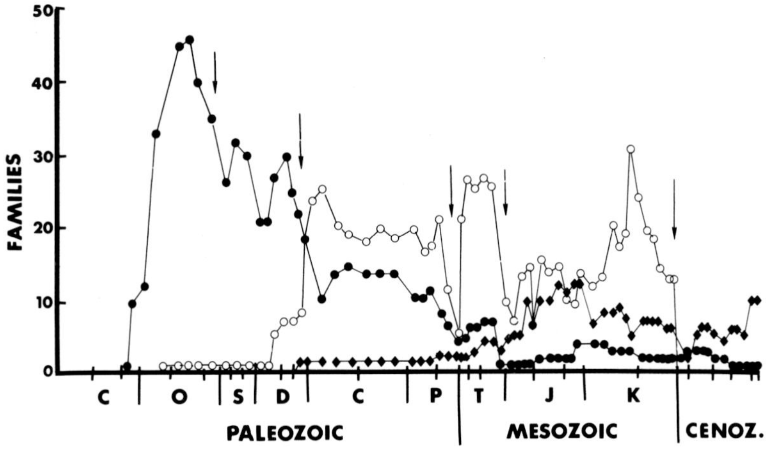


Figure 5. Cephalopod standing familial diversity through time. ●, nautiloids; ○, ammonoids; ◆, endocochleates. Each symbol represents data for one stage; symbol spacing represents stage length. Vertical arrows indicate major Phanaerozoic mass extinctions as follows, from left to right: Ashgillian, Frasnian–Fammenian, Leonardian–Dzhulfian, Rhaetian, Maastrichtian-Danian. Letters denote geological periods.

kinds of diversity decrease appear in the ectocochleate data: (1) rapid diversity crashes associated with mass extinction episodes, as in the Maastrichtian or late Permian; and (2) prolonged, progressive declines, as in the Ordovician, which are not clearly related to mass extinction. Likewise, two kinds of diversity expansions can be seen: (1) rapid radiation immediately following mass extinctions, as in the lower Triassic, and perhaps attributable to the re-establishment of ecologically vibrant groups temporarily decimated by mass extinction phenomena, and (2) prolonged expansion, presumably unrelated to mass extinction, as in the early Ordovician.

One way to seek explanations for these patterns is to evaluate them in terms of locomotor rather than taxonomic groupings. Figure 6 shows a progression in cephalopod locomotor design. The progressions are not meant to replicate phylogenetic sequences, although to some extent they do. Four ectocochleate locomotor configurations are depicted. Locomotor design improves toward the right as suggested by the arrows (which are not intended to have any phylogenetic significance). This improvement is in the form of gains in stability, manoeuvrability, speed and metabolic cost of shell construction. Table 9, which is based on the foregoing analysis of cephalopod locomotion, lists these features in more detail.

Ectocochleate history may be evaluated in terms of these locomotor groupings. This may be done by combining Sepkoski's (1982) biostrati-

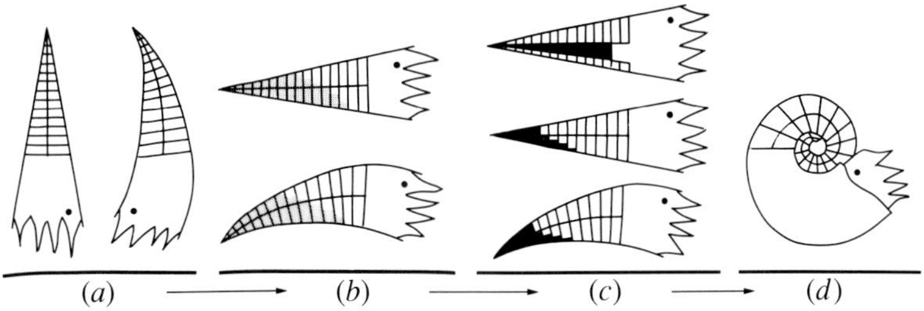


Figure 6. Primary locomotor types for shelled cephalopods. Water ballast indicated by dots, mineralized cameral and siphuncular deposits by solid colour. (*a*) Ballast-absent orthocones; (*b*) water-ballasted orthocones and cyrtocones; (*c*) mineral-ballasted orthocones and cyrtocones; (*d*) coiled forms. Locomotor improvement is from left to right in direction of arrows.

graphic data with morphological information on the taxa involved. I obtained this information from descriptions of nautiloid and ammonoid genera in parts K and L of the *Treatise on Invertebrate Paleontology* (Moore 1957, 1964). Shortcomings of this approach have been noted by Raup (1967), and I found it necessary to exclude from the biostratigraphic listings those genera which were not adequately covered in the *Treatise*. Where a family contained genera from more than one locomotor type, I included it in a specific locomotor category if at least 15% of the genera belonged to that category. Thus, a family can appear on the roster of more than one locomotor category. Also, the fragmentary state in which many Palaeozoic orthocones are preserved precludes separating ballast-absent and water-ballasted categories, so I united them into a single group for this analysis.

Ectocochleate history plotted in terms of these locomotor groupings, under the restrictions noted above, is shown in figure 7. For the most part, the historical pattern of group succession follows the locomotor scenario. Ballast-absent and water-ballasted families appear first (figure 7*a*), then decline as mineral-ballasted families diversify. Finally coiled families dominate.

In its broad features, cephalopod evolution as known from the fossil record conforms to a pattern of improved locomotor design. Better designed groups appear to displace more poorly designed groups, or at least to profit taxonomically from the decline of the latter. Mass extinctions seem to have little effect on this relationship but only temporarily deflect the established trends: a good example is seen in the Triassic. Yet, ectocochleate evolution is more complex than this. Within individual locomotor groups, secondary patterns can be observed. Among coiled forms, for example, families with ornamented shells increase in relative abundance, especially in the Mesozoic (figure 7*b*). Families with smooth, discoidal shells (figure 7*b*) become significant components of total standing diversity from the mid Palaeozoic onwards. Ward (1981, 1983) has interpreted such patterns as a response to intensified predation brought about by the expansion of durophagous predators in the early and mid Mesozoic. This may well be the case, but following my earlier argument about the possible ecological equivalence of deep-bodied (chaetodontiform) fish and coiled cephalopods (Chamberlain 1981), these changes may also reflect increasingly widespread cephalopod accommodation to life in cluttered, topographically complex benthic habitats dominated by high level suspension feeders (e.g. crinoids, corals and bryozoans) characteristic of late Palaeozoic and Mesozoic communities.

Fish can also be categorized with respect to locomotion: Webb (1982, 1984) recognized two broad locomotor categories, specialists and generalists. Specialists have characteristics leading to efficient sustained swimming as in thunniform and carcharhiniform fish, to high manoeuvrability as in chaetodontiform types and to rapid acceleration as in esociform fish. Carangiform generalists develop features promoting reasonably good performance levels in all three areas. These locomotor types are distributed preferentially among the major fish taxa. Ostracoderms, placoderms and acanthodians, chondrosteans and chondrichthians all show an abundance of

Table 9. *Characteristics of ectocochleate locomotor types*
Ballast-absent (BA), water-ballasted (WB) and mineral-ballasted (MB) categories include both orthocones and cyrtocones. Descriptive terms used here should be understood as applying to ectocochleates only. Thus mineral ballasted orthocone drag is low only in the context of ectocochleates; compared to fish, it is actually quite high (table 2).

| Locomotor type | Orientation | Stability | Drag |
|---|---|---|---|
| Ballast-absent | Vertical | High | High in horizontal direction |
| Water-ballasted | Horizontal | Low | Low |
| Mineral-ballasted | Horizontal | Low, but higher than WB in asymmetrically ballasted forms | Low |
| Planispirally coiled | — | Low to high | Low to high |

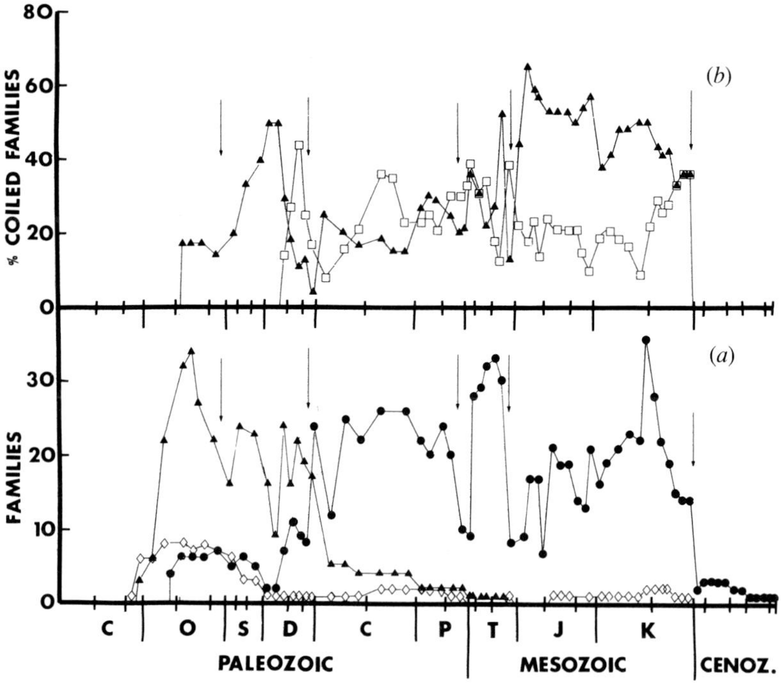


Figure 7. Ectocochleate standing familial diversity through time. (*a*) Diversity of locomotor categories. ◇, Ballast-absent and water-ballasted orthocones and cyrtocones; ▲, mineral-ballasted orthocones and cyrtocones; ●, planispirally coiled forms. (*b*) Diversity of coiled ornate families (▲) and smooth, discoidal forms (□) expressed as a percentage of total coiled families. Ornate forms are those with medium to coarse shell ornament (e.g. ribs, spines, tubercules) as described by Ward (1981, 1983). Discoidal forms are those in which aperture width is less than one third of shell diameter. Vertical arrows indicate mass extinctions as in figure 5. Each symbol represents data for one stage. Symbol staging reflects stage length. Letters as for figure 5.

Table 9. (*cont.*)

| Speed | Manoeuvrability | Metabolic cost of shell |
|---|---|---|
| Near zero, owing to upright shell | Low | Low |
| Low, owing to stability and control problems | Poor, owing to shell elongation | Low but higher than BA because of relatively larger shell |
| Same as WB | Poor, but better than WB | High, owing to expense of secreting ballast |
| High in low drag forms | Good | Low |

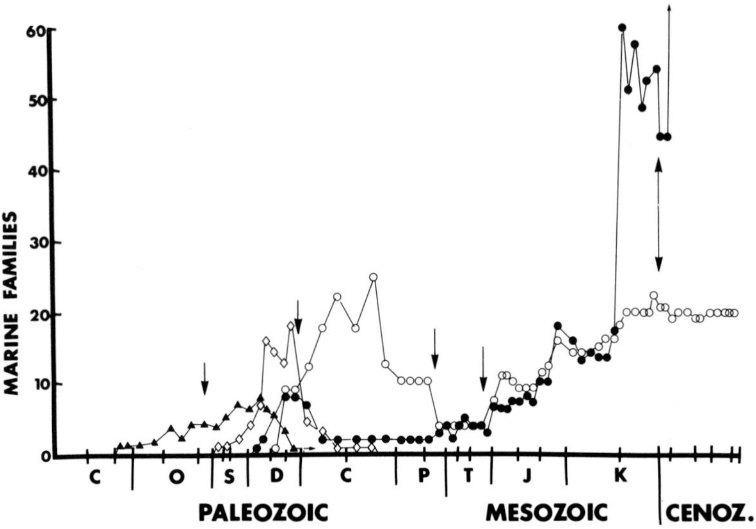


Figure 8. Standing familial diversity through time for major fish taxa. Data for marine families only. ●, Osteichthians; ○, chondrichthians; ◇, placoderms and acanthodians; ▲, ostracoderms. Continuation of ostracoderm plot from upper Devonian to Recent to account for modern agnaths omitted for clarity. Expansion of osteichthians to 186 families in post-Palaeocene omitted for clarity. Vertical arrows indicate mass extinctions as in figure 5. Each symbol represents data for one stage. Symbol staging reflects stage length. Letters as for figure 5.

carcharhiniform types. The other specialist groups and carangiform generalists predominate among the holosteans, and especially the teleosts.

Figure 8 shows standing familial diversity among fish compiled from Sepkoski's compendium (Sepkoski 1982, and 1986 revision). The data are for marine families only and exclude fish that would have no direct interaction with cephalopods. This plot emphasizes the well established succession of these taxa in time (see, for example, Romer 1966), which is generally explained as a consequence of improvement in locomotor (mainly carcharhiniform) design (see, for example, McFarland *et al.* 1979; Carroll 1988) as, for example, in tail shape (Thomson 1971, 1976), or loss of body armour (Webb 1982), and of improvement in feeding anatomy by the development of hinged jaws and teeth (Romer 1966; McFarland *et al.* 1979, Carroll 1988). Not until the holostean, and especially teleost, expansions in the Jurassic and Cretaceous, respectively, do other specialist character complexes and carangiform generalist designs become abundant in the marine fauna. As in cephalopods, mass extinctions do not appear to cause significant, long-term departures from established trends.

As for cephalopods, fish diversity changes can be analysed in terms of

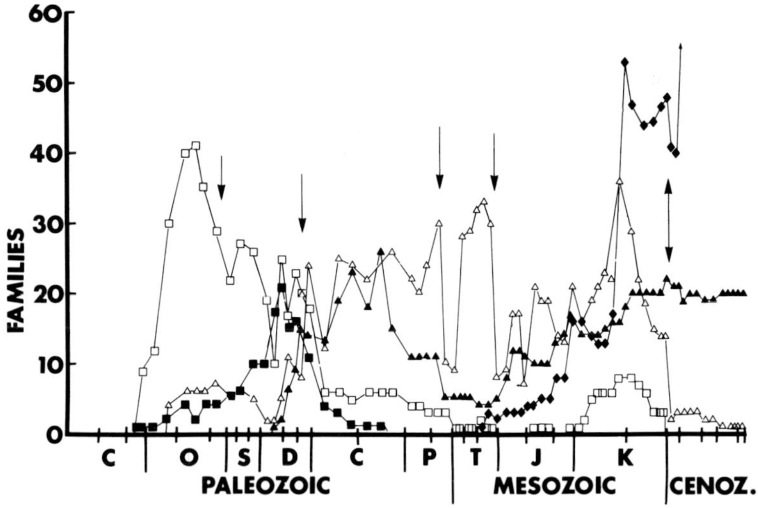


Figure 9. Standing familial diversity of cephalopod and fish locomotor categories. Fish data for marine families only. △, Planispirally coiled 'advanced design' ectocochleates; curve from figure 7*a*. □, Non-coiled 'primitive design' ectocochleates including the various orthocone and cyrtocone morphologies from figure 7*a* plus heteromorphic scaphitid (hook-shaped) and turrelitid (helical) ammonoids. These latter two groups, together with the baculitids, account for the rise in mid Cretaceous diversity noted for non-coiled families. ■, 'primitive design' fish; ▲, 'intermediate design' fish; ◆, 'advanced design' fish. See text for definition of fish locomotor categories. Vertical arrows indicate mass extinctions as in figure 5. Each symbol represents data for one stage. Symbol staging reflects stage length. Letters as for figure 5.

locomotor design. Such a stance also allows us to evaluate interactions between the two groups. Figure 9 shows familial diversity for fish and cephalopod locomotor groups plotted together on the same time axis. In making this plot, I have grouped ectocochleates into two categories: (1) 'primitive design' locomotors (non-ballasted and ballasted orthocones and cyrtocones), and (2) 'advanced design' locomotors (coiled forms). I have also restructured Webb's fish locomotor categories to produce a smaller number of more comprehensive categories: (1) 'primitive design' fish (ostracoderms, placoderms, acanthodians), which are characterized by carcharhiniform body shape, heterocercal tails, extensive body armour, spine-supported fins (which are often not paired), no known buoyancy apparatus and generally small body size; (2) 'intermediate design' fish (chondrichthians and sarcopterygians), which are characterized by carcharhiniform body shape, heterocercal tails, reduced or absent body armour (but with scutes or thickened dermal scales), flexible paired, but not rayed, fins, buoyancy generated by hydroplaning or low density materials, and generally medium to large body size; and (3) 'advanced design' fish (teleosts, especially acanthopterygians), which are characterized by carangiform and non-carcharhiniform specialist body shapes, homocercal tails, thin or no dermal scales, flexible paired and rayed fins often shifted forward on the body, buoyant gas bladder, and small to large body size.

My approach in setting up these fish categories has been far from perfect. For example, by including entire classes of fish within a single category I inevitably include some families that do not qualify for inclusion on the basis of their attributes. In addition, I have neglected entire major taxa, notably the chondrosteans and holosteans. It is important to note, however, that non-conforming families comprise only a small fraction of the total for the higher taxa forming each locomotor category, and that at the family level the omitted taxa were never very diverse in the marine realm. Thus, my simplifications should not materially influence the results. Nevertheless, it is best to consider the fish diversity trends plotted in figure 9 as being less quantitatively founded than the cephalopod curves.

Although evidence of cephalopod-fish interactions are not obvious in figure 9, the data presented in this plot suggest several interesting possibilities. One striking feature appears in the data for the upper Devonian and lower Carboniferous. This was a period during which coiled ectocochleates replaced elongate forms and when chondrichthians become the dominant type of marine fish (figures 7, 8). Although these events may be unrelated, it is worth noting that, because of their hard shell, successful predation of mobile ectocochleates requires a predator with a strong prey grasping and shell-crushing mechanism. Jawless ostracoderms do not have the requisite anatomy to be effective cephalopod predators, and toothless placoderms would not have been much better. Sharks, however, have both teeth and jaws, and it is perhaps no coincidence that shark-bitten cephalopod shells begin to appear at about this time (Mapes & Hansen 1984). Thus, expanding chondrichthian predation beginning in the upper Devonian may have been a factor in reducing the diversity of the more unwieldy, less effective orthocone propulsors, while the more manoeuvrable and compact coiled forms could better relieve this pressure by seeking protection in the crevices and understorey of complex benthic habitats. Thus the greater manoeuvrability of coiled cephalopods may better account for their eventual displacement of orthoconic forms than difference in swimming speed (table 9), since speed would clearly be far lower in all shelled cephalopods than in their chondrichthian predators.

The trends of the Mesozoic are also of interest. Beginning in the Triassic the diversity of swimmers increases markedly. This expansion continues more or less unabatedly to the mid Cretaceous, and involves 'advanced locomotor' ectocochleates as well as 'intermediate locomotor' and 'advanced locomotor' fish. Even 'primitive locomotor' cephalopods, which had not been numerically significant members of the marine fauna since the Carboniferous, make an impressive, although temporary, resurgence in the lower Cretaceous. The parallelism evident in the diversity increases among these groups may imply a single forcing agent, one probably external to any competitive relation among members of the two groups. Perhaps the most obvious candidate for this forcing agent is the Jurassic-Cretaceous expansion of submerged continental shelves and shallow epeiric seas. By increasing the area of shallow marine habitat, these plate tectonic phenomena may have provided an expanded ecological framework in which this diversification could occur. Although widespread shallow seas continue into the upper Cretaceous, ectocochleate diversity enters a long, profound decline which eventually terminates with their virtual extinction in the Maastrichtian. This process is coupled with a dramatic expansion of 'advanced locomotor' fish. The correlation of these two events has long been noted (Packard 1972; Wiedmann 1973), and the interpretation of these trends in terms of the deleterious competitive effect on ectocochleates of the superiority of teleost locomotory design receives added strength from this analysis (specifically table 8). In this regard, it is especially interesting that deep-bodied, chaetodontiform fish, which may be the chief competitors of coiled ectocochleates (Chamberlain 1981), did not become numerically significant in marine ecosystems until the mid Cretaceous diversification of improved suction feeding acanthopterygian teleosts in the Jurassic–Cretaceous (Lauder 1982). Piscine feeding adaptations may thus have had an influential role in this process as well.

Competitive interactions are much easier to identify than to validate for fossil organisms. All of the relevant points I have raised obviously need further analysis: a treatment at the genus level would undoubtedly prove to be helpful. Nevertheless, certain broad patterns emerge. The evolutionary history of fish and ectocochleates follows a similar pattern of replacement on the basis, at least in part, of locomotor design. The impetus for this situation may largely have come from within: that is, cephalopods interacting with cephalopods, fish with fish. The influence of one group on the other is difficult to establish at the family level of analysis, but may occur at several points, particularly in the upper Devonian–Carboniferous and the mid to upper Cretaceous. In both cases, improved piscine feeding mechanisms may have been at least as significant as disparities in locomotor potential between cephalopods and fish. Thus, Packard's (1972) idea of a broad fish–cephalopod relation based on many different areas of competition, and involving diverse aspects of anatomy, may be right.

**Acknowledgements**

I am grateful to Jeremy Rayner of the University of Bristol for inviting me to participate in this symposium and for his forbearance and support in seeing my contribution through to its conclusion. I also thank Jack Sepkoski (University of Chicago) for providing the 1986 update to his fossil compendium and Martin Wells (University of Cambridge) for his advice on matters relating to cephalopod locomotion. I appreciate the help of R.B. Chamberlain in all phases of the work. P. Kennedy and A. Tomo typed the manuscript. This research was supported by grants from the National Science Foundation (PCM-8019064) and the PSC-CUNY Research Award Program of the City University of New York, and by travel grants from the Society of Experimental Biology, Brooklyn College, and the Graduate School of CUNY.

## References

Alexander, R.McN. 1972 The energetics of vertical migration of fishes. *Symposia of the Society for Experimental Biology* **26**, 273–294.

Alexander, R.McN. 1977 Swimming. In *Mechanics and Energetics of Animal Locomotion*, ed. R.McN. Alexander & G. Goldspink, pp. 222–248. London: Chapman & Hall.

Arkell, W.J. 1957 Introduction to Mesozoic Ammonoidea. In *Treatise on Invertebrate Paleontology*, ed. R.C. Moore, part L, *Mollusca*, vol 4, pp. 81–129. Lawrence: University of Kansas Press.

Bainbridge, R. 1958 The speed of swimming fish as related to size and to the frequency and amplitude of the tail beat. *Journal of Experimental Biology* **35**, 109–133.

Bainbridge, R. 1960 Speed and stamina in three fish. *Journal of Experimental Biology* **37**, 129–153.

Baldwin, J. 1987 Energy metabolism of *Nautilus* swimming muscles. In *Nautilus: the Biology and Paleobiology of a Living Fossil*, ed. W.B. Saunders & N.H. Landman, pp. 325–329. New York: Plenum Press.

Bidder, A.M. 1962 Use of tentacles, swimming and buoyancy control in the pearly *Nautilus*. *Nature, London* **196**, 451–454.

Blake, R.W. 1976 On seahorse locomotion. *Journal of the Marine Biological Association of the United Kingdom* **56**, 939–949.

Blake, R.W. 1978 On ballistiform locomotion. *Journal of the Marine Biological Association of the United Kingdom* **58**, 73–80.

Blake, R.W. 1983 Swimming in electric eels and knifefishes. *Canadian Journal of Zoology* **61**, 1432–1441.

Carlson, B.A., McKibben, J.N. & DeGruy, M.V. 1984 Telemetric investigation of vertical migration of *Nautilus belauensis*. *Pacific Science* **38**, 183–188.

Carroll, R.L. 1988 *Vertebrate Paleontology and Evolution*. New York: W.H. Freeman.

Chamberlain, J.A., Jr 1969 Technique for scale modelling cephalopod shells. *Palaeontology* **12**, 48–55.

Chamberlain, J.A., Jr 1976 Flow patterns and drag coefficients of cephalopod shells. *Palaeontology* **19**, 539–563.

Chamberlain, J.A., Jr 1980*a* Motor performance and jet propulsion in *Nautilus*: implications for cephalopod paleobiology and evolution. *American Malacological Union Incorporated Bulletin*, pp. 123–127.

Chamberlain, J.A., Jr 1980*b* The role of body extension in cephalopod locomotion. *Palaeontology* **23**, 445–461.

Chamberlain, J.A., Jr 1981 Hydromechanical design of fossil cephalopods. In *The Ammonoidea*, ed. M.R. House & J.R. Senior. *Systematics Association Special Volume* **18**, pp. 289–336. London: Academic Press.

Chamberlain, J.A., Jr 1987 Locomotion of *Nautilus*. In *Nautilus: the Biology and Paleobiology of a Living Fossil*, ed. W.B. Saunders & N.H. Landman, pp. 489–525. New York: Plenum Press.

Chamberlain, J.A., Jr & Chamberlain, R.B. 1985 Septal fracture in *Nautilus*: implications for cephalopod paleobathymetry. *Lethaia* **18**, 261–270.

Chamberlain, J.A., Jr & Chamberlain, R.B. 1986 Is cephalopod septal strength index an index of cephalopod septal strength? *Alcheringa* **10**, 85–97.

Chamberlain, J.A., Jr & Moore, W.A. 1982 Rupture strength and flow rate of *Nautilus* siphuncular tube. *Paleobiology* **8**, 408–425.

Chamberlain, J.A., Jr & Weaver, J.S. 1978 Equations of motion for post-mortem sinking of cephalopod shells. *Mathematical Geology* **10**, 673–689.

Chamberlain, J.A., Jr & Westermann, G.E.G. 1976 Hydrodynamic properties of cephalopod shell ornament. *Paleobiology* **2**, 316–331.

Collins, D.H., Ward, P.D. & Westermann, G.E.G. 1980 Function of cameral water in *Nautilus*. *Paleobiology* **6**, 168–172.

Crick, R.E. & Teichert, C. 1979 Siphuncular structures in the Devonian nautiloid *Archiacoceras* from the Eifel of West Germany. *Palaeontology* **22**, 747–766.

Daniel, T.J. 1984 Unsteady aspects of aquatic locomotion. *American Zoologist* **24**, 121–134.

Denton, E.J. & Gilpin-Brown, J.B. 1961 The buoyancy of the cuttlefish, *Sepia officinalis* (L.). *Journal of the Marine Biological Association of the United Kingdom* **41**, 319–342.

Denton, E.J. & Gilpin-Brown, J.B. 1966 On the buoyancy of the pearly nautilus. *Journal of the Marine Biological Association of the United Kingdom* **46**, 723–759.

Dunbar, C.O. 1924 Phases of cephalopod adaptation. In *Organic Adaptation to Environment*, ed. J. Thorpe, pp. 187–223. New Haven: Yale University Press.

Flower, R.H. 1955 Trails and tentacular impressions of orthoconic cephalopods. *Journal of Paleontology* **29**, 857–867.

Flower, R.H. 1957 Nautiloids of the Paleozoic. *Geological Society of America Memoir* **67** (2), 829–852.

Furnish, W.M. & Glenister, B. 1964 Paleoecology. In *Treatise on Invertebrate Paleontology*, Part K, *Mollusca*, vol 3, ed. R.C. Moore, pp. 114–124. Lawrence: University of Kansas Press.

Gee, J.H. 1983 Ecologic implications of buoyancy control in fish. In *Fish Biomechanics*, ed. P.W. Webb & D. Weihs, pp. 140–176. New York: Praeger.

Gray, J. 1968 *Animal Locomotion*. London: Weidenfeld & Nicolson.

Greenwald, L., Cook, C.B. & Ward, P.D. 1982 The structure of the chambered nautilus siphuncle: the siphuncular epithelium. *Journal of Morphology* **175**, 5–22.

Greenwald, L., Ward, P.D. & Greenwald, O. 1980 Cameral lipid transport and buoyancy control in the chambered nautilus (*Nautilus macromphalus*). *Veliger* **28**, 356–368.

Hayasaka, S., Oki, K., Tanabe, K., Saisho, T. & Shinomiya, A. 1987 On the habitat of *Nautilus pompilus* in Tanon Strait (Philippines) and the Fiji Islands. In *Nautilus: the Biology and Paleobiology of a Living Fossil*, ed. W.B. Saunders & N.H. Landman, pp. 179–200. New York: Plenum Press.

Hewitt, R.A. 1984 Growth analysis of Silurian orthoconic nautiloids. *Palaeontology* **27**, 671–677.

Hewitt, R.A. & Westermann, G.E.G. 1986 Function of complexly fluted septa in ammonoid shells. I. Mechanical principles and functional models. *Neues Jahrbuch für Geologie und Paläontologie Abhandlungen* **172**, 47–69.

Hoerner, S.F. 1965 *Fluid Dynamic Drag*. Midland Park, N.J.: S.F. Hoerner (author).

Holland, C.H. 1987 The nautiloid cephalopods; a strange success. *Journal of the Geological Society* (*London*) **144**, 1–15.

Jablonski, D. 1986 Causes and consequences of mass extinctions; a comparative approach. In *Dynamics of Extinction*, ed. D.K. Elliot, pp. 183–200. New York: John Wiley.

Jordan, M., Chamberlain, J.A., Jr & Chamberlain, R.B. 1988 Response of *Nautilus* to variation in ambient pressure. *Journal of Experimental Biology* **137**, 175–189.

Jordan, R. 1968 Zur Anatomie mesozoischer Ammoniten nach den Struktur-elementen der Gehause-Innenwand. *Geologisches Jahrbuch Beihefte* **77**, 1–64.

Kanie, Y., Fukuda, Y., Nakayama, H., Seki, K. & Hattori, M. 1980 Implosion of living *Nautilus* under increased pressure. *Paleobiology* **6**, 44–47.

Kanie, Y. & Hattori, M. 1983 Shell implosion depth of living *Nautilus*.

*Kagoshima University Research Centre for the South Pacific Occasional Papers* **1**, 30–35.

Kennedy, W.J. & Cobban, W.A. 1976 Aspects of ammonite biology, biogeography, and biostratigraphy. *Special Papers in Palaeontology* **17**, 1–94.

Kummel, B. & Lloyd, R.M. 1955 Experiments on relative streamlining of coiled cephalopod shells. *Journal of Paleontology* **29**, 159–170.

Lauder, G.V. 1982 Patterns of evolution in the feeding mechanism of actinopterygian fishes. *American Zoologist* **22**, 329–342.

Lehmann, U. 1971 New aspects in ammonite biology. In *Proceedings of the North American Paleontological Convention*, pp. 1251–1269.

Mapes, R.H. & Hansen, M.C. 1984 Pennsylvanian shark-cephalopod predation: a case study. *Lethaia* **17**, 175–183.

McFarland, W.M., Pough, F.H., Cade, T.J. & Heiser, J.B. 1979 *Vertebrate Life*. New York: Macmillan.

Moore, R.C. (ed.) 1957 *Treatise on Invertebrate Paleontology*, part L, *Mollusca*, vol. 4. Lawrence: University of Kansas Press.

Moore, R.C. (ed.) 1964 *Treatise on Invertebrate Paleontology*, part K, *Mollusca*, vol 3. Lawrence: University of Kansas Press.

Mutvei, H. 1957 On the relations of the principle muscles to the shell in *Nautilus* and some fossil nautiloids. *Arkiv för Mineralogi och Geologi* **2**, 219–254.

Mutvei, H. 1964 Remarks on the anatomy of recent and fossil Cephalopoda. *Stockholm Contributions in Geology* **11**, 79–102.

Mutvei, H. 1967 On the microscopic shell structure in some Jurassic ammonoids. *Neues Jahrbuch für Geologie und Paläontologie Abhandlungen* **129**, 157–166.

Mutvei, H. & Reyment, R.A. 1973 Buoyancy control and siphuncle function in ammonoids. *Palaeontology* **16**, 623–636.

O'Dor, R.K. 1982 Respiratory metabolism and swimming performance of the squid, *Loligo opalescens*. *Canadian Journal of Fisheries and Aquatic Sciences* **39**, 580–587.

Packard, A. 1972 Cephalopods and fish: the limits of convergence. *Biological Reviews* **47**, 241–307.

Packard, A., Bone, Q. & Hignette, M. 1980 Breathing and swimming movements in a captive *Nautilus*. *Journal of the Marine Biological Association of the United Kingdom* **69**, 313–327.

Raup, D.M. 1967 Geometric analysis of shell coiling: coiling in ammonoids. *Journal of Paleontology* **41**, 43–65.

Raup, D.M. & Chamberlain, J.A., Jr 1967 Equations for volume and center of gravity in ammonoid shells. *Journal of Paleontology* **41**, 566–574.

Romer, A.S. 1966 *Vertebrate Paleontology*. University of Chicago Press.

Saunders, W.B. 1984 The role and status of *Nautilus* in its natural habitat: evidence from deep-water remote camera sequences. *Paleobiology* **10**, 469–486.

Saunders, W.B. & Shapiro, E.A. 1986 Calculation and simulation of ammonoid hydrostatics. *Paleobiology* **12**, 64–79.

Saunders, W.B. & Spinosa, C. 1979 *Nautilus* movement and distribution in Palau, Western Caroline Islands. *Science* **204**, 1199–1201.

Saunders, W.B. & Ward, P.D. 1987 Ecology, distribution and population characteristics of *Nautilus*. In *Nautilus: the Biology and Paleobiology of a Living Fossil*, ed. W.B. Saunders & N.H. Landman, pp. 137–162. New York: Plenum Press.

Saunders, W.B. & Wehman, D.A. 1977 Shell strength of *Nautilus* as a depth limiting factor. *Paleobiology* **3**, 83–89.

Schmidt, H. 1930 Über die Bewegungsweise der Schalencephalopoden. *Zeitschrfit für Paläontologie* **12**, 194–207.

Sepkoski, J.J. 1982 A compendium of fossil marine families. *Milwaukee Public Museum Contributions in Biology and Geology* **51**, 1–125.

Sepkoski, J.J. & Raup, D.M. 1986 Periodicity in marine extinction events. In *Dynamics of Extinction*, ed. D.K. Elliot, pp. 3–36. New York: John Wiley.

Sturmer, W. 1970 Soft parts of cephalopods and trilobites: some surprising results of X-ray examinations of Devonian slates. *Science* **170**, 1300–1302.

Taylor, G.I. 1952 Analysis of swimming of long, narrow animals. *Proceedings of the Royal Society of London* A **214**, 158–183.

Teichert, C. 1933 Der Bau der actineroiden Cephalopoden. *Paläontographica* A **78**, 111–230.

Teichert, C. 1967 Major features of cephalopod evolution. In *Essays in Paleontology and Stratigraphy*, ed. C. Teichert & E. Yochelson, pp. 162–210. Lawrence: University of Kansas Press.

Teichert, C. & Moore, R.C. 1964 Classification and stratigraphic distribution. In *Treatise on Invertebrate Paleontology*, part K, *Mollusca*, vol. 3, ed. R.C. Moore, pp. 94–106. Lawrence: University of Kansas Press.

Thomson, K.S. 1971 The adaptation and evolution of early fishes. *Quarterly Review of Biology* **46**, 139–166.

Thomson, K.S. 1976 On the heterocercal tail in sharks. *Paleobiology* **2**, 19–38.

Trueman, A.E. 1941 The ammonite body-chamber, with special reference to the buoyancy and mode of life of the living ammonite. *Journal of the Geological Society* (*London*) **96**, 339–383.

Trueman, E.R. & Packard, A. 1968 Motor performances of some cephalopods. *Journal of Experimental Biology* **49**, 495–508.

Vogel, S. 1981 *Life in Moving Fluids*. Boston: Willard Grant Press.

Ward, P.D. 1981 Shell sculpture as a defensive adaptation in ammonoids. *Paleobiology* **7**, 96–100.

Ward, P.D. 1983 The extinction of the ammonites. *Scientific American* **249**, 136–147.

Ward, P.D. 1985 Periodicity of chamber formation in chambered cephalopods: evidence from *Nautilus macromphalus* and *Nautilus pompilius*. *Paleobiology* **11**, 438–450.

Ward, P.D. 1986 Rates and processes of compensatory buoyancy changes in *Nautilus macromphalus*. *Veliger* **28**, 356–368.

Ward, P.D., Carlson, B., Weekly, M. & Blumbaugh, B. 1984 Remote telemetry of daily vertical and horizontal movement of *Nautilus* in Palau. *Nature, London* **309**, 248–250.

Ward, P.D. & Chamberlain, J.A., Jr 1983 Radiographic observation of chamber formation in *Nautilus pompilius. Nature, London* **304**, 57–59.

Ward, P.D., Greenwald, L. & Magnier, Y. 1981 The chamber formation cycle in *Nautilus macromphalus. Paleobiology* **7**, 481–493.

Ward, P.D., Greenwald, L. & Rougerie, F. 1980 Shell implosion depth for living *Nautilus macromphalus* and shell strength of extinct cephalopods. *Lethaia* **13**, 182.

Ward, P.D. & Martin, A. 1978 On the buoyancy of the pearly nautilus. *Journal of Experimental Zoology* **205**, 5–12.

Ward, P.D., Stone, R., Westermann, G.E.G. & Martin, A. 1977 Notes on animal weight, cameral fluids, swimming speed, and color polymorphism of the cephalopod *Nautilus pompilius* in the Fiji Islands. *Paleobiology* **3**, 377–388.

Wardle, C.S. 1975 Limits of swimming sped. *Nature, London* **255**, 725–727.

Weaver, J. S. & Chamberlain, J.A., Jr 1976 Equations of motion for post-mortem sinking of cephalopod shells and the sinking of *Nautilus. Paleobiology* **2**, 8–18.

Webb, P.W. 1971 The swimming energetics of trout. I. Thrust and power output at cruising speeds. *Journal of Experimental Biology* **55**, 489–520.

Webb, P.W. 1975 Hydrodynamics and energetics of fish propulsion. *Bulletin of the Fisheries Research Board of Canada* **190**, 1–158.

Webb, P.W. 1982 Locomotor patterns in the evolution of actinopterygian fishes. *American Zoologist* **22**, 329–342.

Webb, P.W. 1984 Body form, locomotion and foraging in aquatic vertebrates. *American Zoologist* **24**, 107–120.

Webb, P.W. 1986 Kinematics of lake sturgeon, *Acipenser fulvescens*, at cruising speeds. *Canadian Journal of Zoology* **64**, 2137–2141.

Weihs, D. 1972 A hydromechanical analysis of fish turning manoeuvres. *Proceedings of the Royal Society of London* B **182**, 59–72.

Weihs, D. 1973 Mechanically efficient swimming techniques for fish with negative buoyancy. *Journal of Marine Research* **31**, 194–209.

Weihs, D. & Webb, P.W. 1983 Optimization of locomotion. In *Fish Biomechanics*, ed. P.W. Webb & D. Weihs, pp. 339–371. New York: Praeger.

Wells, M.J. 1987 Ventilation and oxygen extraction by *Nautilus*. In *Nautilus: the Biology and Paleobiology of a Living Fossil*, ed. W.B. Saunders & N.H. Landman, pp. 339–350. New York: Plenum Press.

Westermann, G.E.G. 1973 Strength of concave septa and depth limits of fossil cephalopods. *Lethaia* **6**, 383–403.

Westermann, G.E.G. 1977 Form and function of orthoconic cephalopod shells with concave septa. *Paleobiology* **3**, 300–321.

Westermann, G.E.G. 1982 The connecting rings of *Nautilus* and Mesozoic ammonoids: implications for ammonoid bathymetry. *Lethaia* **15**, 373–384.

Wiedmann, J. 1973 Evolution or revolution of ammonoids at Mesozoic system boundaries. *Biological Reviews* **48**, 159–194.

Yates, G.T. 1983 Hydromechanics of body and caudal fin propulsion. In *Fish Biomechanics*, ed. P.W. Webb & D. Weihs, pp. 177–213. New York: Praeger.

Zann, L.P. 1984 The rhythmic activity of *Nautilus pompilius*, with notes on its ecology and behavior in Fiji. *Veliger* **27**, 19–28.

Zuyev, G.V. 1966 Characteristic features of the structure of cephalopod molluscs associated with controlled movements. *Kiev Special Publications*, pp. 14–41; translated *Fisheries Research Board of Canada Translation Series* **1011**.

ROBIN J. WOOTTON & CHARLES P. ELLINGTON

# Biomechanics and the origin of insect flight

## Introduction

There is no insect *Archaeopteryx*. When insect fossils first appear, in the Middle and Upper Carboniferous, they are diverse and for the most part fully winged. There are a few primitively wingless forms, but no convincing intermediates are known. Reconstructing the 'protopterygotes' – the immediate ancestors of winged insects – therefore relies, as it always has, on indirect evidence, drawn mainly from Palaeozoic fossils and from primitive extant forms.

Remarkably, two nineteenth century hypotheses are still in contention. The first, due to Müller (1875), and many times restated and developed (see, for examples, Forbes 1943; Hinton 1963; Rasnitsyn 1981), proposes that protopterygotes bore a segmental series of paired, initially immobile dorsolateral lobes; two pairs of these lobes became enlarged, developed movable articulations, and evolved into flapping aerofoils, while the remainder became reduced and ultimately lost. The second, due originally to Oken (1811), and revived and developed by Wigglesworth (1973, 1976) and Kukalová-Peck (1978, 1983), has wings developing by enlargement of the meso– and metathoracic pairs of a series of already articulated and movable 'winglets'. These winglets would be homologous with the gills of mayfly nymphs (Ephemeroptera; figure 1*a*) and with exites of archaic basal leg segments, since incorporated into the body wall.

These topics have been extensively discussed within the last few years – see reviews by Quartau (1985, 1986) and Wootton (1986) – and we will not cover the ground again. Our purpose is to examine some biomechanical implications of the design of plausible protopterygotes in deciding the pathway by which insect flight evolved.

This subject too has been discussed for decades. The many published hypotheses fall into three broad groups. Two of these are very similar to

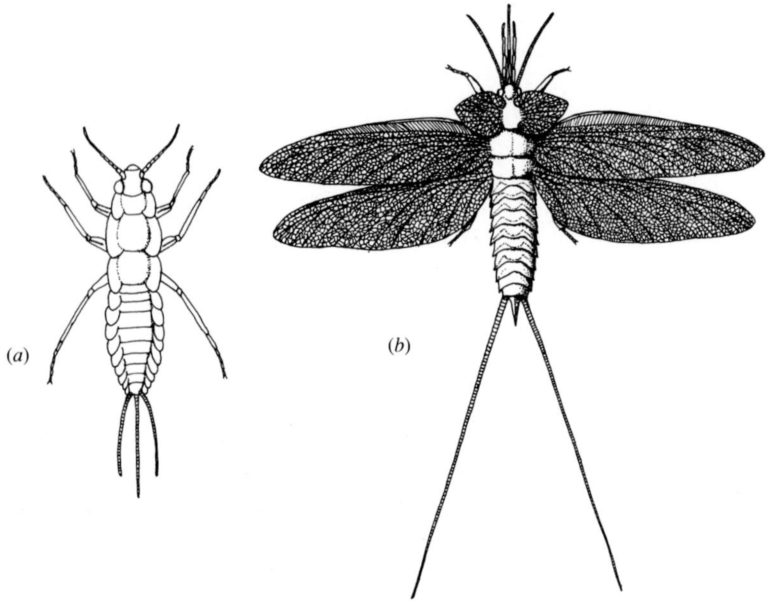


Figure 1. Two relevant Upper Carboniferous insects. (*a*) *Bojophlebia prokopi* Kukalová-Peck (Order Ephemeroptera: a gigantic young mayfly nymph (length 102 mm). Reconstruction, after Kukalová-Peck (1985). Notice the tails (paired cerci and median caudal filament), and the abdominal gills; all are characteristic of the Order. (*b*) *Stenodictya* (Order Palaeodictyoptera): reconstruction based on several species, after Kukalová (1970). Body, minus cerci and mouthparts, around 60–70 mm long. Notice the long cerci, the prothoracic winglets and the specialized piercing mouthparts, an apomorphic character.

those proposed for the evolution of vertebrate flight (see, for example, Rayner, this volume).

(1) *The Paraglider theories*. Protopterygotes, living in tall vegetation, used their lateral lobes for stable parachuting, and later gliding, to the ground or to nearby plants. Improved gliding and powered flight followed, with the enlargement of appropriate winglets, the loss of others, and the development of a basal articulation and an effective flapping mechanism. The precise sequence of events differs from version to version (Forbes 1943; Hinton 1963; Flower 1964; Rasnitsyn 1981).

(2) *The Floating theories*. Protopterygotes were small insects whose lateral winglets, flat or feathery, assisted their passive transport by thermals and winds (Wigglesworth 1963, 1976; Norberg 1972).

(3) *The Running Jump/Flying Leap theories.* Ground-dwelling insects took to gliding or actively flying (Kukalová-Peck 1987) from a running or a jumping start.

The three families of theory were all current when a belief in fixed lateral lobes was in fashion, but each can readily be adapted to take account of the primitively movable winglets for which Kukalová-Peck has now assembled a good deal of evidence. These also raise the question of whether flapping may have contributed to propulsion from an early stage.

In all this discussion few authors have seriously addressed the mechanical aspects of the problem. In 1964 J. W. Flower, an aeronautical engineer, published what is still the most detailed aerodynamic study on the origin of insect flight. He likened insect bodies to circular cylinders, and noted that glide angles of 45° or more could be obtained from such cylinders if they could be held in a suitable attitude. He reasoned that winglets were initially too small significantly to improve on this glide performance, and hence that the original function of the winglets would have been for attitude control, allowing the insect to realize the best glide from its relatively much larger body area.

Such glide angles are not very impressive – indeed they fall within some of the broader definitions of parachuting – but they would allow substantial horizontal movement by a falling insect. The next problem, left unconsidered by Flower, is whether significant reductions in the glide angle could then be achieved by enlarging the area of the winglets. Kingsolver & Koehl (1985) recently investigated this experimentally, by measuring in a wind tunnel the maximum lift-to-drag ratios of model insects with thoracic winglets of different sizes. For small models, 2 cm and 6 cm long, they found no substantial improvement until the wing lengths exceeded 70% and 10% of the body length respectively. Even if the original function of the winglets was to stabilize the gliding body, there would therefore be little aerodynamic advantage in increasing winglet area until this stage was reached. For a larger model (10 cm body length) contrary results were obtained. The lift-to-drag ratio increased rapidly with larger wing lengths, supporting the aerodynamic explanation for wing enlargement. However, Kingsolver & Koehl concluded that the palaeontological evidence did not favour such large protopterygotes, and that non-aerodynamic reasons must be sought for the enlargement of winglets.

They also suggested that the initial function of winglets could not have been attitude control because the rolling moments generated by their models were very small. However, the moment of inertia about the roll axis is also small, and these rolling moments might still be adequate for roll stability. Kingsolver & Koehl did not investigate the moments about the other two axes, pitch and yaw, so the hypothesis that winglets initially served in attitude control remains largely untested.

## Outstanding questions

Leaving aside the matters, which we do not consider here, of the homology and the original, pre-flight functions of the wing precursors, several questions remain unsolved.

### *The functions of the winglets in the early stages of flight development*

There seems little doubt that tiny winglets could not have functioned to improve the initial lift-to-drag ratio of gliding protopterygotes. The most probable aerodynamic role, suggested by Flower, has not been tested thoroughly: did the winglets control the attitude of a parachuting or gliding body? Some active control of gliding may have been possible by using movements of legs, antennae and cerci, but it seems unlikely that these would have cured any problems of gross instability. A degree of inherent stability would have been highly desirable, if not essential.

Our interpretation of Flower's results is that pitch control would have been more important initially than yaw or roll, especially given the small moment of inertia about the roll axis. Problems of pitch arise because the centre of pressure for cylindrical bodies at small angles of attack is located near the front of the cylinder, producing a head-up pitching moment. As the angle of attack increases, the centre of pressure travels towards the middle of the cylinder, reducing the pitching moment (Hoerner & Boerst 1975). Stable equilibrium is reached at an angle of attack of 90°, with the centre of pressure coinciding with the centre of gravity, and the cylinder falling broadside to the flow.

The best glide performance for cylinders is obtained at angles of attack around 30° or less (Flower 1964), but this would obviously be an unstable attitude for simple cylinders. Stable glides at angles of attack less than 90° can be achieved by moving the centre of gravity forward; shifts in the centre of pressure with angle of attack will then provide restoring moments about the centre of gravity. This type of pitch stability can be illustrated by an aeromodelling parlour trick: gliding flat plates with a very carefully positioned centre of gravity. The moment arm for the restoring forces is necessarily small, however, and the stability is marginal. Fossil forms suggest that the centre of gravity in protopterygotes was anterior to the middle of the body, as in modern insects, and they may have gained some stability in this manner. Even so, the large restoring moments about the pitching axis that could be generated by posterior winglets and caudal projections would have been very useful.

Yaw stability might also have posed serious difficulties for the protopterygotes. Lacking a vertical tail fin, they could perhaps have benefited from a weathercock-type of stability from the body if the centre of gravity was far enough forward, but this might conflict with the requirements for pitch stability. Alternatively, the paired cerci and median caudal filament of many primitive insects (figure 1) look promising and purposeful with regard to yaw stabilization.

The evidence that the winglets may have been movable from the first – now possibly the mainstream view – raises several questions as to how this mobility may have assisted in aerodynamic functioning. Two in particular seem important: whether the ability to alter winglet attitude would have been beneficial in gliding or parachuting; and whether flapping could have contributed significant thrust while the winglets were still small.

## *Problems of transition*

If in course of time the meso- and metathoracic winglets became enlarged and those on the abdomen lost, the aerodynamics of transition need to be considered.

Why should the winglets enlarge? The results of Kingsolver & Koehl indicate that the body length must be of the order of 10 cm for aerodynamic benefits to explain the increase. Is this really an unacceptable size based on palaeontological evidence? Many Carboniferous insects were very large indeed (figure 1*a*). Are the results of these experiments convincing? They are presented as a relative lift-to-drag ratio (the ratio for body plus winglets divided by that for the body alone), but the authors comment that the absolute value of the lift-to-drag ratio for the large model is some four or five times worse than that for the smaller models. This radical deterioration at higher Reynolds numbers is hardly to be expected, unless their models ventured into some of the separated flow phenomena which are not uncommon in this range; this possibility is supported by their observations of the airflow patterns.

Furthermore, if Flower's hypothesis that the winglets functioned initially in attitude control proves correct, what would be the consequence of losing all but two pairs? The models of Kingsolver & Koehl have no abdominal winglets, and the work sheds no light on their function or on their dispensibility.

### *The feasibility and relative likelihood of the paragliding, floating and running/jumping scenarios*

Here several questions arise. First, was the vegetation at the time of the origin tall enough for parachuting and/or gliding from plants to be beneficial? There are new indications that the colonization of land and the early radiation of myriapods and wingless insect stock may have taken place earlier than has usually been believed, but there is no real reason to look earlier than the upper Devonian for the origin of flight, by which time tree-like lycopods, sphenopsids, pteridosperms and progymnosperms were becoming established.

Would insects have been liable to damage in falling a few metres from a tree? It seems so. We have experimented by dropping cockroaches (*Periplaneta americana*) of all nymphal instars repeatedly from a height of 5 m onto sheet paper overlying soft plastic foam, and found appreciable damage to antennae and legs at all sizes.

Next, how probable is it that ground-dwelling insects with little or no power of active flight could have made significant use of aerial dispersal? Rainey & Waloff (1951) and Wigglesworth (1963) have stressed that insects up to locust size are carried up by thermals in arid country. However, as Flower (1964) pointed out, selection for passive drifting would favour the evolution of very small insects, though this might perhaps be countered by other pressures independent of dispersal. Selective trends apart, what matters, clearly, is sinking speed: if protopterygotes frequently encountered air which was rising faster than they needed to fall, then passive aerial dispersal would be possible.

A running insect would not benefit from gliding; neither could a protopterygote conceivably generate enough lift and power to fly up from the ground in still air. Whether it could usefully glide from a jump depends on its glide angle, and on whether this could be reduced, and the glide prolonged, by flapping.

## Model protopterygotes

More experiments are needed. The best method of experimental investigation is probably to drop models of possible protopterygotes, and to note their glide performance and stability. Compared with wind tunnel studies of mounted models this approach has one main advantage: instabilities revealed in a wind tunnel may, in free flight, be no more serious than small oscillations about a control axis. Thus a weak pitch instability may cause limited phugoid oscillations, in which the glide path follows a series of swooping arcs, but the average glide angle may still be quite impressive.

It would obviously be desirable for models to be the same size and weight as the protopterygotes, but such small models would be difficult to trim; for example the angle of attack of tiny winglets could not be adjusted accurately. A more practical solution is to construct larger, lightweight models which operate at the same Reynolds numbers. This would ensure that the force coefficients, and hence the glide angle, remain unchanged for stable, equilibrium glides. The larger models would show a relatively faster dynamic response, but this can be helpful when observing stability behaviour; attitude changes, such as recovery from stalls, would occur after relatively short distances.

The larger models must be geometrically similar to the real insects, and constructed so that they glide at the same Reynolds number *Re*, given by $Re = Ud/\nu$, where $\nu$ is the kinematic viscosity of air, $U$ is the gliding speed, and $d$ is the diameter of the cylindrical body. To keep the product $Ud$ constant, the larger model must therefore glide more slowly. The gliding speed for given force coefficients is proportional to $(M/A)^{0.5}$, where $M$ is the mass of the model and A is its area. Since the area of the gliding body is proportional to $d^2$, the product $Ud$ is then proportional to the square root of the mass. The Reynolds number of geometrically similar gliders will therefore remain constant if they have the same mass, and one can calculate the gliding speeds of the real insects by the proportionality

$$U_{\text{insect}}/U_{\text{model}} = d_{\text{model}}/d_{\text{insect}}.$$

After some preliminary tests using variously shaped prototypes, we chose to model our insects as cylindrical balsa tubes, with a length-to-diameter ratio of 8:1 (the ratios of Palaeozoic insects and relevant modern forms mostly lie between 6:1 and 10:1). The cylinder ends were hemispheric shells of paper laminated with PVA adhesive (figure 2).

Two models bore nine pairs of square winglets, each pair mounted on a balsa rod which passed through the cylinder at the diameter of its cross-section, and was held in place by friction only (figure 2*a*). The angle of inclination $\phi$ of each winglet pair relative to the cylinder long axis could be set by simple twisting; measurement was inevitably approximate, but was assisted by small protractors drawn at the base of each winglet on one side.

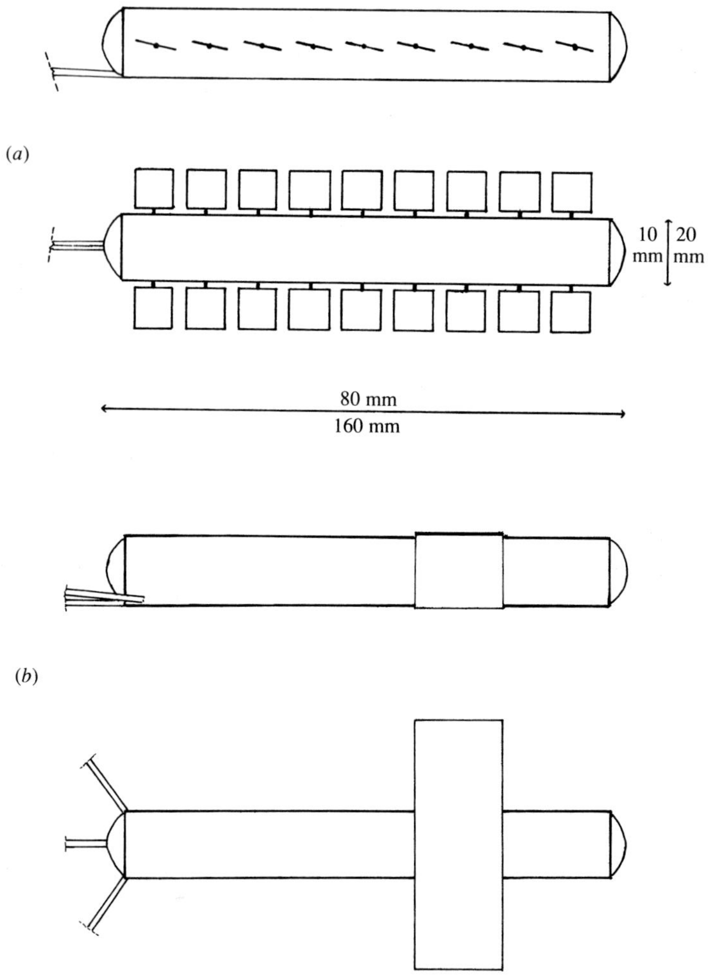


Figure 2. Balsa models: (*a*) with serial winglets, and (*b*) with a single pair of winglets on a movable sleeve. The paper 'tails' are shown truncated. The double scales indicate the sizes of the two models.

A thin, stiff paper 'tail' was glued to the posterior end of the model, and could be adjusted to give yaw stability.

The centre of gravity was adjusted by placing a small blob of Bostik 'Blu-tack' on the nose of the model, and was set between 35% and 45% of the body length from the nose, realistic values for relatively generalized insects.

The models were 80 mm and 160 mm long, with winglets respectively 4 mm and 8 mm square. They were thus isometric in planform, though isometry could not be extended to the thickness of the balsa. The models had mass 368 mg and 4250 mg, respectively; the first was approximately equivalent to a locust hopper of 25 mm body length, the second to a larger similar orthopteran some 70 mm long.

The models were dropped, or gently thrown, from a height of 6.5 m – enough to allow them to reach terminal velocity – and the glides recorded with a Panasonic video recorder NV 8050 for frame by frame analysis. The effects of various angles and combinations of winglet inclinations were noted. Only stable glides were analysed, and the glide angle and glide speed were recorded at the shallowest glide angle obtainable for each model. It was also possible to estimate the angle of attack of the body $\alpha_b$, and to calculate the angle of attack of the winglets $\alpha_w$ from the glide angle $\theta$, the winglet inclination angle $\phi$, and from $\alpha_b$ (figure 3).

After these tests were completed, the effects of removing posterior winglets were examined: it proved necessary to add a pair of paper tails to maintain pitch stability. Finally, the effects of winglet enlargement were tested on similarly proportioned and weighted cylinders with single rectangular wings dorsally mounted on a paper sleeve, so that the centre of lift was adjustable. These models again required three stiff caudal filaments for glide stability (figure 2*b*).

## Results

All the models tested were remarkably stable about the rolling axis. In those with multiple winglets, pitch stability varied with the angles of winglet inclination, but could always be achieved when these were appropriately set. Yaw stability, poor in tail-less models, was easily provided by gluing a slender paper filament, about two thirds as long as the body, in the position of the caudal filament of many apterygotes and mayflies.

Performance, and particularly versatility, were sharply affected by the model size. The smaller (80 mm) model proved to be quite unstable in pitch unless the winglets were inclined in a nose-up sense to the body axis. With the inclination angle $\phi$ set at about 20° the model descended steeply, without turning over, but with $\alpha_b$ and the glide angle oscillating. At $\phi = 30°$ the model consistently parachuted stably in a mushy stall, at glide angles close to 55°. The best recorded angle was 54°, at a glide speed of 1.6 m/s and a sinking speed of 1.3 m/s; $\alpha_w$ was about 85°. The Reynolds number was calculated as 1066, $C_L$ as 0.97 and $C_D$ as 1.34: force coefficients are based on total planform area, that is body plus winglets.

In these tests the glide angles reached were greater than the 45° quoted by Flower for simple cylinders, but the winglets clearly lowered the gliding and

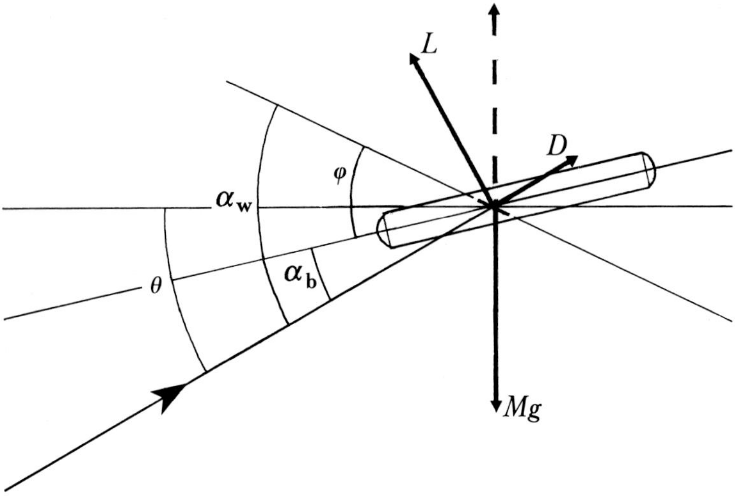


Figure 3. Forces and characteristic angles of a gliding model protopterygote. *L*, lift; *D*, drag; *M*, mass; *g*, acceleration due to gravity; for explanation and other symbols see the text.

sinking speeds, which were equivalent respectively to *ca* 5 m/s and *ca* 4 m/s for 25 mm insects of the same shape.

By contrast, the larger (160 mm) model could readily be made to glide at shallow angles, with most winglets aligned parallel to the body axis or at small negative values of $\phi$. In these tests the winglets were clearly functioning properly as aerofoils, with $\alpha_w$ varying between 20° and 32°. Most interestingly, the performance of this model – and of another, slightly smaller model whose performance was not analysed in detail – could be controlled precisely by the angle of inclination of the most posterior pair of winglets. Changing the inclination by a few degrees could alter the flight path from a shallow glide to a steep nose-dive, and also affected gliding speeds and the angles of attack of body and winglets.

Two good glides illustrate respectively the lowest stable glide angle and the lowest gliding and sinking speeds recorded for this model. In the first, all winglet pairs except the last were set parallel to the body axis; the last pair were inclined slightly upwards, with $\phi$ equal to *ca* 3°. At terminal velocity $\alpha_b$, and hence $\alpha_w$, were 21°; glide angle was 17°, glide speed 5.6 m/s, and sink speed 1.6 m/s. The Reynolds number was calculated as 7470, $C_L$ as 0.37, $C_D$ as 0.11.

In the second glide, all winglet pairs but the last were set at $-5°$; the last was $-3°$. At terminal velocity $\alpha_b$ was 37° and $\alpha_w$ 32°; glide angle was 21°, glide speed 2.6 m/s, and sink speed 0.93 m/s. *Re* was calculated as 3420, $C_L$ as 1.65, and $C_D$ as 0.63.

Equivalent speeds in similar glides for a 70 mm insect of the same shape would be as follows. For the first glide glide speed was 12.8 m/s, and sink speed 3.7 m/s. For the second glide speed was 5.9 m/s, and sink speed 2.1 m/s.

Rotating the winglets to higher values of $\phi$ did not destroy stability, but increased the glide angle: the option of parachuting steeply in a mushy stall was still available.

In both the 80 mm and the 160 mm models the removal of posterior winglets destroyed pitch stability, but this could be restored by attaching lateral paper tails, analogous to the long cerci of mayflies and many Palaeozoic insects (figure 1*b*), at about 45° to the body axis.

Models with bodies identical with those with serial winglets, but bearing instead a single rectangular wing, dorsally mounted, and projecting on each side of the body as a square equivalent in area to four winglets behaved similarly to the equivalent winglet-bearing models, provided they too were given lateral tails, and provided the wing was positioned with its geometric centre just anterior to the centre of mass of the model. The best glide angle recorded from the 160 mm model was 34°, with $\alpha_b$, and hence $\alpha_w$, at 30°, with a glide speed of 4.5 m/s, and sink speed 2.5 m/s, equivalent, respectively, to 10.2 m/s and 5.7 m/s for an isometric insect. This performance was poorer than the equivalent model with complete winglets but with less than half the wing area.

## Discussion

These are simplified models, but the results are informative. First, they confirm Flower's proposal that winglets of reasonable size can confer pitch stability on a simple cylinder falling with an angle of attack less than 90°, and hence allow some horizontal movement. This, together with a reduced terminal velocity and the capability to land the right way up could all have been advantageous to a protopterygote travelling through the air, however it got there, and whatever the speed and direction of the relative airflow.

Second, it is clear that the parachuting/gliding performance is very strongly affected by the angle of inclination of the winglets. In both models, pitch stability is only possible if the winglets are appropriately inclined. Changes in the inclination angle allow and control a wide range of glide angles and speeds in the larger model, with the angle of the last winglet pair acting as a sensitive fine adjustment; the performance of the smaller model was restricted to a much narrower range.

Rotatable winglets would therefore have been far more useful to hopeful protopterygotes than would fixed lateral lobes. The latter, it seems, could at best have served to slow down, and give some roll stability to insects falling broadside to the flow, or to allow fast, uncontrolled gliding at more shallow angles.

Third, it appears that, in insects below a certain size, winglets may only have served to stabilize steep, stalled parachute drops, but that a modest size increase would have made true gliding possible, at a wide range of angles and speeds, sometimes at Reynolds numbers no more than three or four times higher.

Fourth, enlargement of the winglets closest to the centre of mass, simultaneously with the loss of those on the abdomen, could have occurred without loss of aerodynamic functioning, provided that the caudal appendages were large enough to assume a stabilizing role.

Summarizing: if one takes as a starting point an insect with twistable serial winglets on thorax and abdomen, and not – as do Kingsolver & Koehl (1985) – with thoracic winglets only, aerodynamic considerations seem adequately to explain the evolution of wings and flight. While still small, winglets could have begun to contribute stability and drag, and these effects would favour their enlargement. In larger insects, moderately sized winglets would have allowed surprisingly good glides; our results do not reflect Kingsolver & Koehl's discovery that lift-to-drag ratio declined in their largest model. It seems improbable that flapping forces could have had a significant effect at the serial winglet stage of evolution; but loss of abdominal winglets and enlargement of those on the thorax, could have progressed without loss of performance, to the size where flapping would have been effective in prolonging and directing glides, and ultimately in fully powered flight.

Paragliding, floating or jumping? It is easy to oversimplify: a jump from a high rock into rising air would have elements of all three.

Jumping glides from level ground seem unlikely, at least in the early evolutionary stages. The shallow glides reached by our larger models require high speeds, far beyond the 3.4 m/s jumping speed quoted by Brown (1963) for locusts.

Passive transport in thermals can perhaps not be ruled out. The sinking speeds indicated by both our large and small models are inside the figure of 5 m/s given by Rainey & Waloff (1951) for convection currents. Floating, however, favours smaller insects still, which would have been incapable of gliding at decent angles, and which seem to be poor candidates for the development of flapping flight. It is far easier to envisage flapping coming into use in larger gliding insects which could operate at lower frequencies.

Our results, however, underline the strength of the parachuting/gliding scenarios, and add the factor of glide control by variable winglet inclination. Flower (1964) concluded that gliding insects more than some 10 mm long would have dangerously high maximum terminal velocities, and would need to fall a long way before these were reached; but if the insects gained speed and height with a jumping take-off (Norberg 1972), and if glide angle and speed could then be controlled by winglet twist, these problems may disappear. There seems no reason why fast-gliding insects could not have been capable at will of slowing, stalling, and settling down to a soft landing.

Drs Kukalová-Peck, Kingsolver and Koehl have seen this chapter in manuscript and have commented on its conclusions (Kukalová-Peck 1987; Kingsolver & Koehl 1989). The latter authors reviewed the similarities and differences between their previously published results (Kingsolver & Koehl 1985) and our own.

**Acknowledgements**

We thank Roland Ennos for practical help and discussion, and Michael Reddaway, of the Northcott Theatre, Exeter, for the loan of background scenery.

**References**

Brown, R.H.J. 1963 Jumping arthropods. *Times Science Review* Summer 1963, 6–7.

Flower, J.W. 1964 On the origin of flight in insects. *Journal of Insect Physiology* **10**, 81–88.

Forbes, W.T.M. 1943 The origin of wings and venational types in insects. *American Midland Naturalist* **29**, 381–405.

Hinton, H.E. 1963 The origin of flight in insects. *Proceedings of the Royal Entomological Society of London* C **28**, 24–25

Hoerner, S.F. & Boerst, H.V. 1975 *Fluid Dynamic Lift*. Bricktown, New Jersey: Hoerner Fluid Dynamics.

Kingsolver, J.G. & Koehl, M.A.R. 1985 Aerodynamics, thermodynamics, and the evolution of insect wings: differential scaling and evolutionary change. *Evolution* **39**, 488–504.

Kingsolver, J.G. & Koehl, M.A.R. 1989 Selective factors in the evolution of insect wings: response to Kukalová-Peck. *Canadian Journal of Zoology* **67**, 785–787.

Kukalová, J. 1970 Revisional study of the Order Palaeodictyoptera in the Upper Carboniferous shales of Commentry, France. Part III. *Psyche, Cambridge* **77**, 1–44.

Kukalová-Peck, J. 1978 Origin and evolution of insect wings and their relation to metamorphosis, as documented by the fossil record. *Journal of Morphology* **15**, 53–126.

Kukalová-Peck, J. 1983 Origin of the insect wing and wing articulation from the arthropodan leg. *Canadian Journal of Zoology* **61**, 1618–1669.

Kukalová-Peck, J. 1985 Ephemeroid wing venation based upon new gigantic Carboniferous mayflies and basic morphology, physiology and metamorphosis of pterygote insects (Insecta, Ephemerida). *Canadian Journal of Zoology* **63**, 933–955.

Kukalová-Peck, J. 1987 New Carboniferous Diplura, Monura, and Thysanura, the hexapod ground-plan, and the role of thoracic side lobes in the origin of wings (Insecta). *Canadian Journal of Zoology* **65**, 2327–2345.

Müller, F. 1875 Beiträge zur Kenntnis der Termiten. *Jenaische Zeitschrift für Naturwissenschaften* **9**, 241–265.

Norberg, R.Å. 1972 Evolution of flight in insects. *Zoologica Scripta* **1**, 247–250.

Oken, L. 1811 *Lehrbuch der Naturphilosophie*. Jena.

Quartau, J.A. 1985 On some objections to the paranotal theory on the origin of insect wings. *Boletim da Sociedade Portuguesa de Entomologia Suplemento* **1**, 359–371.

Quartau, J.A. 1986 An overview of the paranotal theory on the origin of insect wings. *Publicaçoes do Instituto de Zoologia 'Dr Augustio Nobre', Faculdade de Ciencias do Porto* **194**, 1–42.

Rainey, R.C. & Waloff, Z. 1951 Flying locusts and convection currents. *Anti-locust Bulletin* **9**, 51–72.

Rasnitsyn, A.P. 1981 A modified paranotal theory of insect wing origin. *Journal of Morphology* **168**, 331–338.

Wigglesworth, V.B. 1963 The origin of flight in insects. *Proceedings of the Royal Entomological Society of London* C **28**, 23–24.

Wigglesworth, V.B. 1973 Evolution of insect wings and flight. *Nature, London* **246**, 127–129.

Wigglesworth, V.B. 1976 The evolution of insect flight. In *Insect Flight*, ed. R.C. Rainey, pp. 255–269. *Symposia of the Royal Entomological Society* **7**. Oxford: Blackwell.

Wootton, R.J. 1986 The origin of insect flight: where are we now? *Antenna* **10**, 82–86.

SARAH J. BUNKER & KENNETH E. MACHIN†

# The hydrodynamics of cephalaspids

## Introduction

Pitfalls and traps abound for biomechanicists who venture to ply their trade among extinct animals; it is only too easy for them to rush in where palaeontologists fear to tread. Being used to reliable quantitative data, they may not realize that the reconstruction on which they base their studies is inaccurate, speculative, incomplete or even outdated. Being over familiar with the physiological properties of components of 'modern' animals (for example the specific power output of various sorts of muscle), they may forget that the fossil record has no evidence whatsoever to offer on such matters. Being untutored in sedimentology and phylogeny, they may suggest modes of life or draw analogies which bring down on their heads the scorn of the professionals.

These difficulties might tend to discourage wide, quantitative (if often naïve) biomechanical speculation arising from theorizing and experimenting on fossil material. Yet such speculation must surely be useful: it may tell the palaeontologist what is possible and what is impossible, and even perhaps what is likely. But there is one important proviso: all the assumptions in the biomechanical analysis must be explicitly stated. Only in this way can the reliability and applicability of the analysis be assessed, the accuracy of its quantitative conclusions estimated, and its results brought up to date in the light of further information.

In no sense is this chapter intended to be a review of the hydrodynamics of fossil fish. It is offered as a 'case study' in the application of biomechanical experiment and unfettered speculation to a single reconstruction. Clearly we hope that it may tell palaeontologists something useful about cephalaspids; failing that, perhaps it will tell them something useful about the methods of biomechanics.

† Deceased.

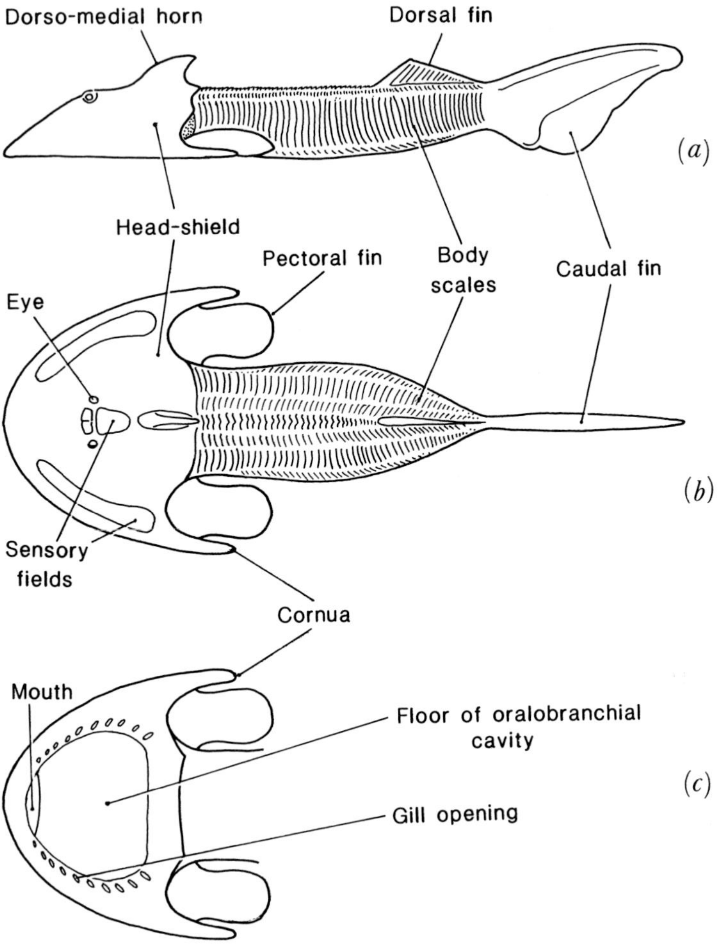


Figure 1. Principal features of cephalaspids: (*a*) side view; (*b*) dorsal view; (*c*) ventral view of head-shield.

## Model experiments and the study of fish hydrodynamics

Harris (1936) was the first to realize that a convenient way of investigating the hydrodynamics of fish was by making measurements on a model fish in a wind tunnel. Provided that the Reynolds number of the model in the tunnel is equal to the Reynolds number of the fish in water, the air flow round the model is dynamically similar to the water flow round the fish. [Reynolds number *Re* is defined as $Vl\rho/\nu$, where $V$ is the fluid velocity relative to the body, $l$ is a characteristic dimension of the body, $\rho$ is the density of the fluid, and $\nu$ its dynamic viscosity.] Wind tunnels are generally provided with balances to measure lift force, drag force and pitching moment directly; the measured forces on the model can then be scaled to give the forces on the fish. Harris showed that, in the dogfish, both the pectoral fins and the heterocercal tail produced lift; together, these lift forces supported the negative buoyancy. But this equilibrium was not stable: if the nose lifted slightly, pitching moments would be produced in such a direction as to lift the nose still further. The delicate balance between the two components of lift necessary for steady swimming must be maintained actively by continuous corrective movements of fins or tail.

Kermack (1943) extended Harris's method to models of fossil fish: he investigated *Pteraspis*, a small, heavily armoured fish with no pectoral fins, but with a hypocercal tail. He showed that the body alone could produce enough lift to support the weight of the fish in water. Although the hypocercal tail was assumed to produce a down-force, Kermack adduced arguments suggesting that equilibrium could still be attained between the body lift, weight and tail down-force. This equilibrium was again unstable, and called for continuous active correction by the tail. Much more recently, Cruickshank & Skews (1980) carried out a very detailed and elegant study of the hydrodynamics of the boomerang-shaped extensions to the skull of certain nectrideans, Palaeozoic amphibians without lifting tails. Once again, the equilibrium between weight and lift was unstable.

## Cephalaspids

Cephalaspids (figure 1) are Devonian fish for which much fossil material is available (for example in the British Museum (Natural History), London). The quality of the material is such that reliable reconstructions can be made, not only of the external appearance, but also of some internal features of the cephalic region (Stensiö 1927, 1932). Fossil specimens cover a range of size from a few centimetres in length to a reconstructed length of almost a metre.

Cephalaspids have an undivided, bony head-shield and a body covered with overlapping, articulating bony scales. On the dorsal surface of the head-shield are various sensory organs, probably including some sensitive to water-borne pressure fluctuations (Moy-Thomas & Miles 1971). Within the angles of the cornua at the posterior corners of the head-shield are attached pectoral fins.

The ventral surface of the entire fish appears to have been quite flat. The floor of the oralobranchial chamber is visible as an array of small polygonal plates, with the mouth at its anterior margin and the gill openings on the lateral margins.

The heterocercal tail is described by Stensiö (1932) as a simple vertical caudal fin, and it is this tail which is illustrated in figure 1. Subsequently Heintz (1967) proposed a different reconstruction with a larger lower lobe and, mounted anteriorly on it, a pair of very small horizontal caudal flaps.

To the eye of a biomechanicist, the large armoured head-shield presaged nose-heaviness; its hydrofoil shape suggested that hydrodynamic lift might be available to counteract the nose-heaviness. These assumptions – the starting-point of the investigation – proved to be oversimple.

## Wind tunnel measurements on a cephalaspid model

A twice-full-scale fibreglass copy was made of the 'generic' cephalaspid model in the Museum of Zoology, Cambridge, which is consistent with the reconstruction of Stensiö (1932). The fibreglass shell was filled with resin loaded with marble dust. Pectoral fins were mounted on stiff wires which could be bent to set the fins at various angles of deflexion. The tail was first shaped in accordance with the model by Stensiö (1932); when all wind tunnel measurements were complete, the tail was rebuilt to accord with the model by Heintz (1967) and the measurements repeated.

The model was mounted in a low-speed wind tunnel in the Department of Engineering, University of Cambridge. Mounting wires were attached to the ends of short, thin metal rods protruding from the head-shield of the model, and to an anchor point near the tail. Errors due to the drag of the attachments or the influence of the walls of the tunnel were negligible. Angles of attack were taken relative to the flat ventral surface of the model.

### *Lift and drag*

The forces read from the aerodynamic balance were converted into the usual dimensionless coefficients of lift and drag, $C_L$ and $C_D$, relative to the whole plan area $S$ of the model; $C_L$ = lift force/$\frac{1}{2}\rho V^2 S$, and similarly for $C_D$. Figure 2 shows the way in which these coefficients vary with angle of attack at a Reynolds number (based on overall length) of $4 \times 10^5$, which corresponds with a 0.4 m long fish swimming at about 1 m/s. The drag can be separated into two components: (1) the drag $C_{D0}$ of the head and body, which is not likely to vary substantially over a moderate range of angle of attack; and (2) the *induced drag*, which is a direct consequence of the generation of lift.

It can be shown theoretically (see, for example, Dommasch, Sherby & Connolly 1967) that

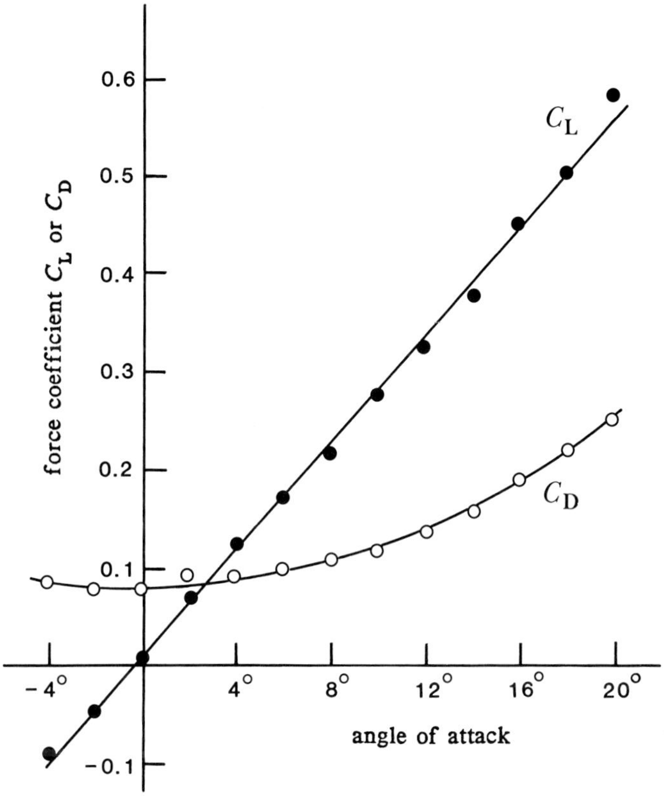


Figure 2. Variation of coefficients of lift $C_L$ and drag $C_D$ with angle of attack. $Re = 4 \times 10^5$; pectoral fins undeflected.

$$C_D = C_{D0} + C_L^2/\pi A,$$

where $A$ is the effective aspect ratio of the model. Figure 3 demonstrates that the theoretical relation holds here. Pennycuick (1971) has drawn attention to the difficulty of reconciling the effective aspect ratio derived from the slope of such a graph with a geometrically determined value, and such difficulties might be expected to be severe with the unusual planform of the cephalaspid. An independent estimate of the effective aspect ratio can, however, be derived from the slope of the relation between $C_L$ and the angle of attack (Dommasch *et al* 1967). The two estimates are here in agreement (0.65 from figure 2, and 0.64 from figure 3) giving confidence that figure 3 does indeed identify the two separate sources of drag.

To compare the drag of a cephalaspid with published figures for the 'dead drag' of other fish, it is necessary to recalculate the value of $C_{D0}$ on the more customary basis of *wetted area* rather than *plan area*. This gives a value of

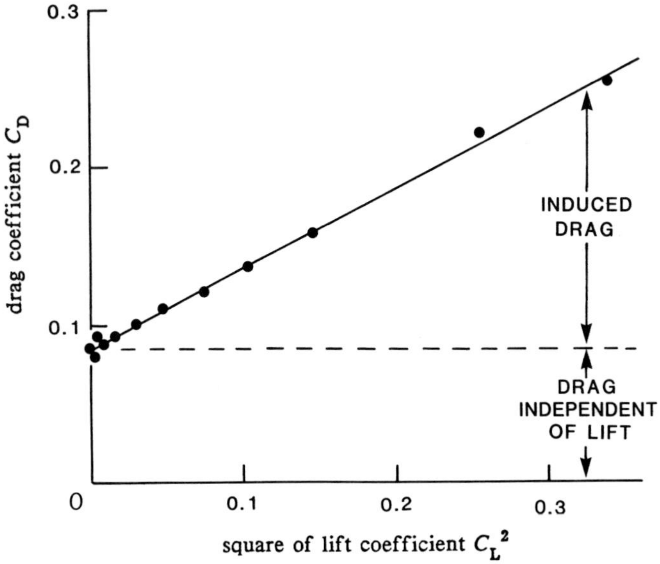


Figure 3. The linear relation between $C_D$ and $C_L^2$.

drag coefficient of 0.033, which may be compared with 0.022 for dogfish (calculated by Webb (1975) from the results of Harris (1936)). We shall mention later the difficulties in equating 'dead drag' figures to the drag actually experienced by a swimming fish.

Some measurements were also taken at the higher Reynolds number of $1.5 \times 10^6$. No change in the behaviour of $C_L$ with angle of attack was expected or observed. Nor was the slope of the relation between $C_D$ and $C_L^2$ (figure 3) changed. The value of $C_{D0}$, the component of drag coefficient independent of lift, was however reduced from 0.088 (at $Re = 4 \times 10^5$) to 0.064 (at $Re = 1.5 \times 10^6$). It is likely that the boundary layer at the lower Reynolds number would be largely laminar, while that at the higher Reynolds number would be largely turbulent; this transition is commonly associated with a significant reduction of drag coefficient (Vogel 1981).

Both lift and drag were unaltered, at all angles of attack, when the tail was changed from 'Stensiö' to 'Heintz'.

### *The effect of pectoral fin deflexions*

Wind tunnel measurements were made with the pectoral fins of the model deflected dorsally (+20°) and ventrally (−20°). Fin deflexion caused a substantial shift in the relation between $C_L$ and angle of attack (figure 4),

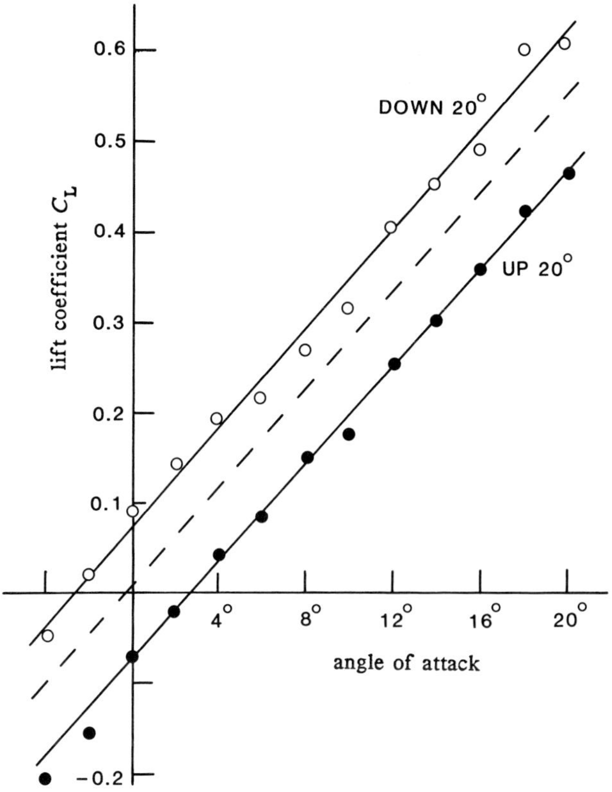


Figure 4. The effect of pectoral fin deflexion on lift coefficient. The dashed line shows the relation between $C_L$ and angle of attack when the pectoral fins are undeflected.

but the values of drag observed with deflected fins did not differ significantly from those observed at the same values of $C_L$ without fin deflexion.

### *Pitching moments*

At any given angle of attack, the wind tunnel balance gives readings not only of lift force and drag force, but also of the pitching moment about the front mounting axis of the model. It is always possible to calculate the position of a point through which the lift and drag forces are taken to act, so that there is no net pitching moment on the model. This point is the *centre of pressure*, and its position relative to the centre of gravity and centre of buoyancy determines the equilibrium and stability of a swimming fish. The measured pitching moments are here presented in terms of a dimensionless

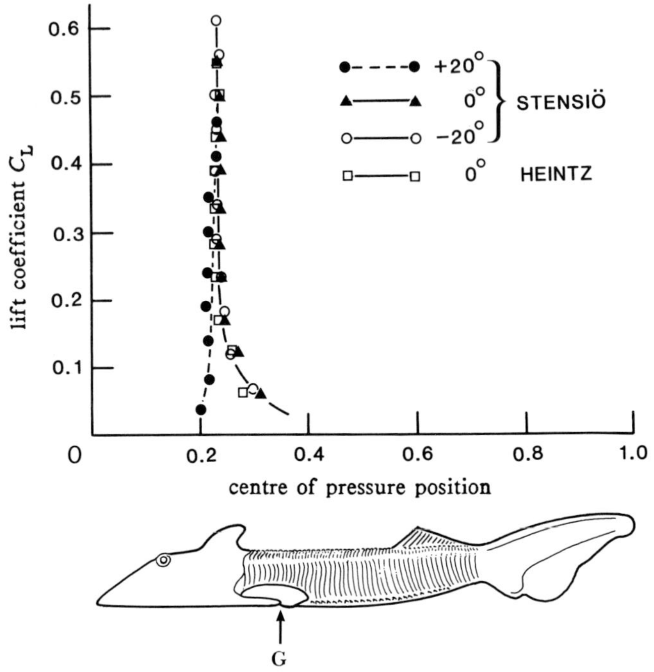


Figure 5. The position of the centre of pressure (as a proportion of the length of the fish) as a function of lift coefficient $C_L$, for various angles of pectoral fin deflexion and for two tail configurations ('Stensiö' and 'Heintz'). G marks the position of the centre of gravity of the fish in water.

position of the centre of pressure: the ratio between the distance from nose to centre of pressure and the length of the model.

The centre of pressure position is shown in figure 5 as a function of $C_L$ for the various configurations of the pectoral fins. The results for fin deflexions of 0° and −20° are indistinguishable: at moderate and high $C_L$ the centre of pressure position is constant, while at low $C_L$ it moves tailwards. This implies passive instability at low angles of attack: a slight nose-down pitch will move the line of action of the hydrodynamical force backwards, causing further nose-down pitching. On the other hand, the position of the centre of pressure for pectoral fin deflexion of +20° moves slightly nosewards at low $C_L$. This passive stability is not unexpected: upward deflexion of the pectoral fins adds to the head-shield a reflexed trailing edge of the type used by designers of tailless aircraft to achieve stability (Pennycuick 1971). For the one fin configuration (0°) for which comparison was made there is no difference between 'Stensiö' and 'Heintz' tails. Heintz (1967) was indeed right when he wrote: 'They [the caudal flaps] may have had some effect as stabilizers similar

Table 1. *Estimated mass and weight for the 'standard' cephalaspid*

| | | | Specific gravity | Weight in water (N) | Mass (kg) |
|---|---|---|---|---|---|
| Volume of head | $1.7 \times 10^{-4}$ m³ | 20% cartilage | 1.1 | 0.03 | 0.04 |
| | | 20% other tissue | 1.0 | 0 | 0.04 |
| | | 60% oralo-branchial chamber | 1.0 | 0 | 0.10 |
| Volume of tail and body | $1.9 \times 10^{-4}$ m³ | 80% muscle | 1.06 | 0.09 | 0.16 |
| | | 20% other tissue | 1.0 | 0 | 0.04 |
| Area of head | 0.024 m² | 1 mm thick bony armour | 2.0 | 0.24 | 0.05 |
| Area of tail and body | 0.025 m² | 1 mm thick bony scales, 30% overlapped | 2.0 | 0.34 | 0.07 |
| Totals: | | | | 0.70 | 0.50 |

to the horizontal small wings in the tail region of an aircraft; but it is hardly likely that these small and narrow flaps could be of any special importance as stabilizers.'

### Mass, weight, centre of gravity, centre of buoyancy

Examination of fossil specimens of cephalaspids in the British Museum (Natural History) revealed that the bony exoskeleton of the head was surprisingly thin: thicknesses between 0.5 mm and 2 mm were measured. Reliable measurements of the thickness and degree of overlap of the body scales were more difficult to obtain. Estimates of the total mass of the fish become a matter of informed guesswork; indeed the weight of the fish in water probably cannot be estimated to better than a factor of 2 either way.

Table 1 presents our best estimate of the mass and weight of a cephalaspid of length 0.38 m (the same length as the model in the Zoology Museum, Cambridge, of which our wind-tunnel model is a twice-full-size copy). The volumes were estimated from the solid models, and the areas of the bony exoskeleton from hollow cardboard models. The point of balance of the solid and hollow models was found, from which it was concluded that the centre of buoyancy and the centre of gravity of the fish in water both lie roughly on the line joining the tips of the cornua, a distance of 0.13 m from the nose. Once again, the position of the two centres can only be estimated very approximately. Indeed the figures presented in this section represent the major uncertainty in the quantitative deductions which follow. Any error should, however, manifest itself only in the numerical part of the conclusions, and should not invalidate the qualitative aspects.

## Implications for the mode of life of cephalaspids

Starting from the experimental data, we can construct a chain of argument about the behaviour of cephalaspids. The reader should be warned again: biomechanics can say that something was 'impossible' or 'possible', but it cannot assert that it happened. Much of the argument will be presented in terms of the 'standard' cephalaspid of 0.38 m length, the fish to which table 1 refers.

### *A cephalaspid stationary on the bottom*

The head-shield is highly cambered, and might therefore be expected to produce lift even at zero incidence. Surprisingly, the results show (figure 2) very small lift at zero incidence with the pectoral fins undeflected; a small upward deflexion of the fins suffices to make the lift negative (figure 4). Furthermore, the pitching moments measured at zero incidence are invariably negative (i.e. nose-down). There is therefore no danger of the fish being lifted off the bottom, even by strong water currents.

### *Steady swimming*

When a 'standard' cephalaspid is swimming horizontally at constant speed, there must be an equilibrium between three vertical forces: (1) the lift $L_H$ due to the head and body (2) the lift $L_T$ due to the oscillation of the heterocercal tail and (3) the weight $W$ of the fish in water. $L_H$ acts through the centre of pressure, which for most values of the lift coefficient is about 0.095 m from the nose (figure 5). $L_T$ will act through some notional 'centre' of the tail fin, say 0.30 m from the nose. $W$ acts through the point which is both the centre of buoyancy and the centre of gravity of the fish in water, estimated earlier as 0.13 m from the nose. For equilibrium both of vertical force and pitching moment, $L_H = 0.83\ W$ and $L_T = 0.17\ W$.

By setting the lift force $L_H$ equal to 83% of the estimated weight of the fish in water (0.7 N) we can determine at what lift coefficients the fish must be operating when swimming steadily at various speeds. These results are shown in figure 6. For each value of lift coefficient, figure 3 allows us to derive a drag

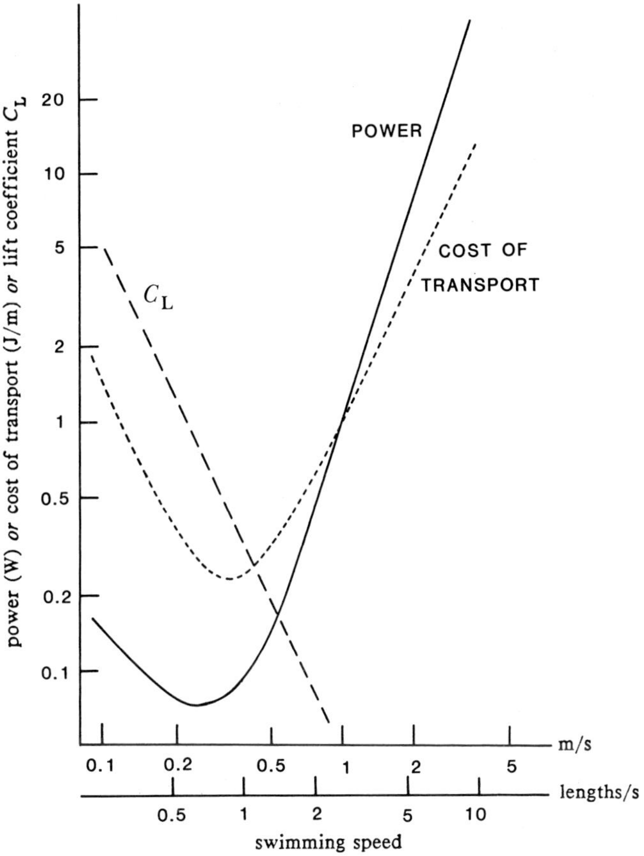


Figure 6. The lift coefficient $C_L$ needed to support the weight of the fish in water (continuous line), the thrust power (full line) and the mechanical cost of transport (dotted line), all as functions of swimming speed on logarithmic scales for a 'standard' cephalaspid.

coefficient, and hence to calculate the drag force for each swimming speed. The product of drag force and speed gives the thrust power required for locomotion, which is also displayed in figure 6.

The power curve involves two separate terms: the downward-going left-hand branch of the curve is the power needed to overcome induced drag, while the upward-going right-hand branch is the power to overcome the drag of the head and body. Each of these branches is subject to considerable uncertainty. Estimates of induced drag are only as accurate as the estimates of weight of the fish in water, about which reservations have already been made. Estimates of head and body drag assume, for simplicity, that the value of $C_{D0}$ measured at $Re = 4 \times 10^5$ is also applicable throughout, and also assume that the drag of a swimming fish is the same as the dead drag of a rigid, immobile body of the same size and shape. This latter assumption is well known to be invalid (Lighthill 1971), swimming drag always being larger than dead drag; the discrepancy has been attributed to the thinning of the boundary layer due to the sideways movement of the body and tail while swimming. For cephalaspids, much of the drag must be attributable to the rigid head; boundary layer thinning must be confined to a relatively short region near the tail. The large drag discrepancies (a factor of 4 or more) observed with streamlined and flexible fish are unlikely with cephalaspids, although it would be prudent to anticipate that the head and body drag may have been underestimated by a factor of 2.

The power curve shows a pronounced minimum. The warnings given in the previous paragraph about the uncertainties in the two branches of the power curve clearly apply to the numerical value of the speed for minimum power, though this value is in fact relatively insensitive to errors in drag. Also plotted in figure 6 is the cost of transport, the energy required to transport the cephalaspid a distance of 1 m. It is possible to find a speed (about 0.3–0.4 m/s) at which both the power requirement and the cost of transport are close to minimal, and at which the required lift coefficient is readily attainable. The conclusion that this is the likely cruising speed used by cephalaspids is virtually irresistible.

We need to be satisfied that, at this cruising speed, the heterocercal tail could indeed provide its share of the required lift. Bainbridge (1958) found, for three species of fish, a simple relation between swimming speed and tail-beat frequency, and conjectured that 'the speed of any fish may be calculated from the formula ...'. If his conjecture is valid even for Devonian fish, then the tail-beat frequency of a cephalaspid swimming at 0.4 m/s would have been about 3 Hz. Assuming that the tail oscillates through an angular range of $\pm 45°$, one can calculate what the average lift coefficient of the tail must be in order that it should carry 17% of the weight of the fish in water. On the basis of the area of the lower lobe of the (Stensiö) heterocercal tail, the value $C_L = 0.75$ is found. Alexander (1965) measured the lift of heterocercal tails from dogfish and tope; his results cannot be expressed in terms of a constant lift coefficient, since the tails flexed to a differing degree at differing speeds. Nevertheless, at the highest experimental speed, the tope tail was attaining (again on the basis of the area of its lower lobe) $C_L \approx 1$, and the dogfish tail, at the lowest experimental speed, $C_L \approx 0.6$. In the light of these figures, the requirement for a cephalaspid tail, beating rather gently, to produce a lift coefficient of 0.75 seems entirely reasonable. The lift from the tail will, of course, have to be adjusted continuously to match the lift from the head and body. Any hope that upwardly deflected pectoral fins might confer overall static stability (figure 5) is therefore illusory.

At cruising speed, over half the total drag is due to induced drag. It might therefore be thought that a cephalaspid could achieve significant economy by swimming close to the bottom, so that 'ground effect' would reduce the induced drag (Withers & Timko 1977). The theoretical treatments of ground effect in the aeronautical literature (see, for example, Reid 1932; Bairstow 1946) presuppose wings of relatively high aspect ratio and with near-optimum lift distribution. On these idealized theories a reduction of induced drag by 30% would require the 'standard' cephalaspid to swim at about 2 cm from the bottom, implying either remarkable flatness of the bottom, or skilled (and energetically expensive) manoeuvring in pitch. Admittedly cephalaspids have low aspect ratio and a very abnormal lift distribution, but this is unlikely to lead to a *greater* reduction of drag than that predicted by the simple theory. Another presumption of the theory – that the ground is rigid – might well not be valid if the bottom over which the fish is swimming is uncompacted mud. All in all, it seems unlikely that any significant improvement in the cost of locomotion would be available from ground effect.

### *Musculature and oxygen consumption*

The power required to propel a 'standard' cephalaspid at cruising speed is probably about 0.1 W; this is chosen somewhat larger than the minimum of figure 6 to allow for the effect of boundary-layer thinning. This figure can be used to predict the amount of swimming muscle needed if two assumptions (both, admittedly, open to debate) are made: (1) the propeller efficiency of the tail is 50%, and (2) the specific power output of the muscle is 80 W/kg (Webb 1975). Only 2.5 g of muscle is needed: this would occupy some 2% of the volume of the body and tail.

A 'modern' fish would probably use, as its power source for cruising, a slow, aerobic red muscle (and indeed the specific power output assumed above is for this type of muscle). The remaining space would probably be filled with fast, anaerobic white muscle, whose specific power is higher (about 400 W/kg), but whose endurance is very limited (Webb 1975). At the risk of our speculations becoming too wild, we proceed on the assumption that cephalaspids were built on the same plan.

Within the assumptions of table 1, there is room for 0.16 kg of white muscle, capable of producing 64 W. Assuming now a propeller efficiency of 70% (appropriate to the higher swimming speed), this would propel the 'standard' cephalaspid at a maximum speed of about 3 m/s. Webb (1975) has pointed out that the proportion of red muscle in fish myotomes correlates with the general level of activity of the fish; thus the very active mackerel and trout have around 25% of red muscle, halibut has 7.5%, and benthic fish less than 1%. On this criterion, cephalaspids are to be classified towards the 'sluggish' end of the activity scale.

Estimates of oxygen consumption for locomotion depend critically on the efficiencies assumed for the various stages of energy conversion (Webb 1975). Assuming 25% for the overall efficiency of the muscles, and taking the propeller efficiency again to be 50%, a cruising cephalaspid would have a specific oxygen consumption of about 400 mg $O_2$/(kg h). For maximum sustained activity at normal environmental temperatures, Brett (1972) quotes a range of 300–950 mg $O_2$/(kg h) for five species of 'streamlined, active' fish. Whether the cephalaspid respiratory system, without the benefit of ram ventilation, and perhaps with reduced availability of oxygen in the water, could fully sustain the estimated specific oxygen consumption is by no means certain. If it could not, then presumably even when the animal was cruising the red muscle would be assisted by the anaerobic activity of a little white muscle. Perhaps, even, there was no red muscle at all, and all locomotion was performed anaerobically.

Beamish (1978) has given the specific metabolic cost of transport and its scaling relation with mass for twelve species of fish, covering a mass range of almost a factor of 1000. With the same assumptions about efficiency that were made in the previous paragraph, the specific metabolic cost of transport for a cephalaspid swimming at cruising speed is 4.5 J/(kg m), which lies a factor of 2 above the line of Beamish's scaling relation. It is perhaps surprising that the cephalaspid, needing to support weight and to overcome large drag, should emerge so creditably in comparison with 'modern', streamlined fish. Once again, though, it must be emphasized that the accuracy of the figures is only as good as the validity of the assumptions.

### *Manoeuvrability*

When the *weight* of a fish is largely compensated by buoyancy, it is easy to forget that its *mass* is not compensated at all; thus manoeuvres involving sudden changes of speed or direction, or even only changes of attitude, may require substantial inertial forces. Indeed, not only must the mass of the fish be considered: the surrounding water contributes an additional 'virtual mass' or 'virtual moment of inertia'. It is not possible to calculate these virtual quantities for shapes other than very simple ones (Bairstow 1946), though rough estimates can be made.

In their study of nectrideans, Cruickshank & Skews (1980) noted that support of the negative buoyancy could be achieved at modest speeds and/or modest angles of attack. They pointed out that increased angles of attack could give rise to upward accelerations: 'the hydrodynamic forces that could be developed are sufficient to make the animal highly manoeuvrable in pitch'. It was envisaged that the nectridean could rise off the bottom in a steep climb to capture its prey.

It is clear that a cephalaspid swimming near its top speed could also call on large reserves of lift by merely increasing its angle of attack. To take a concrete example: suppose that a 'standard' cephalaspid is swimming at 2 m/s, and suddenly reduces to zero the lift generated by the tail. The now uncompensated nose-up pitching moment generated by the lift of the head rotates the fish in pitch so that the angle of attack increases. Crude calculations, based on estimates of the moment of inertia of the fish and associated water, suggest that the angle of attack might increase to 10° within a quarter of a second. The resulting excess of lift would put the fish into a steep climb, so that it could become vertical within another quarter of a second.

The pitching plane is not the only one in which a cephalaspid could execute sharp manoeuvres. The pectoral fins could be used differentially – one up and one down, after the manner of aircraft ailerons – to initiate rolling. From figure 4 one can estimate the rolling moment produced by fin deflexions of +20° on one side and −20° on the other side. The forces opposing rolling are not easy to calculate: the retarding couple arises from hydrodynamic drag forces, from inertial forces due to the fish's mass, and from the forces due to the virtual moment of inertia attributable to the water. Once again, crude estimates can be made; it appears that differential use of the pectoral fins could roll the fish through 90° in less than a quarter of a second. Then a suitable increase in angle of attack would produce excess lift now acting *sideways*, pulling the fish round in a tightly banked turn. For a fish with a relatively stiff body and rather small vertical fin area, turning in this way is likely to be much faster than the 'conventional' fish turn executed entirely in the yawing plane, with the aid of asymmetrical tail movements.

## Conclusions

The results of our biomechanical speculations are consistent with a fish which could (1) rest on the bottom undisturbed by even strong water currents, (2) cruise at a modest speed (around one length per second), (3) swim rapidly (at eight lengths per second or so) for short periods, and (4) during rapid swimming, execute sharp manoeuvres in both the horizontal and vertical planes.

If cephalaspids were to live in fast-moving water, they would thus have the top speed and agility to swim against the current and to avoid obstacles, but probably with only short endurance. Alternatively, if they were to live in relatively still water, they could put to good use their speed and agility (and perhaps their ability to raise an obscuring cloud of mud with the starting vortex of a sharp climb) to escape from a predator who had also discovered that the impressive head and body armour was, in fact, rather thin.

We do not feel competent to judge the likelihood of either of these possibilities.

## Acknowledgements

We received invaluable help and advice from Dr J. A. Clack during the construction of the model cephalaspid. We are indebted to Dr L. C. Squire for allowing us to use the wind tunnel, and to Mr K. Saunders for his patience and assistance during the tunnel experiments. Dr P. Forey of the British Museum (Natural History) kindly granted access to fossil specimens, and provided extremely helpful information on cephalaspid morphology. Our ideas have evolved (and in some cases, become extinct) owing to the selective pressure of discussion with our colleagues, both in Cambridge and at the Manchester SEB/Palaeontological Association meeting. We are most grateful to all of them, but particularly to Dr C. P. Ellington and Dr K. A. Joysey.

## References

Alexander, R. McN. 1965 The lift produced by the heterocercal tails of Selachii. *Journal of Experimental Biology* **43** , 131–138.

Bainbridge, R. 1958 The speed of swimming of fish as related to size and to the frequency and amplitude of the tail beat. *Journal of Experimental Biology* **35**, 109–133.

Bairstow, L. 1946 *Applied Aerodynamics*. London: Longman's, Green.

Beamish, F.W.H. 1978 Swimming capacity. In *Fish Physiology*, ed. W.S. Hoar & D.J. Randall, vol. VII, pp. 101–187. New York: Academic Press.

Brett, J.R. 1972 The metabolic demand for oxygen in fish, particularly salmonids and a comparison with other vertebrates. *Respiration Physiology* **14**, 151–170.

Cruickshank, A.R.I. & Skews, B.W. 1980 The functional significance of nectridean tabular horns. *Proceedings of the Royal Society of London* B **209**, 512–537.

Dommasch, D.O., Sherby, S.S. & Connolly, T.F. 1967 *Airplane Aerodynamics*. New York: Pitman.

Harris, J.E. 1936 The role of the fins in the equilibrium of the swimming fish. I. Wind tunnel tests on a model of *Mustelus canis* (Mitchill). *Journal of Experimental Biology* **13**, 474–493.

Heintz, A. 1967 Some remarks about the structure of the tail in cephalaspids. In *Problèmes Actuels de Paléontologie* (*Evolution des Vertébrés*), pp. 21–35. Paris: Editions du C.N.R.S.

Kermack, K.A. 1943 The functional significance of the hypocercal tail in *Pteraspis rostrata*. *Journal of Experimental Biology* **20**, 23–27.

Lighthill, M.J. 1971 Large-amplitude elongated-body theory of fish locomotion. *Proceedings of the Royal Society of London* B **179**, 125–138.

Moy-Thomas, J.A. & Miles, R.S. 1971 *Palaeozoic Fishes*. London: Chapman and Hall.

Pennycuick, C.J. 1971 Gliding flight of the dog-faced bat *Rousettus aegyptiacus* observed in a wind tunnel. *Journal of Experimental Biology* **55**, 833–845.

Reid, E.G. 1932 *Applied Wing Theory*. New York: McGraw Hill.

Stensiö, E.A. 1927 The Downtonian and Devonian vertebrates of Spitzbergen. I. Family Cephalaspidae. *Skrifter om Svalbard og Nordishavet* **12**.

Stensiö, E.A. 1932 *Cephalaspids of Great Britain*. London: British Museum (Natural History).

Vogel, S. 1981 *Life in Moving Fluids*. Boston, Massachusetts: Willard Grant Press.

Webb, P.W. 1975 Hydrodynamics and energetics of fish propulsion. *Bulletin of the Fisheries Research Board of Canada* **190**, 1–159.

Withers, P.C. & Timko, P.L. 1977 The significance of ground effect to the aerodynamic cost of flight and energetics of the black skimmer (*Rhynchops nigra*). *Journal of Experimental Biology* **70**, 13–26.

JÜRGEN RIESS & E. FREY

# The evolution of underwater flight and the locomotion of plesiosaurs

The evolutionary progression from rowing to underwater flight in tetrapods is discussed, and is illustrated with turtles and plesiosaurs as examples. A biomechanical analysis of the locomotor apparatus of plesiosaurs leads to the conclusion that these animals have two pairs of hydrofoils – a character unique among vertebrates – and that the upstroke is passive owing to the morphology of their girdles.

## Introduction

Swimming vertebrates commonly use one of two forms of locomotion, *axial* or *paraxial* swimming (figures 1 and 2). Axial swimmers, such as most Cetacea, sea snakes and newts, move by either horizontal (figure 1*a*) or vertical (figure 1*b–d*) undulations of the body axis; paraxial swimmers obtain propulsion by movements of their extremities. For a classification of swimming modes in vertebrates see the survey by Braun & Reif (1985). Paraxial swimmers include animals which propel themselves by accelerating water backwards by means of oar-like limbs, which produce small strong 'drag' vortices (Rayner 1985). Such creatures are defined as *rowing*, and include most ducks (Anatidae), frogs (Anura) and polar bears (*Thalarctos maritimus*) (figure 2*a–e*). They may swim either on the water surface or submerged. Other paraxial swimmers move their extremities like wings, producing 'lift' vortices which are large in size but have low circulation (Rayner 1985). These animals are termed *underwater fliers*, and include penguins (Spheniscidae), auks (Alcidae) and sea turtles (Chelonidae and Dermochelidae) (figure 2*f–h*). The propulsive organs of these animals are not oar-like, but form hydrofoils. The characteristic differences between rowing and underwater flying are central to our discussion, and are discussed briefly here.

Rowing paraxial swimmers generate thrust as a reaction to drag on the surfaces of their limbs. A single rowing stroke can be divided into two

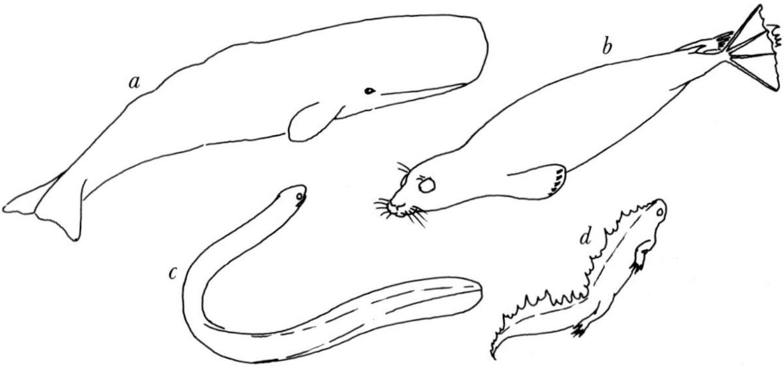


Figure 1. Axial swimming tetrapods. (*a*) Sperm whale (*Physeter catodon*, Odontoceti); (*b*) common seal (*Phoca vitulina*, Phocidae); (*c*) sea snake (*Laticauda laticauda*, Hydropheidae); (*d*) great crested newt (*Triturus cristatus* , Urodela).

Figure 2. Paraxial swimming tetrapods. Rowers with alternating strokes: (*a*) coot (*Fulica atra*, Rallidae); (*b*) polar bear (*Thalarctos maritimus*, Ursidae); (*c*) freshwater turtle (*Pseudemys* sp., Emydidae). Rowers with synchronous strokes: (*d*) African clawed toad (*Xenopus laevis*, Anura); (*e*) darter (*Anhinga rufa*, Anhingidae). Rowing fliers: (*f*) California sealion (*Zalophus californianus*, Otariidae). Underwater fliers: (*g*) leatherback turtle (*Dermochelys coriacea*, Dermochelidae); (*h*) penguin (*Spheniscus* sp., Spheniscidae).

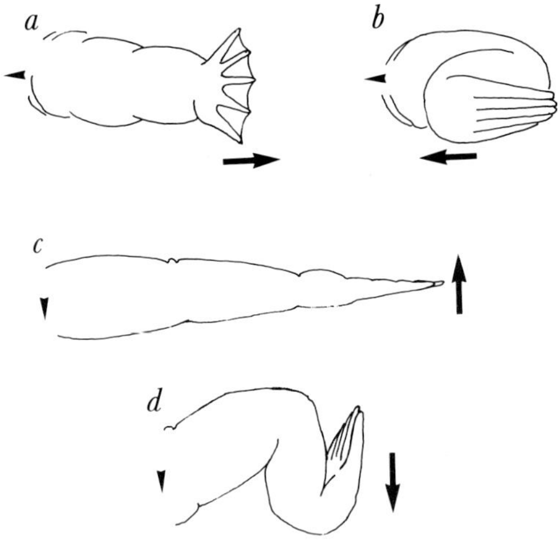


Figure 3. Principles of rowing illustrated with a generalized extremity. The triangular arrows mark the direction of locomotion, the other arrows the direction of movement of the limb. (*a*) and (*b*) Lateral view of stroke cycle (*a*, power stroke; *b*, recovery stroke). (*c*) and (*d*) Dorsal view of stroke cycle (*c*, power stroke; *d*, recovery stroke). The direction of movement of the limb is fore-and-aft.

components: the power stroke and the recovery stroke (figure 3). During the power stroke (figure 3*a*, *c*) the limb is pushed posteriorly relative to the body, with maximum drag to accelerate a mass of water backwards and give the body a forward impulse. The recovery stroke (figure 3*b*, *d*) moves the limb forwards again with minimum drag. The resulting locomotory force is discontinuous, but this can be made steadier by alternating power strokes between limbs on either side of the body (figure 2*a*–*c*) or between fore- and hindlimbs. Rowing efficiency can be improved by minimizing drag during the recovery stroke, since this limits mean forward speed of the animal. This limit is also related to the absolute size of the animal: the greater the size, the less favourable is the relation between maximum power stroke drag for propulsion and the hydrodynamic resistance of the body. With increasing mass (and hence inertia) thrust discontinuity also becomes a problem: during the power stroke the body is accelerated, during the recovery stroke the animal loses speed. This means that with each power stroke a rowing animal has to accelerate its body mass to a velocity greater than the mean forward speed. This is no problem for small creatures such as protozoa or swimming insects, but larger animals require considerable energy to maintain even a moderate speed by rowing. For this reason subaqueous rowers among vertebrates swim only short distances. Polar bears, the largest recent rowing

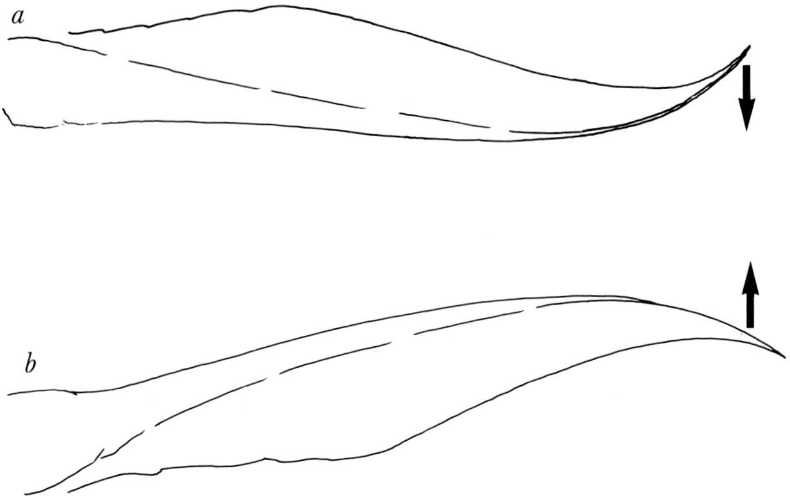


Figure 4. Principles of underwater flight shown at a generalized hydrofoil seen from the front. (*a*) Downstroke with the hydrofoil pronated; (*b*) upstroke with the hydrofoil supinated. The arrows mark the up– and downstrokes of the hydrofoil.

swimmers, prefer to swim at the water surface where drag is considerably less than when fully submerged; they try to avoid swimming longer distances or diving.

Underwater flight is achieved by means of pairs of hydrofoils moving synchronously. Unlike the rowing stroke cycle, the main direction of movement of the hydrofoils is perpendicular to the swimming axis (figure 4). The hydrofoils can produce propulsive force throughout the stroke cycle, and therefore the locomotion may be more continuous. Compared to rowing, underwater flight is a more highly derived mode of swimming in tetrapods. In the following discussion we use a biomechanical model to explore how underwater flight may have evolved from rowing, under the premise that no 'disadvantageous' biological forms should occur as an 'intermediate' state.

## Taxa with both rowers and underwater fliers

The question of the origin of underwater flight can be answered only by considering animal groups in which both forms of locomotion are realized. Although birds include both rowers and underwater fliers, they are inappropriate for this purpose, since their prime locomotor mode is aerial flight; underwater flight is essentially the same form of movement as this (Clark & Bemis 1979), and it is probable that underwater flight in birds did not evolve from rowing. This is demonstrated clearly by the auks (Alcidae), which fly with equal proficiency in both water and air; the largest of the family, the extinct great auk *Alca impennis*, was an underwater flier but was flightless in air.

We therefore concentrate on two groups: the Sauropterygia and the Chelonia. The Sauropterygia comprise the Nothosauria, which were rowers, and the Plesiosauria, which were underwater fliers. The monophyly of the group has been confirmed by Schmidt (1985, 1986, 1987), who also considered the locomotor modes of nothosaurs. Schmidt clearly demonstrated that the common ancestor of plesiosaurs and nothosaurs was a rowing animal. There is equally little doubt concerning the monophyly of the Chelonia (Gaffney 1975, 1984). The most primitive of the aquatic forms, the Pleurodira, and the more derived Emydidae (figure 2*c*) swim by rowing, while *Dermochelys* (figure 2*g*) and the Chelonidae (marine turtles) use underwater flight with the forelimbs. *Carettochelys* and the Trionychidae are able to use both modes of swimming (Davenport, Munks & Oxford 1984; Walker 1971).

There are of course limitations to any attempt to reconstruct the functional morphology or behaviour of extinct taxa. For instance, muscles can be reconstructed only in the most general way: since individual muscles seldom have identifiable insertion marks on the bone, and within conspecific living animals there is a considerable degree of variability in the musculature, any attempt at muscle reconstruction must be conjectural. The resulting model may give a general idea of what the animal looked like, but it is unlikely to be realistic enough to form a base for calculations. There is a similar problem with the physical properties of bio-materials. All that can be said realistically is that the materials were much like those found today; but we can only guess at the special modifications that an animal might have evolved to meet its particular needs. Calculations can only indicate a range of feasible values, based on those appropriate for contemporary animals; but even this can be wrong. From our point of view there is no question that when the calculations disagree with the documented biology, then it is the biological evidence which should be trusted: the experience of biologists shows that even in living organisms many ‘impossible’ constructions are realized.

To provide a basis for the biomechanical consideration of underwater flight, we give first a short general outline of the skeletons of chelonians and plesiosaurs (figure 5*a*, *b*).

In the Chelonia the modes of locomotion are restricted by the carapace. The shoulder girdle is situated inside the rib cage, and movements of the humerus are confined by the borders of the shell. For this reason the

a

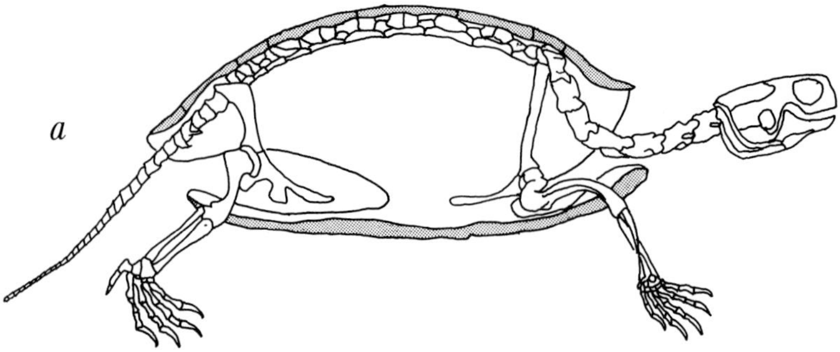

b

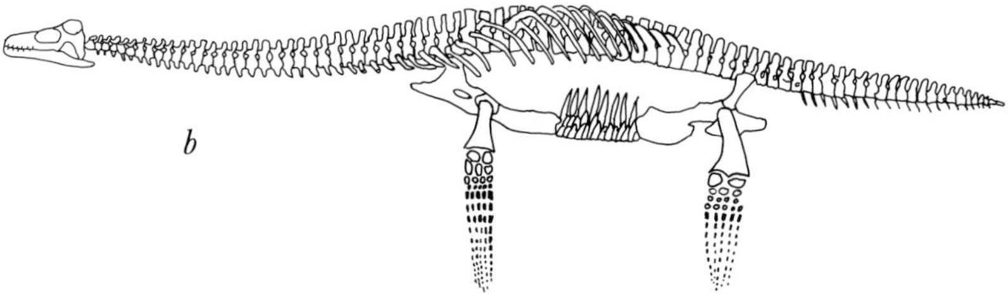

Figure 5. Schematic drawings of the skeletons of (*a*) a typical chelonian and (*b*) a typical plesiosaur.

functional underwater wing in turtles is confined to the portion of the limb distal to the humerus: the wing moves partly around the distal end of the humerus, and the humerus acts as a lever arm that connects the girdle with the wing.

Plesiosaurs, which lived in Jurassic and Cretaceous oceans, ranged from 2 to 18 m in length. Two constructional types can be distinguished: the plesiosaurs *sensu stricto*, which are characterized by very long necks and small heads, and the pliosaurs, which had long and broad heads and relatively short necks. In plesiosaurs, as in chelonians, the trunk is stiffened, but by an 'internal carapace' (for general descriptions see Romer (1959, 1976), Williston (1914); figure 5*b*). The two-headed ribs are strong, broad bones extending mainly in the lateral direction from the vertebral column to stiffen the dorsal part of the thorax. On the ventral side the dorsally reduced girdles form large, bony plates, which leave only a small gap between shoulder girdle and pelvis. This gap is filled by very well developed gastral ribs, which are situated inside the rectus muscles, and form a massive and hardly movable plastron. Most reconstructions of the skeleton show the distance between ribs and plastron too large: figure 5*b* attempts to correct this. As an adaptation to marine life all plesiosaurs have no bony connection

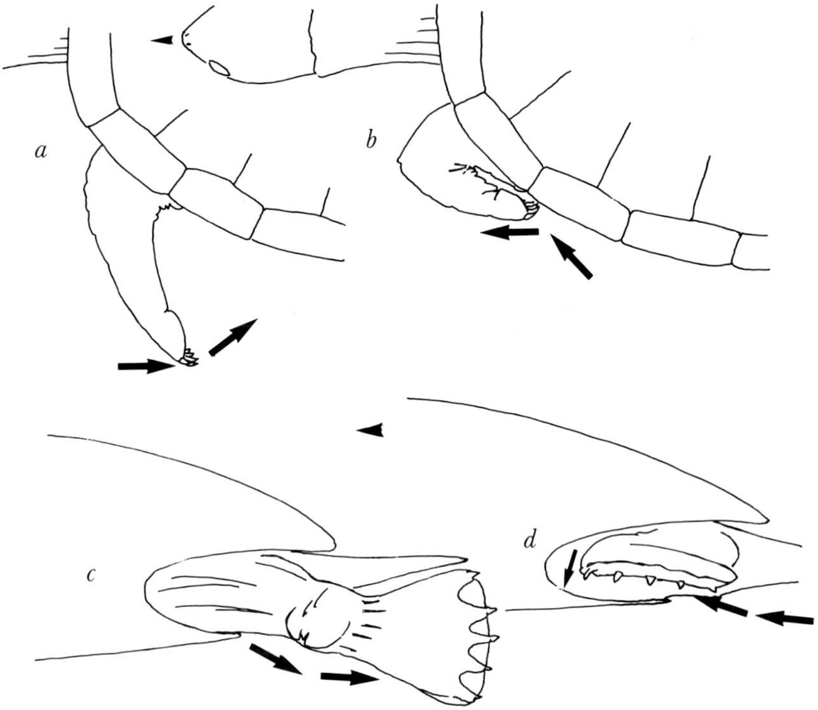


Figure 6. The principles of the recovery stroke. The triangular arrows mark the direction of locomotion. (*a*) and (*b*) Dorsal view of recovery stroke of an emydid front limb; the arrows show how the limb is (*a*) folded towards the body and then (*b*) is swung anteriorly. (*c*) and (*d*) lateral view of recovery stroke of an emydid hind limb; the limb supinates from the posteriormost position (*c*) into the horizontal plane and then is brought forward nearly fully extended (*d*); the power stroke is induced by a pronation.

of the ilium to the sacral rib. Altogether the body outlines of turtles and plesiosaurs are very similar, with the obvious exception of the neck and head region.

**From rowing to underwater flight**

As we have discussed above, rowing can be optimized by reducing the drag of the recovery stroke. Paraxial swimming vertebrates do this in two essentially different ways. The first can be observed in the Emydidae (figure 6*a*, *b*): the front limb is pulled towards the body and the webbed toes are folded close together. All joints of the extremities are in action during the recovery stroke until the limbs are unfolded at the beginning of the power stroke, and strong muscles are needed to move and to control them. As a

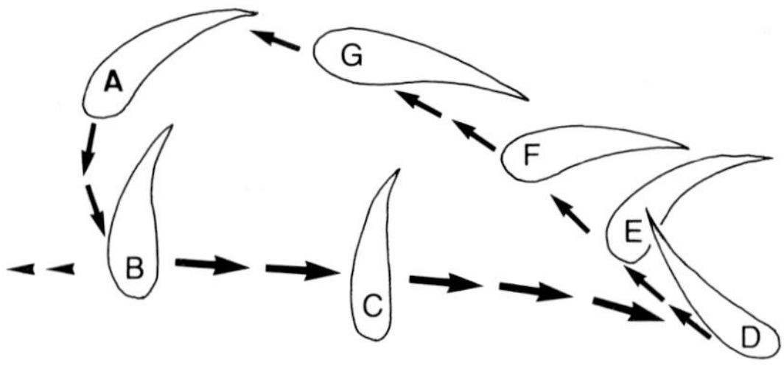


Figure 7. Principle of rowing flight shown in schematized cross-section. The direction of movement is shown by triangular arrows. The power stroke (broad arrows) starts at B and is completed at D. Then the hydrofoil is supinated and sweeps antero-dorsally (E–A shows the recovery stroke, thin arrows). From A the power stroke is induced by pronation and a downward movement, whereby the lift produced by the hydrofoil is inclined into the direction of locomotion (A–B).

consequence it is necessary to accelerate large limb masses during locomotion. However, owing to their primitive tetrapod construction, the extremities remain useful for terrestrial locomotion and for bottom walking as well as for swimming, and they confer a high degree of manoeuvrability in water. This form of rowing requires neuromuscular control similar to that of terrestrial locomotion; as in swimming, it is advantageous to fold the autopodia and to bring the limb towards the body in walking to avoid friction through grass and other irregularities of the soil surface. In short, the limb is of a general locomotion type. Owing to the high mobility of the limb segments relative to each other it is hard to imagine how underwater flight may have evolved from rowing with this type of recovery stroke (Davenport *et al.* 1984; Marlow 1985).

A second form of recovery stroke is less widespread among aquatic vertebrates; emydids use it only in the hindlimb (figure 6*c*, *d*), and it is seen occasionally in the trionychids, in *Carettochelys* and in juvenile *Caretta* (Davenport & Clough 1986). Only the otariids (figure 2*f*) adopt it frequently (English 1976; Godfrey 1984). At the posterior-most point of the power stroke the extremity is supinated and swept forward without being folded (figure 6*c*, *d*; figure 7, positions *D*–*G*). This creates considerable drag along the surface of the limb, but the lift which also is produced can compensate for the drag, despite its different direction relative to the movement of the limb. If the limb is shaped like a hydrofoil, with a rounded leading edge and a sharp, flexible trailing edge (figure 7), the lift can exceed the drag in magnitude. In hydrofoil swimming the mobility of the limb elements can be restricted, and this facilitates reduction of the distal muscle masses and their substitution by tendons and/or connective tissue, and allows limb mass to be reduced. When the limb is moved anteriorly, the lift delivered by this type of recovery stroke provides a dorsally directed force component (figure 7). The power stroke is induced by pronation of the entire limb in the glenoid, with a simultaneous initiation of the downstroke (figure 6*d*; figure 7, positions A, B). During this movement the lift of the hydrofoil-shaped limb acts parallel to the direction of locomotion. After it has been pronated into a vertical plane, the limb is pulled back with maximum drag (figure 7, positions B–D). This can lead to a form of locomotion which is best described by the term 'rowing flight', because components of both rowing (drag based fore-and-aft) and 'flying' (lift based up-and-down) movements can be observed in a single stroke cycle. Similar movement patterns in portunid crabs and some fossil eurypterids are considered by Briggs *et al.* (this volume).

It is obvious, however, that modifications of the limbs which are effective in the recovery stroke in rowing flight constrain the utility of the extremity on solid ground; they therefore can be regarded as an adaptation to a more pelagic way of life.

From rowing flight it is relatively straightforward to imagine how underwater flight might have evolved as the fore-and-aft (drag) component of the stroke cycle is gradually reduced without any overall loss of thrust or any malfunction. Among the Chelonia, form and function of the extremities clearly demonstrate this: the intermediate rowing flight stage is represented by the Trionychidae and in the genus *Carettochelys*.

## Plesiosaurs: advanced underwater fliers

According to the predictions of the model outlined above, plesiosaurs descended from a rowing animal which used a horizontal recovery stroke with fully extended extremities. Such a creature might have been very similar to the nothosaur *Ceresiosaurus* from the Upper Muschelkalk (Landin) of Monte St Georgio in Italy (Peyer 1931). The bones of the Stylo-, Zeugo- and Autopodia are flattened and indicate a hydrofoil-like cross-section through the limb. The glenoid is a well-developed ball-and-socket joint, as is necessary for limb rotation during locomotion. With the exception of those between carpus and tarsus, the remaining joints are reduced in function.

The final result of the tendency to transform the oar-like limb into an effective hydrofoil are seen in the underwater wing of the plesiosaurs. The Stylo- and Zeugopodia are similarly flattened, but also reduced in length, while the bones of the Autopodia are enlarged in both number and relative size. Apart from the humero-glenoidal ball-and-socket joint, there are no functional joints in the extremities. A remarkable flexible trailing edge to the hydrofoil is well documented in the fossil record, for instance in the specimen of *Plesiosaurs brachypterygius* from the Liassic of Holzmaden now preserved at Tübingen (Robinson 1975). The plesiosaur hydrofoil is ideally designed for underwater flight (see figure 9*a*, below). However, some unusual anatomical features prevented former palaeontologists from reconstructing plesiosaurs as underwater fliers, and these must be considered in any functional reconstruction of the animal. The most serious problem is the lack of a dorsal expansion to the scapula or ilium to serve as an attachment area for the upstroke muscles. The ilium is small, and the scapula, with the exception of a minute dorsal process, is turned into an ante-glenoidal position. The ventral elements of the girdles are tremendously enlarged to deliver attachment areas for strong muscles. Plesiosaurs possessed two pairs of hydrofoils, and this also prevented their reconstruction as underwater fliers, since all other recent and extinct tetrapods using this form of locomotion have a single pair only.

The extension of the ventral girdle plates anterior and posterior to the basal limb joints finally led to an accepted reconstruction of plesiosaurs as rowers. For a long period this seemed the only plausible solution, until consideration of the Reynolds number of swimming plesiosaurs set an end to the hypothesis. Rowing is simply not sufficiently efficient for an active predator of this size (Nachtigall 1977). Curiously, Nachtigall describes plesiosaurs as inefficient rowing animals, and attributes their extinction to this.

More recently, a few authors have tried to model plesiosaurs as flying through the ancient seas. However, all hypotheses which have been advanced have experienced problems with the apparent lack of upstroke muscles, and none has hitherto solved the problem of the four hydrofoils (Robinson 1975, 1977; Godfrey 1984). However, plesiosaur locomotion can be explained plausibly without the need to invoke migration of muscles (S. Tarsitano, personal communication), or without requiring that upstroke muscles originate from the interior aspect of the ventral girdle plates, which is in fact anatomically impossible (Robinson 1977).

The simplest part of the plesiosaurian wing stroke is the downstroke. The large masses of muscle attached to the ventral plates of the girdles indicate that this was certainly powerful and made the dominant contribution to thrust.

The cross-section of a plesiosaur shoulder girdle (figure 8, after Andrews 1895) clearly shows the presence of a ventral keel. There is a similar keel in some ichthyosaurs (Riess 1986) and in the Amazon dolphin *Inia* (Klima, Oelschläger & Wunsch 1980), and it is a further indication of the presence of

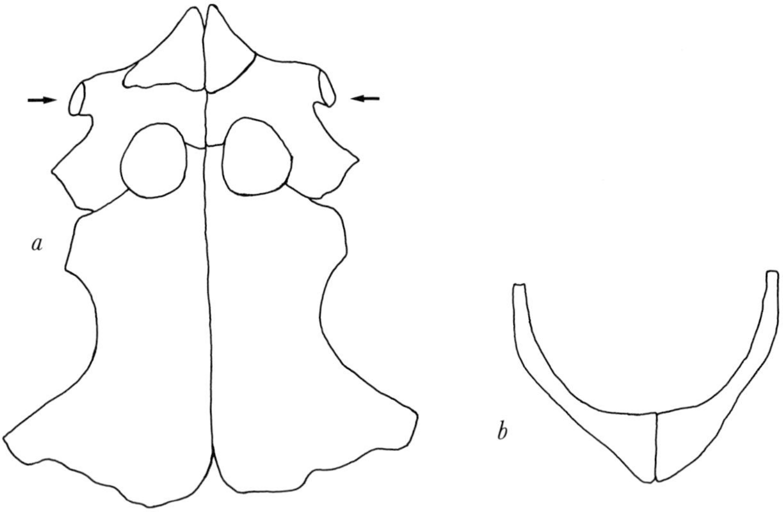


Figure 8. (*a*) Dorsal and (*b*) cross-sectional views of the shoulder girdle of a plesiosaur. The arrows mark the area of the cross-section. (After Andrews 1895.)

powerful downstroke musculature. The keel improves the lever arm of the muscles and increases the diameter of the body and the area for muscle attachment, and it also stabilizes the bones of the girdle during the downstroke. The comparable stabilizing structure necessary for a powerful *upstroke*, as in the dermal girdle in ichthyosaurs and the sternum–scapula connection in *Inia*, is absent in plesiosaurs.

The absence of dorsal muscles means that the animal was unable to develop an active (muscle-powered) upstroke to elevate the wing, but rather that the wing was elevated passively in the upstroke by hydrodynamic forces (figure 9*a*; Frey & Riess 1982). (Rayner (1985 and this volume) uses the terms 'active' and 'passive' in the opposite sense relating to the aerodynamic action of the avian wing in flight.) This passive elevation can be possible only with sufficient forward motion to provide the necessary lift. We have suggested that this type of wingbeat cycle requires the up– and downstrokes of the two limb pairs to be alternate: one pair moves ventrally in the downstroke providing thrust; the other pair is supinated into the water flow with a positive angle of attack. This pair is swept up passively, reaching the dorsal extreme of the wingbeat cycle when the other wing pair is at the ventral extreme; the downstroke is then initiated by pronation of the wing.

It is possible that the co-ordination of the limbs of plesiosaurs might have been similar to that in terrestrial movement (Ball, *in litt.*). This would have

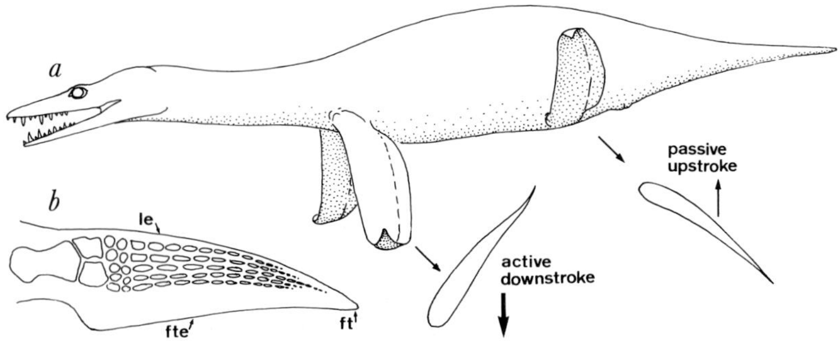


Figure 9. Locomotion of a plesiosaur (*a*) and the construction of its hydrofoil (*b*). le, leading edge; fte, flexible trailing edge; ft, flexible tip. See text for further explanation.

avoided alternation of lift at the front and the back of the trunk. This pattern of limb movement would require no additional steering or ballasting devices (the latter function was suggested for the gastroliths by Darby & Ojakangas (1980), Taylor (1981) and Rayner (1985)). Horizontal forces associated with the alternating limb beats could readily be counteracted by the stiffness of the plesiosaur trunk. It would not be surprising if the plesiosaur body was stabilized by skin folds along the body, although no structures of this kind have been preserved in the fossil record.

When plesiosaurs were moving slowly or starting it is most likely that fore-and-aft (largely drag) movements of the flippers dominated. (So in this sense rowing would have initiated underwater flight!) When moving slowly plesiosaurs might have been able to propel themselves backwards as well as forwards; whether this ability might have had any ecological significance cannot be said.

## Summary

By comparative biomechanical studies of turtles and plesiosaurs a model for the transition from rowing to underwater flight in swimming vertebrates has been developed. It is shown that such a transition is possible only if in the rowing recovery stroke the extremity is swept forward fully extended. For this movement the flattening of the limb to a hydrofoil is advantageous, and provides additional propulsion by thrust when the limb is pronated and moved ventrally at the beginning of the power stroke. Extremities acting in this intermediate way are documented from some turtle groups and are assumed for some nothosaurs. Gradually the fore-and-aft rowing movement can be reduced and the dorso-ventral 'flying' movement begins to dominate. This intermediate stage between rowing and underwater flight can be observed in otariids, in trionychid and emydid turtles (hindlimbs only) and in *Carettochelys*, and has been termed 'rowing flight'. With the complete reduction of the rowing components and the immobilization of the distal limb joints the full transition to underwater flight is achieved, with no need to invoke evolutionary jumps or interim maladapted phases. The hypothesized ancestry of plesiosaurs appears to support this model.

Nothosaurs seem primarily to have been rowers, and it is likely that the common ancestor of nothosaurs and plesiosaurs swam in this way also. Future research is required to explore how far plesiosaur phylogeny reflects this hypothetical transition to underwater flight.

The unique construction of the plesiosaur locomotor apparatus with two pairs of wings and minimal dorsal expansion of the girdles can be explained by a passive upstroke induced by supinating the hydrofoil into a positive angle of attack relative to the water flow. Since the thrust produced by one pair of wings is discontinuous, a second pair of hydrofoils acting alternately is necessary to provide sufficient water velocity for the wing pairs to be elevated passively, and to ensure continuous locomotion.

## References

Andrews, C.W. 1895 On the development of the shoulder-girdle of a plesiosaur (*Cryptocleidus oxoniensis*, Phillips, sp.) from the Oxford Clay. *Annals and Magazine of Natural History* (6) **15**, 333–346.

Braun, W.E. & Reif, W.-E. 1985 A survey of aquatic locomotion in fishes and tetrapods. *Neues Jahrbuch für Geologie und Paläontologie Abhandlungen* **169**, 307–332.

Clark, D.B. & Bemis, W. 1979 Kinematics of swimming of penguins at the Detroit Zoo. *Journal of Zoology, London* **188**, 411–428.

Darby, D.G. & Ojakangas, R.W. 1980 Gastroliths from an upper Cretaceous plesiosaur. *Journal of Paleontology* 54, 548–556.

Davenport, J. & Clough, W. 1986 Swimming and diving in young loggerhead sea turtles (*Caretta caretta* L.). *Copeia*, pp. 53–57.

Davenport, J., Munks, S.A. & Oxford, P.J. 1984 A comparison of the swimming of marine and fresh-water turtles. *Proceedings of the Royal Society of London* B **220**, 447–475.

English, A.W. 1976 Limb movements and locomotor function in the California sea lion (*Zalophus californianus*). *Journal of Zoology, London* **178**, 341–364.

Frey, E. & Riess, J. 1982 Considerations concerning plesiosaur locomotion. *Neues Jahrbuch für Geologie und Paläontologie Abhandlungen* **164**, 193–194.

Gaffney, E.S. 1975 A phylogeny and classification of the higher categories of turtles. *Bulletin of the American Museum of Natural History* **155**, 387–486.

Gaffney, E.S. 1984 Historical analysis of theories of the chelonian relationships. *Systematic Zoology* **33**, 283–301.

Godfrey, S.J. 1984 Plesiosaur subaqueous locomotion: a reappraisal. *Neues Jahrbuch für Geologie und Paläontologie Monatshefte*, pp. 661–672.

Klima, M., Oelschläger, H.A. & Wunsch, D. 1980 Morphology of the pectoral girdle in the Amazon dolphin *Inia geoffrensis* with special reference to the shoulder joint and the movement of the flippers. *Zeitschrift für Säugetierkunde* **45**, 288–309.

Marlow, R.W. 1985 Chelonian locomotion: limited options. *Konzepte SFB*230 (*Natürliche Konstruktionen*) **4**, 139–143.

Nachtigall, W. 1977 Zur Bedeutung der Reynoldszahl und der damit zusammenhängenden strömungsmechanischen Phänomene in der Schwimmphysiologie und Flugbiophysik. *Fortschritte der Zoologie* **24**, 13–56.

Peyer, B. 1931 Die Triasfauna der Tessiner Kalkalpen. IV. *Ceresiosaurus calcagni* nov. gen. & nov. spec. *Abhandlungen der Schweizerischen Paläontologischen Gesellschaft* **51**, 3–68.

Rayner, J.M.V. 1985 Vorticity and propulsion mechanics in swimming and flying vertebrates. *Konzepte SFB*230 (*Natürliche Konstruktionen*) **4**, 89–117.

Riess, J. 1986 Fortbewegungsweise, Schwimmbiophysik und Phylogenie der Ichthyosaurier. *Paläontographica* A **192**, 93–155.

Robinson, J.A. 1975 The locomotion of plesiosaurs. *Neues Jahrbuch für Geologie und Paläontologie Abhandlungen* **149**, 286–332.

Robinson, J.A. 1977 Intracorporal force transmission in plesiosaurs. *Neues Jahrbuch für Geologie und Paläontologie Abhandlungen* **153**, 86–128.

Romer, A.S. 1959 *The Vertebrate Story*, 4th edition. University of Chicago Press.

Romer, A.S. 1976 *Osteology of the Reptiles*, 3rd edition. University of Chicago Press.

Schmidt, S. 1985 Biomechanics of nothosaurs. *Konzepte SFB*230 (*Natürliche Konstruktionen*) **4**, 193–197.

Schmidt, S. 1986 *Nothosaurier*. Inaugural dissertation, Universität Tübingen.

Schmidt, S. 1987 Phylogenie der Sauropterygier (Diapsida; Trias-Kreide). *Neues Jahrbuch für Geologie und Paläontologie Abhandlungen* **173**, 339–375.

Taylor, M.A. 1981 Plesiosaurs: rigging and ballasting. *Nature, London* **280**, 628–629.

Walker, W.F. 1971 Swimming in sea turtles of the family Chelonidae. *Copeia*, pp. 229–233.

Williston, S.W. 1914 *Water Reptiles of the Past and Present*. University of Chicago Press.

KEVIN PADIAN

# Pterosaurs: were they functional birds or functional bats?

## Introduction

Pterosaurs, the flying reptiles of the Mesozoic, were the first of three vertebrate groups to evolve active flight (the others being birds and bats). They first appeared in the Late Triassic, about 220 million years ago, and persisted to the end of the Cretaceous, about 65 million years ago. They were almost exact contemporaries of the great Mesozoic dinosaurs, and quite closely related to them. Pterosaurs ranged in size from sparrows to small aircraft; the largest, *Quetzalcoatlus*, had a wingspan of about 11 m. Pterosaurs are generally divided into two groups (figure 1). The earlier 'Rhamphorhynchoidea', a paraphyletic group, had long tails, short metacarpals and long fifth toes, and a 'dinosaurian' carriage in which the head was held in front of and slightly above the torso. The later Pterodactyloidea, which presumably evolved from a Late Jurassic 'rhamphorhynchoid' like *Rhamphorhynchus*, lost the tail and the fifth toe, elongated the metacarpus into the longest element of the wing, and adopted a more 'bird-like' posture in which the neck was curved in an S shape to hold the head higher above the torso, entering the cranium from below rather than from behind (Wellnhofer 1978; Padian 1983*b*).

Pterosaurs have always been recognized as archosaurs, the group that includes crocodiles and birds, but they have been traditionally interpreted as bat-winged gliders with the physiology of a lizard. Empirical evidence from their skeletons and anatomical comparisons with birds, bats, gliders and lizards suggest a very different interpretation. Pterosaurs are more likely to have been active flappers with bird-like posture, wing planform and physiology. They seem to have been quite capable terrestrial bipeds. A re-analysis of their wing area, structure and planform suggests much better and more versatile flight performance than was previously calculated. If this new model is correct, it raises the question of why previous models were not, and

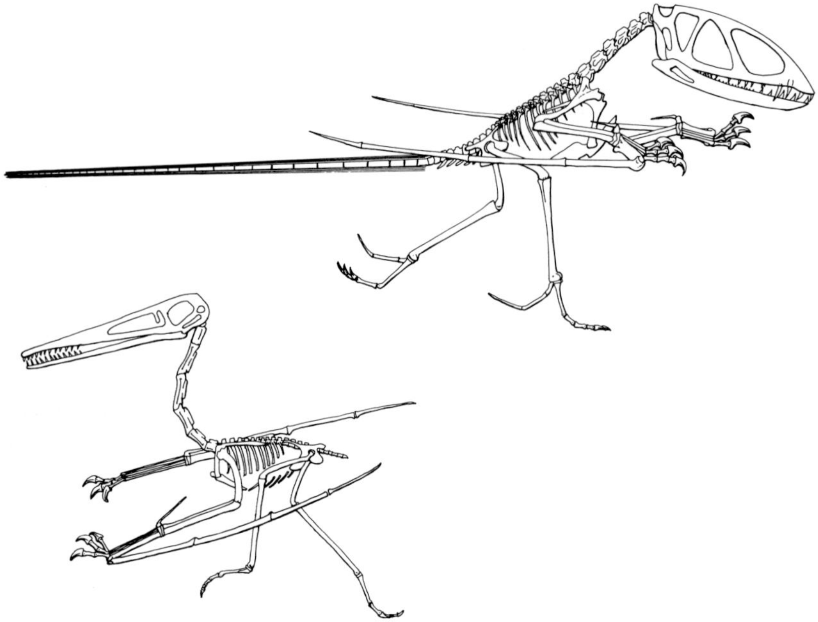

Figure 1. Two generalized forms of pterosaurs: the early 'rhamphorhynchoid' *Dimorphodon* (skull length about 20 cm) and the later, more derived pterodactyloid *Pterodactylus* (skull length about 7 cm). Left forelimbs shown extended, right forelimbs retracted.

thus provides an entry to the analysis of method in biomechanics, especially as applied to extinct organisms.

To begin with, I should clarify that I take a relatively broad view of biomechanics, which I would loosely define as the analysis of physical forces in living organisms. This can range from structural properties of tissues to the physical consequences of locomotion. The relationships between the latter and general functional morphology is refined by the analysis (and frequently quantification) of forces. One point I want to make in this chapter is that, when dealing with extinct organisms, some understanding of functional morphology is important to the application of purely biomechanical analyses; and a correct understanding of functional morphology, in turn, depends on getting the anatomy and evolutionary history right. With extinct organisms we have to make many assumptions about functional morphology, but the possibilities can be constrained to some degree by knowing the evolutionary history of the forms (see, for example, Lauder, this volume).

Influences of history and of several approaches to functional morphology have shaped the interpretations of pterosaurs through time. This chapter can recount some of those influences only in an introductory way; fuller explanations can be found in the literature cited. In the context of this volume, it is of interest to explore the rules for the application of biomechanical analysis to extinct organisms, for two reasons. First, pterosaur flight lends itself naturally to the kinds of biomechanical analyses used on living birds and bats. Second, the history of pterosaurs is also a history of 150 million years of flight evolution. Biomechanical analyses must be set within an understanding of the origin, phylogenetic relationships and evolution of pterosaurs if they are to be useful and valid.

**Four influences on the understanding of pterosaurs**

The first influence on the scientific perception of pterosaurs is historical. This influence, long understood and studied in the field of artistic studies and symbols (Gombrich 1961), amounts to the axiom that first impressions tend to hold sway, regardless of their accuracy and even despite substantial empirical evidence to the contrary. The oldest restoration of a pterosaur in the scientific literature (figure 2) pictures it with bat-like wings. This image, though now regarded as incorrect on the basis of substantial fossil evidence to the contrary (figure 3), was adopted by other artists, and eventually influenced a chain of palaeontologists across two continents and more than 150 years. The details of this historical chain are recounted elsewhere (Desmond 1975; Padian 1983*b*, 1987*b*; Wellnhofer 1983). The point is that to be accepted a pictorial image need not be accurate, but need only be plausible to the mind's eye. This historical dimension bears an important lesson for biomechanical studies of fossil organisms, which often rely on analogies with other animals or with machines to strike comparisons.

A second influence is philosophical, and often involves an historical dimension. The biological philosophy that underlies biomechanics avers that it is appropriate and useful to consider the mechanical design of organisms and the properties of their materials. But this was not always in fashion. Richard Owen's pre-Darwinian, transcendental philosophy of the Archetype exerted tremendous influence in classifying pterosaurs automatically as cold-blooded reptiles, and therefore not capable of the physiological exertion necessary for a bird-like metabolism (Desmond 1975, 1979, 1982; Padian 1980, 1987*b*). This image proved to be very difficult to shake, as it did for the dinosaurs. Pterosaurs are among the closest relatives of dinosaurs, a group that phylogenetically includes birds. Like dinosaurs, pterosaurs evolved from small, active, upright, bipedal ancestors (Padian 1984). Many clues from fossils, including pneumatic foramina, insulatory coverings, skeletal mor-

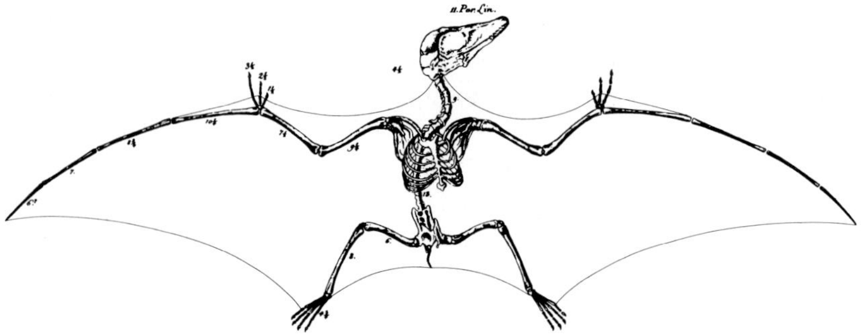

Figure 2. von Soemmerring's (1817) restoration of a bat-like juvenile pterosaur. Note dotted outline of presumed extent of wings, which were not preserved in this specimen. Wingspan about 15 cm.

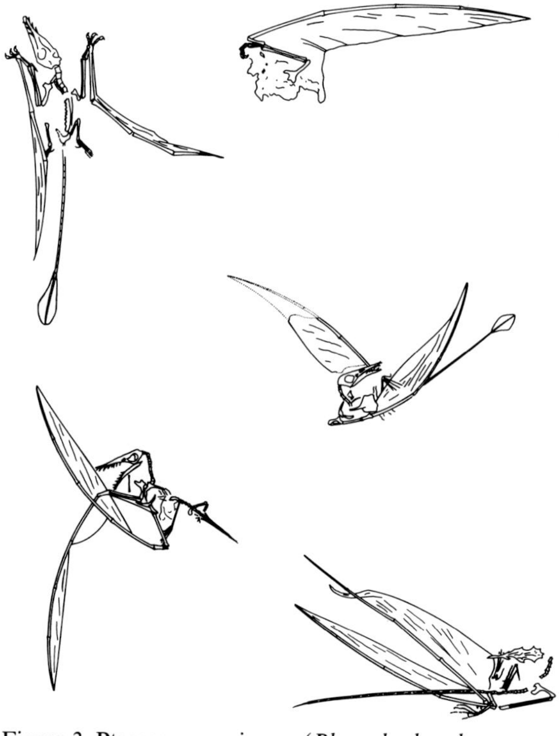

Figure 3. Pterosaur specimens (*Rhamphorhynchus*, average wingspan about 100 cm) showing narrow wing profiles. (After Padian 1983*b*.)

phology and kinematics of the limbs, do not support physiological or functional comparisons between pterosaurs and lizards or gliding vertebrates (Wellnhofer 1978; Padian 1983*b*, 1985). But all evidence must be interpreted within a philosophical matrix. Owen (1870) argued that the pneumatic foramina of pterosaur bones could have served merely to lighten the skeleton, an adaptation useful to any aerial animal and therefore not indicative of powered flight. But H. G. Seeley (1870), who held views very different from those of Owen, replied that air spaces are ubiquitous in animals, occurring even in the skulls of elephants, whereas pneumatic foramina were otherwise known only in birds. 'To discard this fact', wrote Seeley, 'would be to discard the only known clue to their interpretation', namely to cool the blood, not to lighten the skeleton. Owen viewed animals in a non-transmutationist, archetypal context, and therefore considered the reptilian (as opposed to avian or mammalian) morphological characters of pterosaurs to be an unshakable guide to their physiological abilities. Seeley was more an adaptationist who reasoned function closely from structure. His view of phylogenetic determinism was more fluid than that of Owen, but Owen's was philosophically more comprehensive, and this may partly explain the later predominance of Owen's view of pterosaurs (Padian 1987*b*). A biomechanicist of today might prefer Seeley's analogical approach to Owen's rigid archetypal principles, but both views were acceptable to Victorian science.

A third influence on the perception of pterosaurs has been the choice of the analogies used to shed light on the function and flight of these extinct animals. At one time or another, pterosaurs have been compared in some respects to birds, bats, lizards, mammals, fishes, kites, gliders (sailplanes), hang-gliders and to many other animals and machines. In fact, M. J. S. Rudwick (1964) began his classic exposition of the paradigm method with the example of pterosaurs as perfect paradigms for animal gliders. On the basis of inferences about their reptilian physiology, their thin wings and their poorly developed flight apparatus, Rudwick concluded that they had no capability for powered flight. This is curious in view of several anatomical facts, notably a sternum and delto-pectoral crest comparatively larger than in any bird or bat of similar size, a strongly co-ossified pectoral girdle buttressed to the sternum, and a forelimb that is virtually restricted to the motion of flight (Padian 1983*b*; see below).

The assumption made by many workers (again by historical accident, not in accordance with empirical fact; Padian 1983*b*, 1987*b*) that the wings of pterosaurs were attached to the ankles led to natural comparisons of structure and function with those of bats. Clearly, if the wings involved the ankles, the legs would have had to splay in bat-like fashion instead of being tucked under the body in bird-like fashion. Otherwise the position of the legs would have frustrated the aerodynamic effectiveness of the wings. But the hindlimbs of pterosaurs are organized like those of birds, not like those of bats (Padian 1983*b*, 1985), from the medially (not laterally) offset femoral head to the distally (not proximally) reduced fibula, to the parasagittal (not fully flexible) tarsus, to the elongated, asymmetrical (not short and parallel) metatarsals and phalanges, to the short, comparatively straight (not long and recurved) unguals (figure 4). Manipulation of the joint surfaces of articulating hindlimb elements shows that the legs of even the earliest pterosaurs, such as *Dimorphodon* (Padian 1983*a*), *Dorygnathus* and *Campylognathoides*, cannot be placed in bat-like position without dislocating them from the hip sockets.

The wings of bats are supported by the fingers, and each individual feather in birds has a shaft for structural support. Because pterosaurs had no such obvious structure behind their elongated wing bones, it has often been assumed that their wing membranes lacked strong support and were poorly controlled. This, in addition to the low flight speeds usually calculated for pterosaurs, has suggested to some recent workers (e.g. Brower 1983) an analogy to Rogallo-winged hang-gliders (figure 5). There are several problems with this analogy. First, the pterosaur wing was invested with a system of closely spaced, parallel, inflexible fibres, possibly of keratin, which are visible on several well-preserved specimens (von Zittel 1882; Wellnhofer 1975; Padian 1983*b*). These radiate in the same pattern as the fingers of bats and the feather shafts of birds (figure 6), and this pattern cannot be seen as fortuitous. It may be presumed that these structural elements, in addition to their natural stiffness, were under some muscular control, and could have been manipulated to provide arch and camber to the wing, as well as to change the effective area of the aerofoil; thus the wings did not billow like those of hang-gliders. Second, Rogallo wings have a very low aspect ratio, and this type of hang-glider presents a broad, inverted kite-like triangle to the airstream. But the pterosaur wing was long and narrow, with a high aspect ratio. If pterosaur wings were attached to the ankles, as many workers have presumed without evidence, and if the wing fibres are not accorded any structural role, it would be natural to consider an analogy to hang-gliders. But given the evidence to the contrary, pterosaurs only resemble hang-gliders (as well as high-aspect soaring planes) in the low flight speeds that have been calculated for them. Whether these speeds are realistically calculated is another question, touched on below.

Analogies are useful tools in both functional morphology and biomechanics, but their blades can be double-edged: it is difficult to know which possible analogic comparison is most appropriate, and difficult to know when it ceases to be appropriate. The reasons, once again, are partly historical. Of course, there is the historical primacy of early analogical comparisons that are difficult to unseat, as discussed above. But there is a

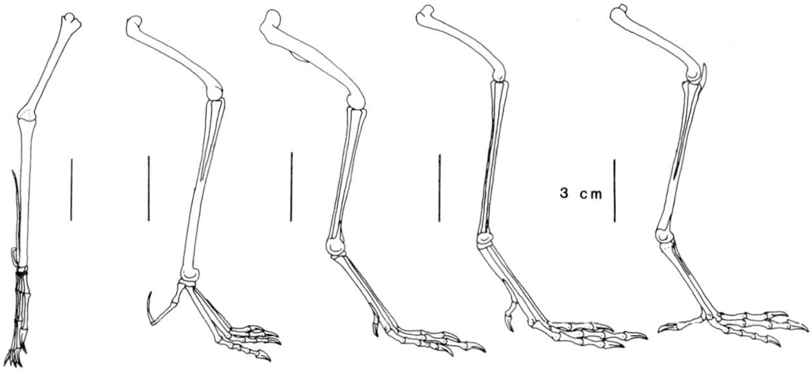


Figure 4. Comparative figures of the hindlimbs of bats (*Pteropus*), pterosaurs (*Dimorphodon*) and dinosaurs (*Compsognathus*, and the birds *Archaeopteryx* and *Columba*), in articulated position. (After Padian 1985.)

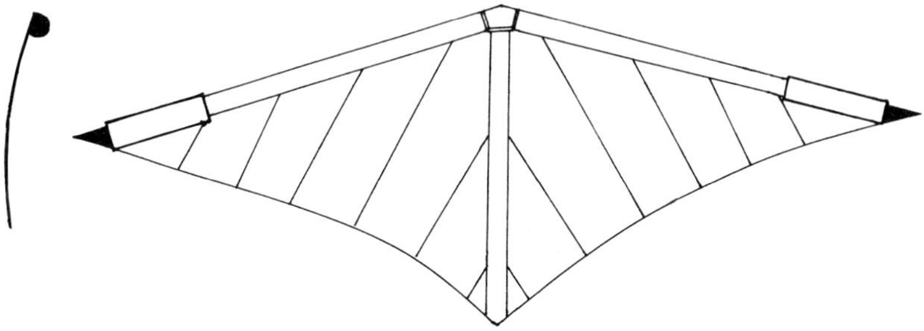

Figure 5. Outline and cross-section of a Rogallo-winged hang-glider. Note large leading-edge spar and stiffening battens in the aerofoil. (Modified from Brower 1983.)

more important historical problem, and that is the artefact of biological history. Not all possible structures or adaptive forms that ever existed are still extant, including transitional evolutionary forms (Bock (1986) gives a divergent viewpoint). The living world provides us with only a limited range of organisms for comparison, and this range is essentially a random historical artefact. Equally random is the available range of mechanical analogies that is provided by man-made machines and objects. For some biological mechanisms there are no good engineering models (ornithopters are far less efficient than flapping animals), whereas some man-made structures (e.g. wheels) are seldom if ever observed in nature, regardless of their utility. Even this is in part an historical artefact. Therefore the choice of suitable analogies, and the recognition of their constraints and limitations, is to me one of the most salient problems in the application of biomechanical analyses to extinct organisms.

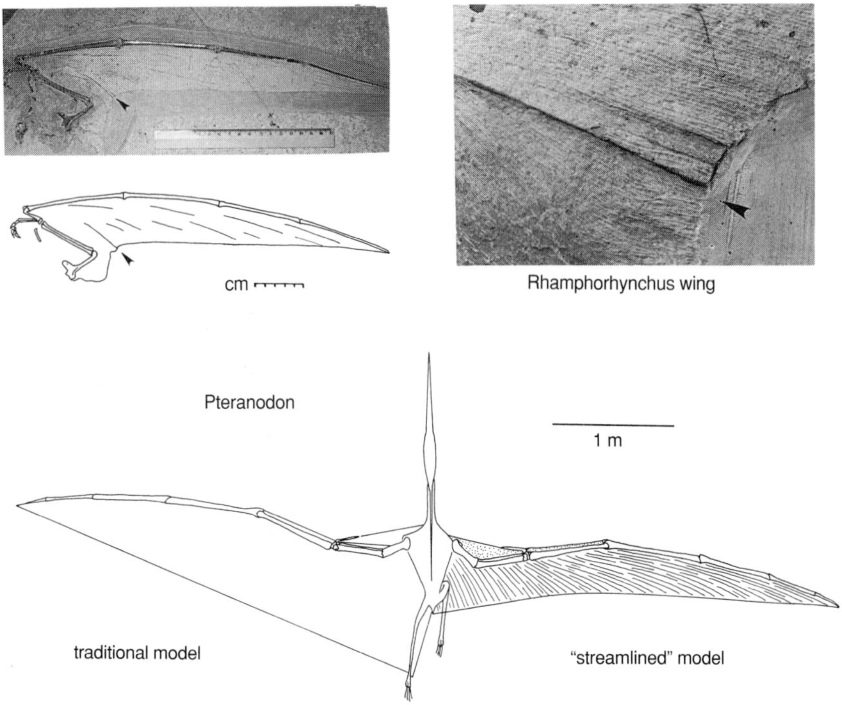


Figure 6. Upper left, photo and diagram of the isolated *Rhamphorhynchus* wing described by von Zittel (1882); note the narrow outline of the wing, and how it is gathered into folds (note arrow) that parallel the stiffening structural 'fibres'; arrow is in same position as in previous photo. Below, diagram of the Cretaceous pterodactyloid *Pteranodon*, comparing the 'traditional' model of the wing (left side) with a newer revision (right side). (After Padian 1983*b*, 1985.)

The fourth influence on the perception of pterosaurs relates more directly to biomechanics, and in this context deserves more extensive consideration in some of its aspects. This fourth influence is the use of engineering analyses to assess the mechanical capabilities of pterosaurs. As I noted earlier, in many respects such analyses have been based on unsubstantiated models of wingspan and wing attachment, and these seem to have underestimated both flight performance and kinematics of the wings (Padian 1985). This is not a necessary flaw of biomechanical analysis; it proceeds from the non-evolutionary (or perhaps 'anevolutionary' is a better term) analytic structure of biomechanics. I will suggest, as I have for the previous three 'influences', that historical factors – this time the historical factor represented by phylogeny – must be taken in hand with biomechanical analysis of organisms if the latter is to have any evolutionary context and validity.

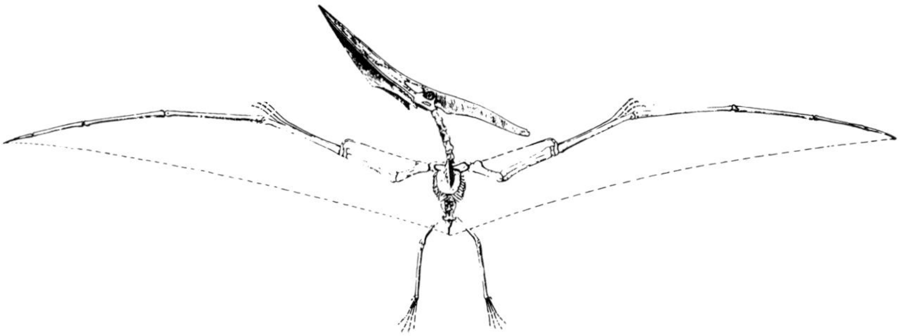

Figure 7. *Pteranodon*, restoration by Eaton (1910), modified to restrict the wings to the extent preserved in other pterosaurs; Eaton connected the wings to the ankles, but without evidence. Original specimen about 5 m in wingspan.

**Some biomechanical issues concerning pterosaurs**

The pioneering biomechanical work on pterosaurs was done by the aerodynamicist E. H. Hankin and the palaeontologist D. M. S. Watson. Their study of the biomechanics of the Cretaceous pterosaur *Pteranodon* (figure 7; actually not this genus, but related forms from the Cambridge Greensand) determined the range of motion available to each joint in the forelimb, and considered how these motions would have influenced flight (Hankin & Watson 1914). In the following decades a number of other studies appeared (reviewed by Padian 1985), but in 1974 a true classic emerged with the publication of Bramwell & Whitfield's study of the biomechanics of *Pteranodon* (again the American and British Cretaceous forms were treated as one genus, but for functional and biomechanical purposes the differences appear to be negligible).

Bramwell & Whitfield (1974), working from the same material that Hankin & Watson (1914) had used, refined the earlier analysis by applying newer insights from aerodynamics. After confirming ranges of motion and articulation in the wing joints, they calculated weight by breaking the animal's planform into a series of geometrical solids. This weight, divided by the estimated wing area, yielded the wing loading that enabled them to calculate polar curves that describe optimal gliding performance. From these measures Bramwell & Whitfield were also able to calculate stalling speed, turning radius, safety factors of the wing and many other parameters.

The straight biomechanical engineering approach was rationalized by functional and analogical comparison of pterosaur structure with those of other fliers, notably bats. This comparison was necessary because biomechanical factors and quantifications cannot by themselves tell us much about the actual workings of an extinct organism; they must be interpreted in a functional and often behavioural or ecological context. In other words, for example, one cannot simply put a strain gauge on a pterosaur's leg bone to determine the forces that it could withstand. Fossilized bone does not have the same resilience as living tissues, and virtually all information about the latter has vanished from preserved fossil remains. One can, however, make judicious use of computed tomography, thin sections of bone and vector analysis to determine the most likely directions by which forces were transmitted through the skeleton. This evidence may prove to be useful in choosing among alternative models of posture or gait. However, it is only one step in the process.

According to Bramwell & Whitfield, *Pteranodon* (wingspan 5 m) flew best at about 7 m/s, was not very tolerant of winds, could lock its arm in the shoulder socket to save energy while it coasted, had severe difficulties in terrestrial locomotion, and preferred to hang upside down from cliffs. Despite the elegance of their study, none of these statements is likely to be correct, and I would suggest that the general reason stems from a problem in appreciating the historical, evolutionary basis for pterosaur design. Bramwell & Whitfield assumed that the wing membrane extended to the foot, for which there has never been any evidence in any pterosaur; on the contrary, there is good evidence in several pterosaurs that the wing extended only to the hip (figure 3). As a result, their estimated wing area is too great by a factor of more than 2, and this substantially alters every aerodynamic calculation in their monograph (Padian 1985; Smith 1986). *Pteranodon* does not seem to have had any special modification to lock its arm in the shoulder socket: the 'locking mechanism' interpreted by Bramwell & Whitfield is merely the suture between the scapula and the coracoid, which all tetrapods have, and which would have been covered by cartilage in life. The proximal humerus had no 'notch' to lock into the shoulder joint: the feature they described, which is present in all other well-preserved pterosaurs, merely appears to mark the discontinuity between epiphyseal surfaces, which in life were covered by tendon and cartilage. The hindlimbs of pterosaurs were built like those of dinosaurs and birds, not of bats (figure 4), and the foot claws were not recurved for grasping. Bramwell & Whitfield concluded that pterosaurs might have become extinct because the average wind speed changed at the end of the Cretaceous. This conclusion is curious because it gives less credit to 150 million years of animal evolution than to the calculated limitations of simple machines (cf. Maynard Smith 1952).

Given the versatility of wing control provided by the structural fibres in the pterosaur wing, it is quite likely that pterosaurs were more capable aerodynamically than are hang-gliders and soaring aircraft. For one thing, if they could alter their wing area and planforms by flexing their joints while maintaining the aerodynamic integrity of their aerofoil, then their speeds could increase markedly over those of the fixed-wing planform (Padian 1985). Although the largest pterosaurs, which are those most readily analysed, were (like the largest birds) mostly confined to soaring, all must have retained some flapping capability for functions such as launching and avoiding danger in the air.

As large pterosaurs (such as *Quetzalcoatlus*) become better known, and as three-dimensional material becomes available for biomechanical study, the application of aerodynamic analyses to this material might enable us to answer questions about some of the more complex operations of pterosaur flight. However, such analyses will be most realistic if they take into account the phylogenetic history of pterosaurs. As the above considerations illustrate, analogical comparisons have their limits, but those limits are often difficult to recognize. To answer the question posed in the title, pterosaurs are neither functional birds nor functional bats. Like birds, pterosaurs have pneumatic foramina, expanded sterna, upright posture and a bipedal, parasagittal gait. But they do not have feathers, and their joints are functionally similar but anatomically different from those of birds. Therefore, many comparative aspects of aerodynamic efficiency and flight stroke kinematics, such as the ability to generate lift on the upstroke, cannot be quantified with confidence. Like bats, pterosaurs had a membraneous wing. But this wing was free from the hindlimbs, and so many aspects of aerodynamics and control, to say nothing of the kinematics of terrestrial locomotion, are not comparable.

What is important from a strictly functional viewpoint is the number of features that pterosaurs share with both birds and bats. For example, there are certain kinematic features of the down-and-forward flight stroke of birds in slow flight that produce the thrust necessary to flight. Other, more vertical motions apply to the generation of thrust in faster modes of flight (Rayner, this volume). Analysis of the wing joints of pterosaurs (e.g. *Dimorphodon*; Padian 1983*a*,*b*) shows that the motion of the humerus at the shoulder is virtually limited to depression-levitation and long-axis rotation. In other words, the humerus could be raised and lowered, and as it did so it could also be rotated forwards and backwards (figure 8). This joint is not strictly comparable to those of birds or bats; rather, it recalls the saddle-shaped, rotatory joints seen in moles, echidnas and dicynodont synapsids, among other vertebrates. It demonstrates that pterosaur forelimbs were capable of the same flight stroke seen in birds and bats, but of little else. Knowing these functional limitations can give us an appreciation for the more strictly

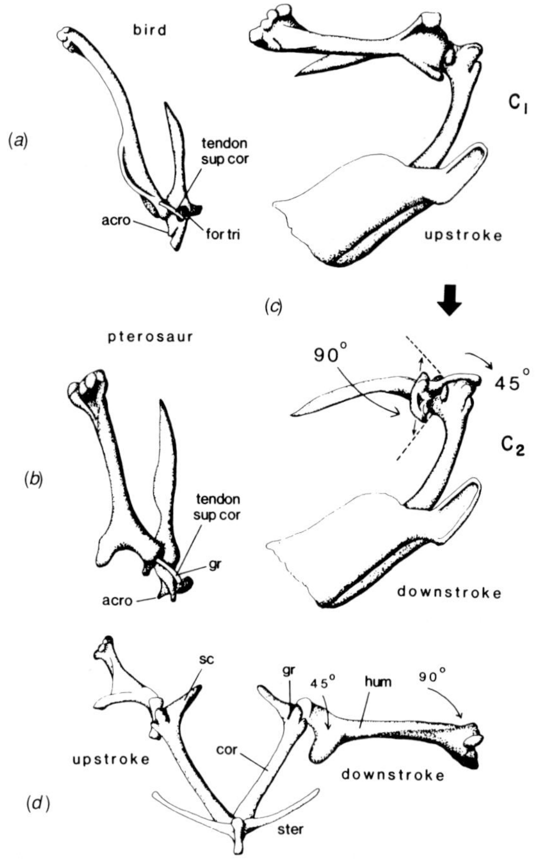


Figure 8. Flight kinematics of the pterosaur forelimb, after Padian (1983*b*). (*a*) Right scapulocoracoid and humerus of a bird in dorsal view. (*b*) The same of a pterosaur (*Dimorphodon*). (*c*) Pterosaur right pectoral girdle and humerus at end of upstroke (above) and downstroke (below). (*d*) Same as (*c*), with right side showing upstroke and left side showing downstroke. Note rotation and depression of humerus during the downstroke. Abbreviations: acro, acrocoracoid process; cor, coracoid; for tri, foramen triosseum; gr, groove for the supracoracoideus tendon; hum, humerus; sc, scapula; ster, sternum; tendon sup cor, tendon of the supracoracoideus muscle, the principle elevator of the wing.

biomechanical aspects of their limb design, such as the shape and orientation of bony crests and pneumatic foramina, the internal struts and stringers of bone and the distribution of cortical and trabecular bone.

On the other hand, with pterosaur hindlimbs we face a quandary. Most workers have compared them to bat hindlimbs, yet as figure 4 shows they are articulated like those of birds and other dinosaurs. Bramwell & Whitfield

(1974), like many other workers, looked at the apparently spindly legs of *Pteranodon* in comparison to its great wings and crested skull, and concluded that the legs were all but useless on the ground. But, as I have suggested above, the evolution of the group and its features needs to be considered. The earliest pterosaurs were smaller, and their skulls and wings accordingly smaller. As pterosaurs achieved phyletic growth, their wing areas had to increase proportionately faster than their masses to maintain a similar order of wing loading; they could not maintain the same geometric shape in relation to the body. The only direction in which a pterosaur wing could grow was by elongating the wing; it could not widen past the pelvis. (One could certainly argue that pterosaurs could have widened the wing by lengthening the body, but they never did. I suspect that the reason may be that it would have unbalanced the bipedal stance, a functional argument; further, archosaurs generally have short trunks, a phylogenetic factor to be considered.) *Pteranodon* and other large pterosaurs appear to have enormous wings because the wing bones must support a proportionately longer wing as mass increases (figure 7). However, the hindlimbs were under no such pressure, and did not enlarge as disproportionately as did the forelimbs. This gives the impression that the hindlimbs were spindly, but if they are compared to the torso they will be seen to be of expected size (Padian 1983*b*, 1985). The wing bones look large and heavy, but they are of course hollow, with their bone walls 'no thicker than a visiting card', as Watson said (Hankin & Watson 1914). The same is true of the bizarre skull and cranial crest characteristic of some large pterosaurs. Therefore the appearance of great mass in these elements is illusory, as is the apparent frailty of the hindlimbs, which have much thicker bone walls, 'honeycombed' lattices near the bone ends and internal struts for strength.

## Conclusion

The point of the foregoing remarks is to suggest that animals such as *Pteranodon* cannot be studied in an evolutionary vacuum without the risk of misplaced analogy and underestimation of biological abilities. The history of pterosaurs shows their evolution from small, active, bipedal, terrestrial predecessors closely related to the ancestors of dinosaurs. The evolution of their wing planforms over 150 million years shows several interesting trends in the proportional change of individual elements (Padian, in progress), the loss of features such as the long tail which suggest a move from inherent stability to dynamic instability (Maynard Smith 1952), and several adaptive radiations in size and ecology (Wellnhofer 1978). Perhaps this is the major drawback with paradigms, analogies, functional morphological analyses and standard biomechanics: they operate without an evolutionary context. They are static approaches, proceeding from 'givens' about form and function. This is perfectly useful when the form of a living organism is considered, because biomechanics serves to describe a system that is already known, often at several levels. But nothing in the biomechanical approach can be claimed to elucidate the evolution of functional complexes, whether in living or extinct forms. Furthermore, in dealing with wholly extinct forms, some assumptions must be made about the operation of a system before biomechanical analyses can be interpreted. These assumptions are necessarily based on paradigm and analogy to living forms and structures. But as the example of *Pteranodon* shows, even these questions can have misleading answers if the course of evolution is not considered.

The course of evolution is best shown by phylogenetic analysis, and elsewhere (Padian 1987*a*) I have suggested an approach to co-ordinate functional-analytic methods with cladistic (historical) analysis. Especially when phylogenies are built on several different kinds of characters – cranial, dental, postcranial, microstructural or biomechanical – the logic of these phylogenies can be quite robust. The proposed approach involves five stages:

(1) Define the adaptation and the groups that possess it.
(2) Perform phylogenetic analysis on each group to determine (*a*) the evolution of functional character complexes from outgroups that lack the adaptation under study and (*b*) the further evolution of the functional complex within each group.
(3) Establish the sequence of evolution of functional characters for each group, and the progression of functional stages.
(4) Construct and compare the functional evolutionary sequences among groups.
(5) Test each scenario with the adduction of additional lines of evidence. (For a fuller explanation of this approach, see Padian 1987*a*.)

Should we frame hypotheses of function based on phylogeny? I would prefer to analyse function based on structure, and then to test different hypotheses of function against the phylogenetic hypothesis (Cracraft 1981; Padian 1982). The ones that agree are most likely to be robust explanations. This is certainly not to say that analogies, paradigms and biomechanics are inherently flawed approaches. To the contrary: we simply need an independent means to evaluate their use in evolutionary studies. Phylogeny, the story of evolutionary history, gives us this independent context (Lauder 1981 and this volume). By combining phylogenetic and mechanical approaches, different lines of evidence and alternative explanations can be tested against others, and the contribution of biomechanics to evolution can be more fully realized.

**Acknowledgements**

I am most indebted to Jeremy Rayner and David Unwin for useful and instructive comments on the manuscript, and to many participants in the SEB/Palaeontological Association symposium on 'Biomechanics in Evolution' for advice and discussion. This work was supported by the National Science Foundation and the Committee on Research of the University of California, Berkeley.

**References**

Bock, W. 1986 The arboreal origin of avian flight. *Memoirs of the California Academy of Sciences* **8**, 57–72.

Bramwell, C.D. & Whitfield, G.R. 1974 Biomechanics of *Pteranodon*. *Philosophical Transactions of the Royal Society of London* B **267**, 503–581.

Brower, J.C. 1983 The aerodynamics of *Pteranodon* and *Nyctosaurus*, two large pterosaurs from the Upper Cretaceous of Kansas. *Journal of Vertebrate Paleontology* **3**, 84–124.

Cracraft, J. 1981 Patterns and process in palaeobiology: the role of cladistic analysis in systematic paleontology. *Paleobiology* **7**, 456–468.

Desmond, A.J. 1975 *The Hot-Blooded Dinosaurs: a Revolution in Palaeontology*. London: Blond & Briggs.

Desmond, A.J. 1979 Designing the dinosaurs: Richard Owen's response to Robert Edmond Grant. *Isis* **70**, 224–234.

Desmond, A.J. 1982 *Archetypes and Ancestors: Palaeontology in Victorian London*, 1850–1875. University of Chicago Press.

Eaton, G.F. 1910 Osteology of *Pteranodon*. *Memoirs of the Connecticut Academy of Arts and Sciences* 2.

Gombrich, E.H. 1961 *Art and Illusion*, 2nd edn. Princeton University Press.

Hankin, E.H. & Watson, D.M.S. 1914 On the flight of pterodactyls. *Aeronautical Journal* **72**, 1–12.

Lauder, G.V. 1981 Form and function: structural analysis in evolutionary morphology. *Paleobiology* **7**, 430–442.

Maynard Smith, J. 1952 The importance of the nervous system in the evolution of animal flight. *Evolution* **6**, 127–129.

Owen. R. 1870 *A Monograph on the Fossil Reptilia of the Liassic Formations*. III. London: The Palaeontographical Society.

Padian, K. 1980 Studies on the structure, evolution, and flight of pterosaurs. PhD thesis, Yale University.

Padian, K. 1982 Macroevolution and the origin of major adaptations: vertebrate flight as a paradigm for the analysis of patterns. *Proceedings of the 3rd North American Paleontological Convention*, vol. 2, pp. 381–392.

Padian, K. 1983*a* Description and reconstruction of new material of *Dimorphodon macronyx* (Buckland) (Pterosauria: Rhamphorhynchoidea) in the Yale Peabody Museum. *Postilla* **189**.

Padian, K. 1983*b* A functional analysis of flying and walking in pterosaurs. *Paleobiology* **9**, 218–232.

Padian, K. 1984 The origin of pterosaurs. In *Third Symposium on Mesozoic Terrestrial Ecosystems: Short Papers*, ed. W.-E. Reif & F. Westphal, pp. 163–168. Tübingen: Attempto Verlag.

Padian, K. 1985 The origins and aerodynamics of flight in extinct vertebrates. *Palaeontology* **28**, 413–433.

Padian, K. 1987*a* A comparative phylogenetic and functional approach to the origin of vertebrate flight. In *Recent Advances in the Study of Bats*, ed. M.B. Fenton, P.A. Racey & J.M.V. Rayner, pp. 3–22. Cambridge University Press.

Padian, K. 1987*b* The case of the bat-winged pterosaur: typological taxonomy and the influence of historical representation on scientific perception. In *Dinosaurs Past and Present*, ed. E.C. Olson & S.M. Czerkas, vol. 2, pp. 64–81. Los Angeles County Museum of Natural History.

Rudwick, M.J.S. 1964 The inference of function from structure in fossils. *British Journal of the Philosophy of Science* **15**, 27–40.

Seeley, H.G. 1870 Remarks on Professor Owen's monograph on *Dimorphodon. Annals and Magazine of Natural History* **6**, 129–152.

Smith, M.K. 1986 The aerodynamics of pterosaurs. M.A. thesis, Department of Paleontology, University of California, Berkeley.

von Soemmerring, Th. 1817 Ueber einen *Ornithocephalus brevirostris* der Vorwelt. *Denkschriften der koenigliche Bayerische Akademie der Wissenschaften* (*Mathematische-Physische Klasse*) **6**, 89–104.

Wellnhofer, P. 1975 Die Rhamphorhynchoidea (Pterosauria) der Oberjura Plattenkalke Süddeutschlands. *Palaeontographica* A **148**, 1–33, 132–186; **149**, 1–30.

Wellnhofer, P. 1978 Pterosauria. *Handbuch der Paläoherpetologie*, Teil 19, ed. O. Kuhn. Stuttgart: Gustav Fischer Verlag.

Wellnhofer, P. 1983 Cuvier and his influence on the interpretation of the first known pterosaur. In *Actes du Symposium Paleontologique Georges Cuvier*, ed. J.-M. Mazin & E. Salmon, pp. 525–538. Montbeliard.

von Zittel, K.A. 1882 Über Flugsaurier aus dem lithographischen Schiefer Bayerns. *Palaeontographica* **29**, 47–80.

DAVID B. NORMAN & DAVID B. WEISHAMPEL

# Feeding mechanisms in some small herbivorous dinosaurs: processes and patterns

Feeding mechanisms in some small (approximately commensurate) herbivorous dinosaurs exhibit a remarkable variety of adaptations. Dinosaurs were highly active creatures; some exhibit essentially lizard-like jaw mechanisms, and may by inference have been specialist feeders, while others have sophisticated mechanisms for chewing fibrous plant material. In their range and variety of feeding adaptations these herbivorous dinosaurs seem more comparable to Recent mammals than to living reptiles.

An attempt is made to review the fossil record of some herbivorous dinosaurs and to compare it with that for major plant types to identify coincident, possibly macroevolutionary, patterns. The rise and diversification of herbivorous sauropodomorphs appear to coincide with that of the gymnosperms of the Jurassic Period, while in the Cretaceous Period the rise of the angiosperms mirrors the rise of the ornithopods and ceratopians. Such coincident patterns pose a number of questions and problems concerning the co-evolutionary history of dinosaurs and plants during the Mesozoic Era.

## Introduction

The study of structural adaptations associated with herbivory in fossil vertebrates has a long history in the disciplines of comparative anatomy, functional morphology and palaeontology. This is particularly true of mammals, where the intricacies of tooth shape and wear characteristics, along with details of jaw and skull morphology, can be analysed in such a way that both diet and the evolutionary history of a specimen can be analysed with a measure of confidence.

In fossil mammals, and also to some extent in their extinct relatives the synapsid or mammal-like reptiles (Kemp 1982), this type of research has distinct advantages over that in some other groups. For example, there has been a great deal of work on the physiological and functional adaptations for herbivory in living mammals, and there are often extant species which provide useful structural and functional analogues for the interpretation of observed structural features in fossil forms.

Herbivorous reptiles are, by contrast, much less well studied and understood, particularly with respect to their feeding adaptations. There would appear to be a number of good reasons why this should be so. The most important are that relatively few living reptiles are obligate herbivores; moreover, of those reptiles that are, the range and variety of morphological adaptations in their skulls, jaws and teeth specifically for feeding upon plant tissues are slight when compared to those of mammals (Throckmorton 1976). This limitation of range and variety among living herbivorous reptiles has severely restrained active research programmes in this area.

However, if we take an historical perspective, living reptiles are a rather pale shadow of a group which appears to have reached a peak both in terms of abundance and diversity during the Mesozoic Era (245–64 million years before present (MYBP)). Dinosaurs were abundant throughout most of the Mesozoic (210–64 MYBP) and included not only predatory forms, but also an enormous variety of small (30+ kg), medium and very large sized (10000+ kg) herbivores; indeed herbivorous dinosaurs show a diversity of form which is paralleled only by that seen in modern ungulate mammals (Norman & Weishampel 1985). Very fortunately, dinosaurs not only have a very reasonable fossil record thanks to the intense collecting activities of research workers in many parts of the globe, but also are in many cases known from surprisingly well preserved skeletal remains. This combination of circumstances permits surprisingly detailed investigations of the anatomy and functional morphology of some dinosaurs in relation to an herbivorous diet (Norman 1984*a*; Norman & Weishampel 1985, 1987; Weishampel 1983, 1984*a*). This review is intended to complement our earlier studies by looking at the remarkable variation in observable, and deducible, feeding adaptations in a range of very similarly sized bipedal herbivorous dinosaurs throughout the Mesozoic.

Despite the current disputes concerning the systematic status of dinosaurs within the Archosauria (Bakker & Galton 1974; Benton 1984; Charig 1980, 1982; Norman 1984*b*; Paul 1984; Sereno 1984), it is beyond dispute that they may be divided into two quite distinct orders: Ornithischia and Saurischia (Seeley 1888). Both of these groups have herbivorous representatives. The saurischian herbivores are collectively referred to as sauropodomorphs (following Charig, Attridge & Crompton 1965); these are further subdivided here, for general convenience, into prosauropods and sauropods. The ornithischians were, so far as can be determined, all herbivorous and are by common consent subdivided into a number of distinctive groups: orni-

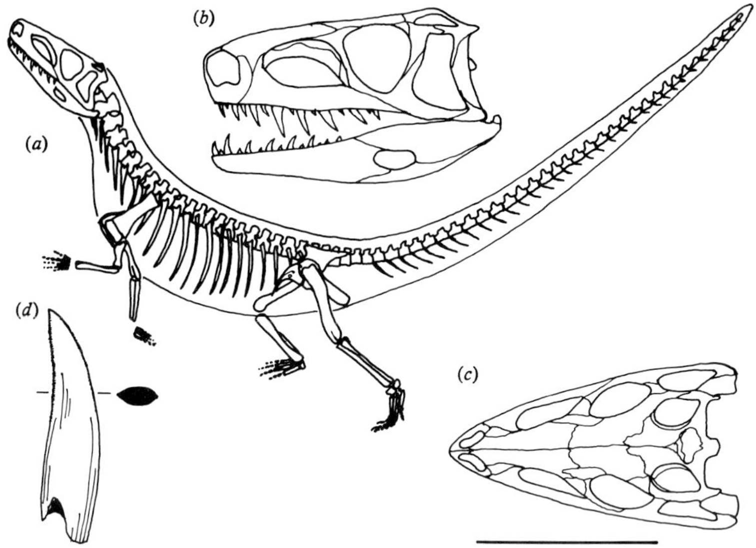


Figure 1. *Euparkeria*, an example of a generalized carnivorous archosaur. (*a*) Reconstruction of skeleton in high speed gait; body length *ca* 70 cm; (*b*) skull in lateral view; (*c*) skull in dorsal view; (*d*) typical archosaur tooth in lateral and sectional view. Scale bar represents 5 cm. ((*a*)–(*c*) From Ewer 1965.)

thopods, stegosaurs, ankylosaurs, pachycephalosaurs and ceratopians. The comments in this chapter concern small representatives of the prosauropod saurischian, ornithopod and ceratopian ornithischian groups alone.

## Herbivorous adaptations in dinosaurs

To provide an evolutionary perspective on herbivory in these selected groups of dinosaurs, it is necessary to review the principal cranial features that can be associated with an herbivorous diet.

Dinosaurs undoubtedly inherited a type of skull which is common to a variety of archosaurian reptiles (Romer 1956). Typically, such a skull is large in comparison with the size of the body (figure 1); it is also narrow and deep, with a moderately vaulted palate, and is notably fenestrated. There are large apertures in the mandible and snout, in addition to the usual ones for nostrils, eyes and the jaw muscles. The braincase is both simplified (by loss of various bones from the skull roof) and strengthened by ossification of the front and side walls. Both of these features produce a rigid unit at the centre of the skull which anchors the framework of facial bones, so that the skull is well adapted to withstand the stresses associated with feeding on active prey. The jaws are straight and have a simple hinge that permits a scissor-like

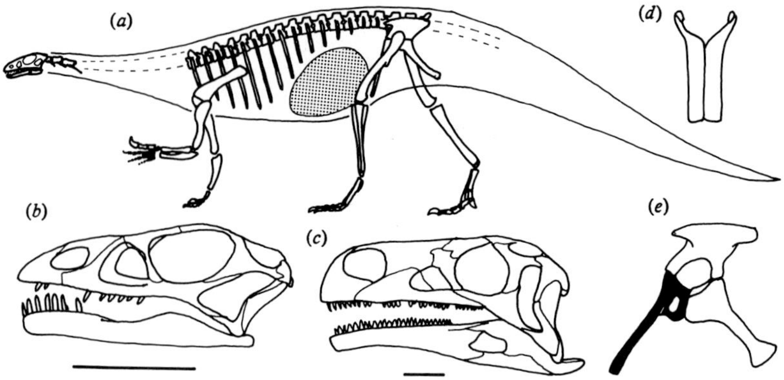


Figure 2. Prosauropods. (*a*) *Anchisaurus*, reconstruction of skeleton; body length *ca* 2 m; stippled area represents approximate position of stomach; (*b*) *Anchisaurus*, restoration of skull in lateral view; (*c*) *Plateosaurus*, reconstruction of skull in lateral view; (*d*) *Anchisaurus*, pubic apron in anteriodorsal view; (*e*) *Plateosaurus* pelvis in lateral view (pubis in black). Scale bars represent 5 cm. (After Galton 1976, 1985.)

closure. The teeth are invariably deeply rooted in the jaws and have trenchant, laterally compressed and recurved crowns with finely serrated margins (figure 1*d*). This combination of features fits well with the active predatory life style proposed for early archosaurs and many types of dinosaur (Norman 1985; Romer 1956).

Herbivorous dinosaurs undoubtedly inherited the main design features of the 'typical' archosaurian skull described above but, as we shall see, have modified this to varying degrees to cope with the very different mechanical and physiological problems associated with plant rather than animal tissues.

### Prosauropods

The small prosauropod *Anchisaurus* (figure 2*a*, *b*) is reviewed, with supplementary information from *Plateosaurus* (figure 2*c*, *e*) and *Massospondylus* (figure 3*a*, *b*). Note also that further information on prosauropods and their diets is provided by Galton (1985).

The skull retains the general archosaur features outlined above: well fenestrated, and a high, narrow snout (especially *Plateosaurus*; see Galton 1984). As a general feature, however, the skull is small relative to body size (compare figures 1 and 2). The snout is long in proportion to the total length of the skull – to accommodate an enlarged number of teeth – and the teeth are also quite distinctive, being small, lanceolate and closely packed in an array that tends to overlap posteriorly in the jaw (figure 3*a*, *b*). In addition, the crowns of prosauropod teeth have obliquely directed coarse

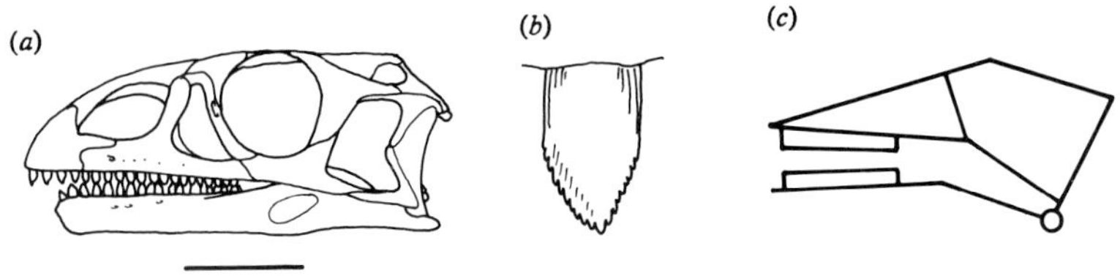


Figure 3. Prosauropods. (*a*) *Massospondylus*, skull in lateral view; (*b*) enlarged view of maxillary tooth crown; (*c*) kinematic abstraction of *Anchisaurus* skull showing simple hinge mechanism for lower jaw. Scale bar represents 5 cm. ((*a*) and (*b*) After Attridge, Crompton & Jenkins 1985.)

marginal serrations (figure 3*b*), while those of carnivorous forms (figure 1 *d*) have minute, perpendicularly oriented serrations.

Another feature common to these dinosaurs is the form of the jaw articulation, which is off-set ventral to the level of the tooth row; furthermore, the lower teeth bite inside the uppers in such a way that there is no regular tooth-tooth contact, since no wear facets have been reported in these species.

Postcranially, small prosauropods are lightly built facultative bipeds (figure 2). The long tail helped to counterbalance the neck and trunk region. However, the necessarily bulky gut lay anteriorly to the pubic apron (figure 2*a*, shaded) and was evidently laden with gastroliths (Bond 1955; Raath 1974; J. Attridge, personal observations on *Massospondylus*). The downward pitching moment of the gut mass anterior to the hip joint necessitated a primarily quadrupedal gait.

### *General comments*

*Diet and jaw mechanism.* Despite claims to the contrary (Cooper 1981), there is strong evidence in favour of a primarily herbivorous diet for prosauropods. The arrangement of overlapping, small, coarsely serrated teeth, with the lower teeth biting inside the uppers, is precisely analogous to the arrangement found in modern herbivorous lizards (Galton 1985; Montanucci 1968; Throckmorton 1976; our personal observations). The ventrally off-set articulation does not seem to have an obvious parallel among modern herbivorous lizards (although the jaw hinge of *Iguana* sp. is slightly off-set). This arrangement is analogous to the dorsally off-set jaw hinge of herbivorous mammals. It is widely recognized that there are two main features associated with off-setting the jaw hinge (Moore 1981): (1) the teeth tend to meet simultaneously along the entire length of the jaw, providing for a uniform, rather than guillotine-like, bite, and (2) the lever-arm of the main jaw-closing muscles is improved, thus increasing the force that can be exerted upon the teeth. Precisely why the jaw hinge of herbivorous reptiles should be ventrally rather than dorsally off-set as in mammalian herbivores has not been satisfactorily explained, yet this is a persistent theme among the groups considered here. The most simple explanation, and the one suggested here, comes from a consideration of the jaw musculature. Among herbivorous reptiles, the main jaw-closing muscles are those of the temporal region. The effect of ventrally off-setting the jaw hinge is to increase the depth, and therefore the volume, of the skull in the region that accommodates these muscles, as well as to improve the lever arm mechanics of these muscles. By contrast, the powerful jaw-closing muscles of the herbivorous mammal are the masseters and internal pterygoids, which are situated lateral and medial to the jaws, respectively. The temporal muscles tend to be greatly reduced. Thus in mammals the dorsal migration of the jaw hinge into the temporal region of the skull probably reflects differences associated with the size and mechanics of the masseter-pterygoid sling and a decrease in the role of the temporal muscle in mastication.

*Food-processing abilities*. The general implication from the arrangement of jaws and teeth is that prosauropods were perfectly capable of tearing off portions of vegetation caught between the teeth, assisted no doubt by jerks of the head to break unsevered fibres. As in modern lizards the skull was essentially rigid (figure 3*c*) and the jaw hinge allowed simple isognathic (bilateral) jaw closure. No oral food processing was likely, given the absence of wear facets on the teeth. After ingestion, the essentially indigestible plant material was prepared for enzymatic degradation by symbiotic gut microbes after pulverization in a stone-laden stomach or discrete gizzard.

'Stomach grinding' would have greatly increased the rate of assimilation compared to similar-sized reptiles not possessing gastric mills, although the efficiency of digestion may not have differed significantly (Hamilton & Coe 1982).

**Ornithopods**

Ornithopod ornithischian dinosaurs are remarkably uniform in their body proportions, even though they range in size from body lengths of 1 m up to an apparent maximum of 13 m. The trunk and neck are relatively short compared to prosauropods, and are counterbalanced by a large tail which permits a predominantly bipedal gait in this group (figures 4, 6, 8). The main variation in form occurs within the skull, and appears to be of direct relevance to this review. For the purposes of a *comparative* review we shall concentrate on three types of skull design, represented by a fabrosaur (*Lesothosaurus* (= *Fabrosaurus* of Thulborn 1970)), a heterodontosaur

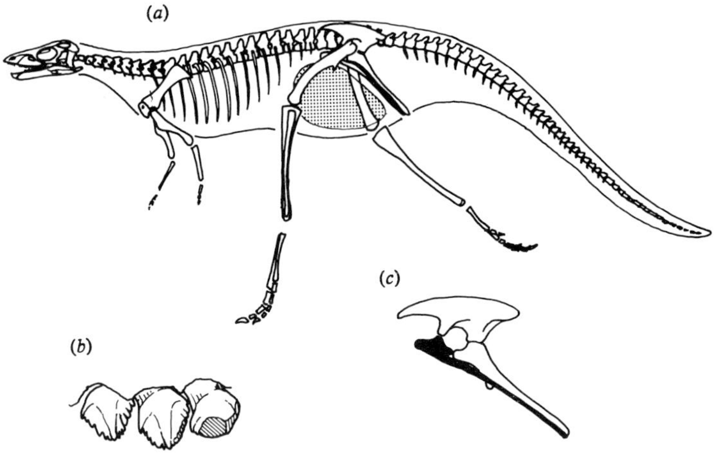


Figure 4. Fabrosaur. *Lesothosaurus*: (*a*) skeletal restoration; body length *ca* 1.5 m; stippled area represents approximate position of stomach; (*b*) enlargement of maxillary tooth crowns showing wear facets (cross-hatching); (*c*) pelvis in lateral view (pubis in black). ((*b*) After Crompton & Attridge 1986.)

(*Heterodontosaurus* (Charig & Crompton 1974)) and a hypsilophodontid (*Hypsilophodon* (Galton 1974)). These three genera are roughly comparable in size (1–3 m body length) – this circumvents the problem of size-related constraints on design – and fall within the size range of *Anchisaurus*.

### *The fabrosaur*

The skull of *Lesothosaurus* is small, lightly built and somewhat reminiscent of that of some small prosauropods (figures 2*b* and 5*a*). The snout is not so prominently fenestrated as in prosauropods or generalized archosaurs (figures 1 and 2), and the facial region is somewhat shorter and more sloped. The lower jaw is quite robustly constructed, again compared to prosauropods which have slender, laterally compressed jaws; the teeth are inset along the inner edge of the mandible, which has a marked external shelf for a cheek pouch. In addition, there is a single, median, toothless predentary bone at the tip of the lower jaw (figure 5*a*, pd), which lies above, and stabilizes, the dentary symphysis; this bone is found only in ornithischian dinosaurs.

The crowns of the teeth of *Lesothosaurus* have, as in the prosauropod, coarsely serrated margins; however, the crowns, although fewer in number,

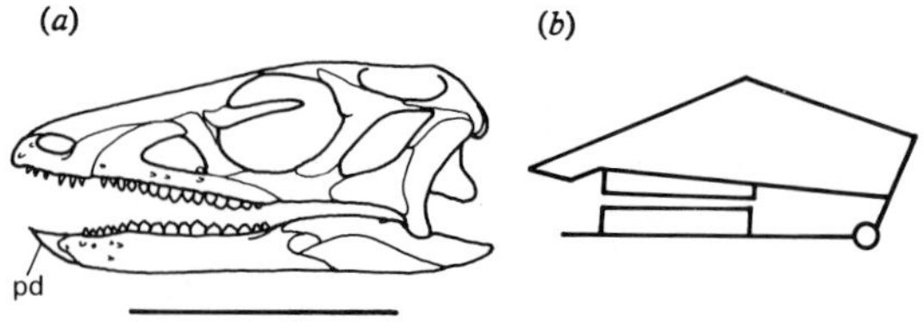


Figure 5. Fabrosaur. *Lesothosaurus*: (*a*) skull reconstruction (from Crompton & Attridge 1986); (*b*) kinematic abstraction of the jaw mechanism indicating simple hinge, as in figure 3 (modified from Weishampel 1984*a*). Scale bar 5 cm. pd, predentary bone.

are broader, more leaf-shaped, and tend to overlap one another posteriorly in the jaw (figure 4*b*). More anteriorly the teeth are reminiscent of those of prosauropods, being smaller, narrower and more lanceolate; those at the front of the upper jaw are differentiated as recurved, semi-conical crowns. Tooth wear is also exhibited, albeit sporadically within the tooth row (figure 4*b*). Two wear facets are sometimes observed (Thulborn 1971), which suggest that the teeth met during jaw closure in a reasonably precise way, with the upper teeth biting outside the lower.

Looking briefly at the postcranial organization of ornithopods, *Lesothosaurus* has a gracile skeleton (figure 4). The tail counterbalanced the trunk over the hips so that these animals were bipedal, and the hindlimb proportions are similar to those of cursorial mammals; taken together with the small size and light build of these creatures, this suggests an energetically demanding life style. The key innovation that made this type of life style possible was, no doubt, the reorganization of the pelvic bones. In ornithischian dinosaurs the pubes are long and slender (figure 4*c*) and have rotated backward to lie parallel with the ischium. Unlike prosauropods (figure 2*e*), where the pubes are directed obliquely forward and downward and form a long but broad apron, the ornithischian arrangement with slender, rotated pubes permits the gut to lie beneath the pelvis and between the legs (figure 4*a*, shaded) so that the centre of balance would have a more posterior position between the hind feet (Alexander 1985). This configuration combines agility and fleetness of foot within a body design that has to incorporate a relatively bulky herbivore gut. Once this pattern had evolved in ornithopods it seems to have been retained throughout the lineage, the only significant feature being in the assumption of a facultatively bipedal gait in some of the largest ornithopods (Norman 1980). Curiously, no gastroliths have ever been reported associated with any ornithopod skeleton.

*General comments*. The general configuration of the skull of *Lesothosaurus* is reminiscent of that of *Anchisaurus*. However, the lower jaw is somewhat more robust and the teeth are inset, leaving a cheek recess, have broader crowns and develop wear facets at intervals. It would seem likely that the skull frame was essentially rigid and jaw action provided a simple isognathic bite (figure 5*b*). Little significant oral food processing (chewing) is likely to have occurred in the absence of well defined and regular wear facets. The only other significant innovation in jaw design is the presence of a specialized toothless, horn-covered, cropping beak.

From these general observations it is proposed that *Lesothosaurus* was a highly active herbivorous dinosaur. In view of the absence of significant chewing or stomach grinding abilities, its diet may have been restricted to highly nutritious fructifications and shoots, rather than to mature foliage, which is much less readily digested. Such a specialized diet would accord not only with the probable high activity levels of these dinosaurs, judged from their body size and limb proportions, but also from snout shape. Jarman (1974) has noted that small, scrub-dwelling duikers have narrow snouts (reminiscent of that of *Lesothosaurus*) for plucking off energy-rich food such as new shoots and fruits.

### *The heterodontosaur*

Superficially, the skull of *Heterodontosaurus* is similar to that of *Lesothosaurus* (figure 6). However, the similarity in proportions belies significant differences in organization. The main differences are associated with the jaws and teeth. The lower jaw is very stoutly constructed, with the teeth inset along the jaw, and behind them a very prominent coronoid process. The jaw hinge is off-set ventrally. The upper jaw is similarly robust, with the bones of the side of the face firmly interlocking with one another in a rigid framework.

The teeth are differentiated into small incisiform types at the extreme anterior of the upper jaw, followed by large recurved caniniforms and then a series of robust, closely packed cheek teeth. Before coming into wear, replacement crowns of the upper teeth have coarsely serrated margins, but these are rapidly worn down to columnar, almost prismatic, crowns (figure 6*c*). The wear surfaces are continuous along the tooth row and steeply inclined: the upper wear surfaces face medio-ventrally, and the lowers dorso-laterally (figure 7*a*). The outer edges of the upper crowns have thickened enamel, as do the inner edges of the lower teeth.

Postcranially, the skeleton of *Heterodontosaurus* is that of a lightly built cursor, similar in most respects to that of *Lesothosaurus*, although the forelimbs are unexpectedly large, powerful and grasping.

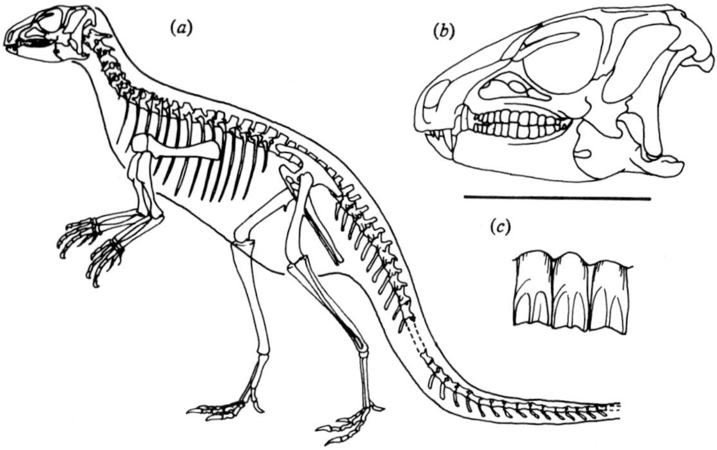


Figure 6. Heterodontosaur. *Heterodontosaurus*: (*a*) skeletal reconstruction, body length *ca* 1 m (after Santa-Luca 1980); (*b*) skull reconstruction; (*c*) maxillary teeth enlarged. Scale bar represents 5 cm. ((*b*) and (*c*) After Charig & Crompton 1974.)

*General comments.* Heterodontosaurus represents a radical departure from the skull design of *Lesothosaurus* and *Anchisaurus*. The skull is small and yet very rigidly constructed (Weishampel 1984*a*; Charig & Crompton in preparation). Occlusion is isognathic and produces low angle (30°–45°) shear between both sets of teeth; this would have imposed both vertical and transverse reaction forces in the jaws. The vertical reaction was resisted by the rigid construction of the skull roof and massive lower jaw. The rotatory forces imposed on the skull roof by the wedging effect of the lower teeth biting between the upper teeth (figure 7*a*) were less easily resisted. The reinforcement of the bones of the skull roof undoubtedly served to prevent lateral displacement of the upper jaw. However, it is claimed (Weishampel 1984*a*; Crompton & Attridge 1986) that transverse movements were possible in the lower jaw (figure 7*b*, *c*).

Thus, in *Heterodontosaurus* jaw action may have been comparatively sophisticated by normal reptilian standards. Jaw closure was bilateral and undoubtedly powerful, resulting in the well defined, continuous, blade-like wear facets along the tooth row. Reaction forces at the occlusal surface induce rotation or adduction of each ramus in the lower jaw (figure 7*b*, *c*). Movement of the lower jaw would have produced a grinding action during occlusion, which can be considered to be broadly analogous to the transverse grinding during the power stroke in anisognathic herbivorous mammals (Norman & Weishampel 1985, 1987).

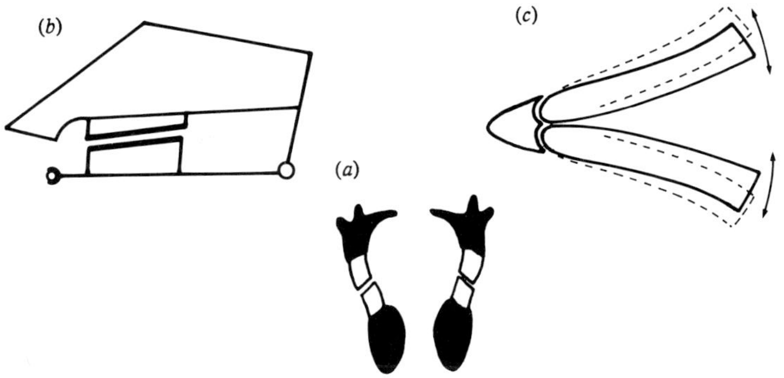


Figure 7. Heterodontosaur. (*a*) Schematic cross-section of the jaws of *Heterodontosaurus* to show oblique angulation of tooth wear facets; (*b*) kinematic abstraction of jaw mechanism showing mandibular rotation model of jaw action. The thick line representing the cutting edges of the teeth indicates that the wear facets of the teeth are continuous (after Weishampel 1984*a*). (*c*) Lower jaw adduction–abduction model suggested by Crompton & Attridge (1986).

Therefore the combination of powerful jaw closure, cheeks and a grinding jaw action suggests that heterodontosaurs may well have been ecologically partitioned from fabrosaurs in their ability to process orally more fibrous, mature plant tissues in preparation for relatively rapid digestion associated with the active life style of these animals.

*The hypsilophodontid*

As in the two previous examples, the outward appearance of the skull of *Hypsilophodon* is very similar to that of fabrosaurs and heterodontosaurs (figure 8*e*). The teeth are differentiated into smaller incisiform types at the front of both upper and lower jaws, while the cheek teeth show well developed oblique, and continuous, wear facets (figure 8*b*–*d*). Unlike heterodontosaurs, there are no enlarged caniniform teeth. Postcranially, *Hypsilophodon* does not differ significantly in its general anatomy from either of the two previously described forms.

*General comments*. *Hypsilophodon* provides a third range of feeding adaptations. As has been argued by Norman (1977, 1984*a*), Norman & Weishampel (1985, 1987) and Weishampel (1984*a*), *Hypsilophodon* had a *pleurokinetic* skull (figure 9*b*). The pleurokinetic hinge was illustrated by Galton (1974; see figure 9*a*): it forms a line of sutural weakness running

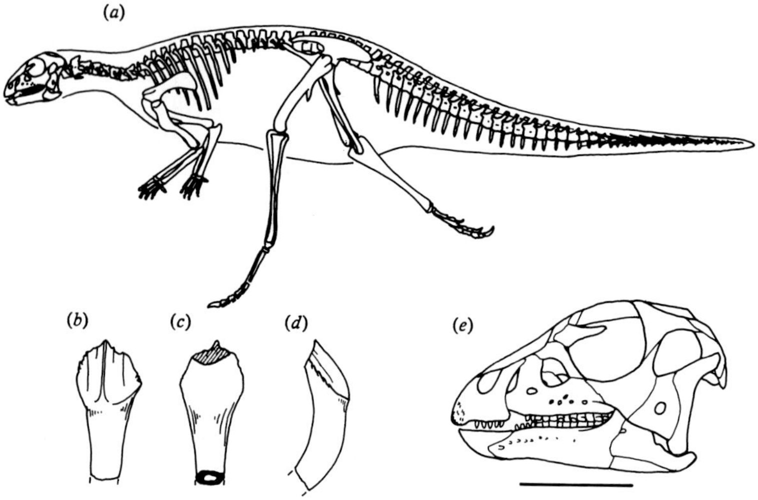


Figure 8. *Hypsilophodon*: (*a*) skeletal reconstruction, body length *ca* 1.5 m (after Galton 1974); (*b*)–(*d*) dentary tooth in lingual (medial), buccal (lateral) and distal (posterior) views respectively (after Galton 1974); (*e*) skull reconstruction (from Galton 1974). Scale bar represents 5 cm.

across the side of the skull. During jaw closure vertical and rotatory forces acted on the jaws, as in heterodontosaurs. Unlike *Heterodontosaurus* the skull roofing bones are neither rigidly fused together to prevent lateral displacement of the upper jaw nor braced internally by a bony secondary palate, as in mammals. The oblique hinge running down the side of the facial region permits maxillary rotation during the power stroke. The essentially passive displacement of the maxillae was controlled by the probably tendinous remnant of the palate musculature running between the margins of the palate and the braincase (Norman 1984*a*; Weishampel 1984*a*). This tendinous system served to maintain tooth-food pressure and provided a recoil system, in which the basipterygoid processes acted as 'spacers' to ensure the accurate separation of the tooth row in the upper jaw.

Thus the jaw action in *Hypsilophodon* relies upon powerful isognathic jaw closure and a grinding mechanism that is caused by rotation of *upper* jaw, rather than rotation or medial displacement of the lower jaw as was seen in heterodontosaurs. The pleurokinetic system is found in all advanced ornithopods investigated to date, and provided ornithopods with an important innovation in oral food processing that permitted them to radiate to a significant extent in the medium size range (Norman & Weishampel 1985, 1987; Weishampel & Norman 1989).

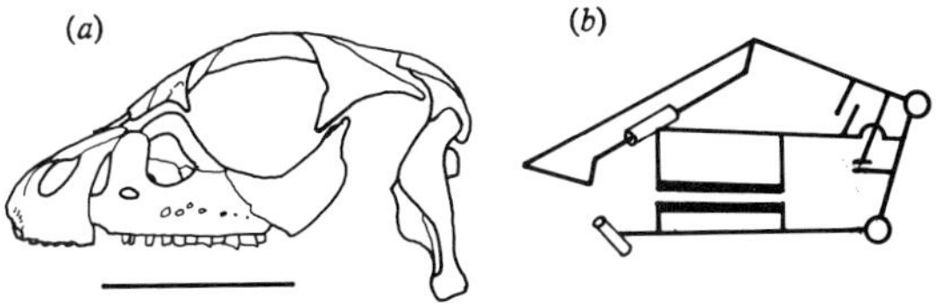


Figure 9. *Hypsilophodon*: (*a*) partial skull in lateral view, simplified version (from fig. 4*a* of Galton 1974) showing the separation of the tooth-bearing cheek region from the central skull roof along the line of the pleurokinetic hinge; (*b*) kinematic abstraction of the jaw mechanism in *Hypsilophodon* (modified from Weishampel 1984*a*).

**Ceratopians**

Although the great majority of ceratopian dinosaurs are large, rhinoceros-like quadrupeds, there is a group of small, primitive ceratopians known as psittacosaurs which bear a superficial similarity to the bipedal ornithopods considered earlier (figure 10*a*). Psittacosaurs do not appear to exhibit the remarkable feeding adaptations seen in later ceratopians (Ostrom 1964), but rather show their own peculiar feeding adaptations.

The skull (figure 10*b*, *c*) is remarkably narrow and deep anteriorly, terminating in a narrow, hooked beak formed by a unique rostral bone; the skull roof is flat, while the sides of the face are practically vertical and diverge strongly, producing a very wedge-shaped head in dorsal view (figure 10*b*). The bones of the skull roof and facial region are very firmly locked together by fusion or interdigitating sutures. The jaw hinge is slightly ventrally displaced and the jaw is very robust. The predentary is firmly attached to the front of the lower jaw by means of an interdigitating suture, while the main body of the jaw is stout, with a wide cheek recess and a large coronoid process.

The teeth (figure 10*d*, *e*) are coarsely serrated, have leaf-shaped crowns, and apparently are closely packed and develop oblique wear facets (P. Sereno, personal communication).

In addition, one notable feature of the postcranial skeleton associated with feeding is the observation of large numbers of gastroliths in an articulated skeleton in the American Museum of Natural History, New York.

*General comments*. Psittacosaurs would seem to be somewhat paradoxical in their jaw mechanism in view of the considerations above: they combine teeth that possess oblique wear facets (implying a transverse grinding stroke during jaw closure) with a rigid skull roof and facial region

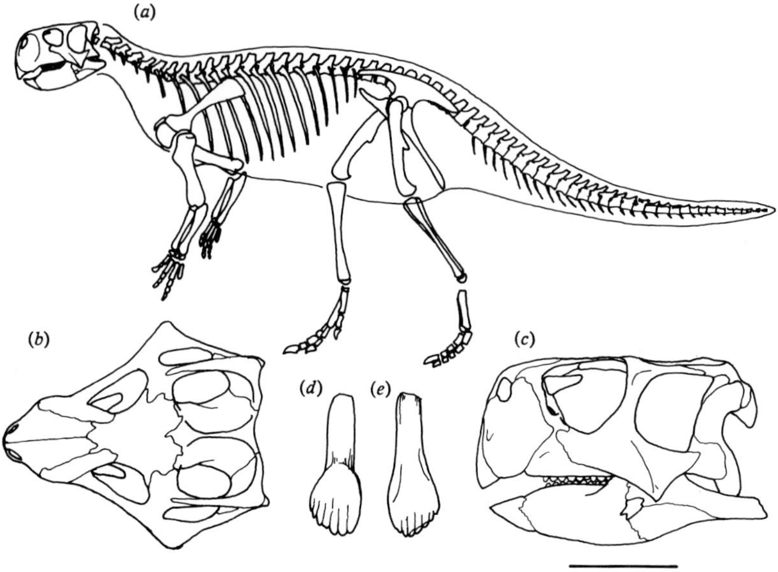


Figure 10. Ceratopian. *Psittacosaurus*: (*a*) skeletal reconstruction, body length *ca* 1.5 m; (*b*) skull restoration in dorsal view; (*c*) skull restoration in lateral view; (*d*)–(*e*) a maxillary tooth in buccal (lateral) and lingual (medial) views respectively (courtesy of P. C. Sereno). Scale bar represents 5 cm.

(preventing pleurokinetic motion, as in ornithopods) and a lower jaw that was rigidly clamped by a complex and extensive dentary-predentary suture (preventing mandibular movements as in heterodontosaurs).

According to Norman & Weishampel (1985) there appear to be only three developmental pathways for producing an orthal grinding system by retaining isognathy and mobilizing (1) the upper jaw (mainly ornithopods) or (2) the lower jaw (heterodontosaurs only), or (3) by becoming anisognathic (as in most herbivorous mammals). Since psittacosaurs could, so far as we can tell, mobilize neither the upper nor the lower jaws, if they had a grinding jaw action, one of the few remaining possibilities would seem to be that they were anisognathic (figure 11*a*). This latter possibility is, however, very unlikely given the generally reptilian biological status of these creatures. For an anisognathic jaw frame to operate, the lower jaw has to be controlled by an equivalent of the mammalian masseter-internal pterygoid sling of muscles, so that the jaw can be moved precisely from side to side during the chewing cycle (figure 11*a*). There is no good evidence in psittacosaurs for the requisite jaw muscles. Consideration of the structure of the jaw joint does, however, offer an alternative explanation to this problem. The jaw hinge region of the lower jaw is rather unusual in that it consists of a shallow, cup-shaped

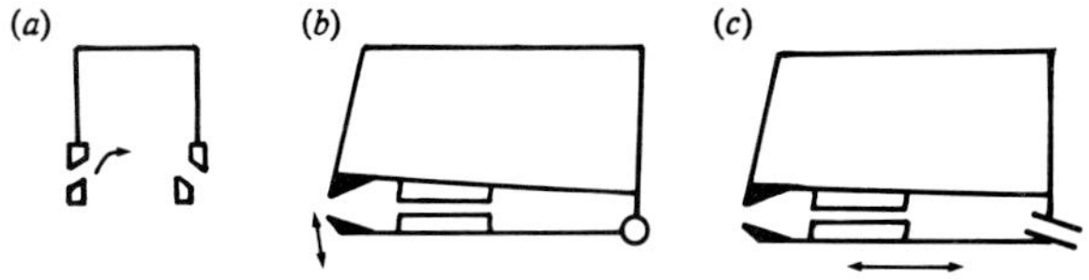


Figure 11. *Psittacosaurus*: potential jaw mechanisms. (*a*) Anisognathic jaw frame used to generate rotatory jaw motion and a grinding stroke; (*b*) simple jaw closure with no significant grinding stroke as in prosauropods and fabrosaurs (figures 3 and 5); (*c*) propalinal (fore–aft) sliding motion accompanying jaw closure as a means of generating a grinding stroke and as a way of using the beak.

depression, anterior to which is a sloping articular pad; this arrangement provides the potential for propalinal (fore-aft) jaw motion (figure 11*c*).

Although tentative at this stage, the jaw mechanism of psittacosaurs may well have involved either (1) anisognathy or (2) propalinal motion of the lower jaw. The latter model is suggested here, and this may have been linked with the cropping function of the peculiarly narrow, parrot-like beak of these dinosaurs as well as serving to generate oblique wear facets on the cheek teeth. With the lower jaw in its fully protracted position, the predentary would engage the rostral for cropping. Retraction of the lower jaw would disengage the beak, and the divergent rows of lower cheek teeth would have been drawn backward against the uppers during jaw closure.

Our model of psittacosaur jaw mechanics is limited by the general lack of biological and anatomical knowledge of these animals, not least by the absence of well preserved tooth rows in the few skulls so far studied. If tooth wear does not produce integrated, angled wear facets then jaw action may have been relatively simple (figure 11*b*). Much new material from Russian, Mongolian and Chinese collections now being studied by Paul Sereno (Chicago) may help to clarify this issue.

## Discussion

The various groups of small herbivorous dinosaur reviewed above display a remarkable range of mechanical adaptations to overcome the problems associated with a diet of plants.

So far as dinosaurs are concerned, but unlike living reptiles, there would appear to be a broad spectrum of adaptive modes associated with herbivory. For convenience these can be viewed as two rather extreme categories: those which provided minimal oral food treatment and presumably employed either extensive passage time for gut fermentation or used a gastric mill to accelerate digestion and assimilation ('gut processors'), and those which provided for extensive oral food processing to increase the rate of assimilation ('mouth processors').

### *Gut processors*

Fabrosaurs were small, bipedal, fleet-footed creatures which probably fed on low-fibre nutritious plants to support their high activity levels, despite their relatively simple dentition, jaw mechanism and presumably unsophisticated guts. Looking at the fossil record of these forms (figure 12), fabrosaurs seem to have persisted as small specialist feeders at low diversity throughout the reign of the dinosaurs.

Prosauropods, represented by small forms such as *Anchisaurus*, were presumably moderately active quadrupeds. Their simple dentition and jaw mechanics rules out oral food processing; however, they may not have been nutritional specialists – as were the fabrosaurs – because they possessed stone-lined gastric mills which would have greatly improved the rate of digestion of high-fibre plants.

As a strategy for dealing with plants the 'gut processing' employed by prosauropods, and their descendants the much larger sauropods, was evidently successful, because they became diverse and numerically abundant in the late Triassic and Jurassic Periods (figure 12), although both these measures, particularly abundance, declined during the succeeding Cretaceous Period (Weishampel & Norman 1989).

### *Mouth processors*

Heterodontosaurs, small and agile like the fabrosaurs, developed complex cheek teeth and an ability to mobilize the lower jaw to generate a grinding jaw action and so process plant food orally. This conferred upon these dinosaurs the ability to maintain the high rates of digestion and assimilation necessary for an active life style while on a high-fibre diet, without the necessity for an excessively large stone-laden gut.

Despite the apparent advantages of mobilizing the lower jaw, the heterodontosaur system was very short lived. After a brief radiation in the early Jurassic, no heterodontosaur is known from the middle Jurassic onwards (figure 12).

*Hypsilophodon* represents an alternative means of generating a grinding jaw action by mobilizing the upper jaw through the development of complex teeth and pleurokinesis. As in heterodontosaurs, the advantages of this system combine an active life style with high rates of digestion on a high fibre diet and a relatively small gut.

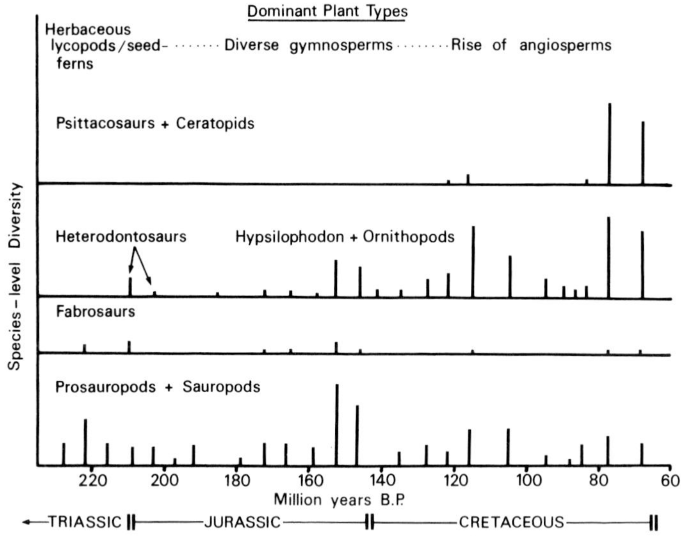


Figure 12. A representation of the fossil record of some herbivorous dinosaurs and contemporary plant types during the Mesozoic Era.

After this system arose in ornithopods in the mid-late Jurassic, the group became increasingly diverse and abundant (figure 12).

Psittacosaurs, as small (1–2.5 m long) bipedal predecessors of the ceratopians, present a number of contrasting features; they were capable of mobilizing neither upper nor lower jaws in the ways described above, and yet have been reported to have obliquely worn, closely packed cheek teeth. Several solutions to this paradox have been proposed. The additional fact that gastroliths are known in *Psittacosaurus* provides evidence that suggests that, whatever the efficiency of the jaw mechanism in these dinosaurs, considerable 'gut processing' was also possible.

After their appearance in the early Cretaceous Period, the psittacosaurs were succeeded by the generally larger, abundant and more diverse quadrupedal ceratopian dinosaurs of the late Cretaceous (figure 12). As Ostrom (1964) has shown, the later ceratopians retained the curious parrot-like beak and developed a powerful vertical shearing dentition totally unlike that seen in either psittacosaurs or any other group of herbivores.

## Summary

Reviewing feeding adaptations in a variety of small herbivorous dinosaurs, where size-related mechanical and physiological constraints are minimized, has revealed a remarkable variety of adaptations when compared to living herbivorous reptiles. Broadly speaking these can be divided into two areas of adaptation, modifications to the gut or to the mouth, both designed to improve rates of digestion.

Looking at the fossil record of dinosaurs from the Mesozoic from an evolutionary perspective, it appears that certain feeding adaptations were 'popular' at particular times. For example, in the late Triassic and Jurassic Periods, 'gut processors' in the form of prosauropods and sauropods are prevalent, while the Cretaceous Period marks the rise and diversification of various ornithischian 'mouth processors'.

These tentative *macroevolutionary* observations pose a number of questions which need to be answered to be able to understand better the long term co-evolutionary relationship between the plants and herbivorous animals during the Mesozoic Era. For example:

(1) Can coincident changes in floral type be identified during the Mesozoic which may explain the apparent shift from one feeding strategy to another?
(2) Can specific feeding adaptations be correlated with specific plant groups, particularly in such cases as are seen in the highly specialized late Cretaceous ceratopians?
(3) Is there significant regional or geographical variation in the patterns of abundance and diversity of herbivores and plants that is masked by the rather general measures of diversity?
(4) To what extent may complex chewing mechanisms be linked to varying degrees of complexity of gut anatomy, as in modern mammals?

Although at present none of these questions can be answered satisfactorily, several areas of research currently being covered should clarify some of the points raised. Data are being collected on the functional morphology, spatial and temporal distribution of all Mesozoic herbivores (Weishampel & Norman 1989). In addition, further work on the types and distributions of plants is being done (see Weishampel 1984*b*) in an attempt to tie jaw mechanics and foraging abilities to specific plant types.

**Acknowledgements**

We thank Jeremy Rayner for the invitation to contribute this chapter, and him and Robin Wootton for editorial support. Much of this work would not have been possible without the support of the following institutions and people: the Royal Society of London and the National

Science Foundation (Washington), Drs A. J. Charig and A. C. Milner (British Museum (Natural History)), Dr E. Gaffney (American Museum of Natural History), P. C. Sereno (Chicago), Professor A. W. Crompton (Museum of Comparative Zoology, Harvard) and Dr H.-D. Sues (Smithsonian Institution, Washington). All have our sincere thanks for their help and co-operation. The manuscript was typed by Miss Sophie Hindley (Department of Zoology, Oxford).

## References

Alexander, R.McN. 1985 Mechanics of posture and gait in some large dinosaurs. *Zoological Journal of the Linnean Society of London* **83**, 1–25.

Attridge, J., Crompton, A.W. & Jenkins, F.A. 1985 The southern African Liassic prosauropod *Massospondylus* discovered in North America. *Journal of Vertebrate Paleontology* **5** (2), 128–132.

Bakker, R.T. & Galton, P.M. 1974 Dinosaur monophyly and a new class of vertebrates. *Nature, London* **248**, 168–172.

Benton, M.J. 1984 Consensus on archosaurs. *Nature, London* **312**, 599.

Bond, G. 1955 A note on dinosaur remains from the Forest Sandstone (Upper Karoo). *Occasional Papers of the National Museums and Monuments of Rhodesia* **2** (20), 795–800.

Charig, A.J. 1980 Differentiation of lineage among Mesozoic tetrapods. *Mémoires de la Société Géologique de France Nouvelle Série* **239**, 207–210.

Charig, A.J. 1982 Problems in dinosaur phylogeny: a reasoned approach to their attempted resolution. *Géobios Mémoire Spécial* **6**, 113–126.

Charig, A.J., Attridge, J. & Crompton, A.W. 1965 On the origin of the Sauropoda and the classification of the Saurischia. *Proceedings of the Linnean Society of London* **176**, 197–221.

Charig, A.J. & Crompton, A.W. 1974 The alleged synonymy of *Lycorhinus* and *Heterodontosaurus*. *Annals of the South African Museum* **64**, 167–189.

Charig, A.J. & Crompton, A.W. *in preparation* The skull of the Lower Jurassic ornithischian *Heterodontosaurus tucki*.

Cooper, M.R. 1981 The Prosauropod dinosaur *Massospondylus carinatus* Owen from Zimbabwe: its biology, mode of life and phylogenetic significance. *Occasional Papers of the National Museums and Monuments of Rhodesia (Zimbabwe) Series B Natural Sciences* **6**, 689–840.

Crompton, A.W. & Attridge, J. 1986 Masticatory apparatus of the larger herbivores during late Triassic and early Jurassic times. In *The Beginning of the Age of Dinosaurs*, ed. K. Padian, pp. 223–236. Cambridge University Press.

Ewer, R.F. 1965 The anatomy of the thecodont reptile *Euparkeria capensis* Broom. *Philosophical Transactions of the Royal Society of London* B **248**, 379–435.

Galton, P.M. 1974 The ornithischian dinosaur *Hypsilophodon* from the Wealden of the Isle of Wight. *Bulletin of the British Museum (Natural History) Geology* **25**, 1–152.

Galton, P.M. 1976 Prosauropod dinosaurs of North America. *Postilla* **169**, 1–98.

Galton, P.M. 1984 Cranial anatomy of the prosauropod dinosaur *Plateosaurus* from the Knollenmergel (Middle Keuper, Upper Triassic) of Germany. I. Two complete skulls from Trossingen with comments on diet. *Geologie und Paläontologie, Marburg* **17**, 207–243.

Galton, P.M. 1985 Diet of prosauropod dinosaurs from the late Triassic and early Jurassic. *Lethaia* **18**, 105–123.

Hamilton, J. & Coe, M.J. 1982 Feeding, digestion and assimilation in a population of giant tortoises (*Geochelona gigantea* Schweigger) on Aldabra atoll. *Journal of Arid Environments* **5**, 127–144.

Jarman, P.J. 1974 The social organisation of antelope in relation to their ecology. *Behaviour* **48**, 215–266.

Kemp, T.S. 1982 *Mammal-like Reptiles and the Origin of Mammals*. London: Academic Press.

Montanucci, R.R. 1968 Comparative dentition in four iguanid lizards. *Herpetologica* **24**, 305–315.

Moore, W.J. 1981 *The Mammalian Skull*. Cambridge University Press.

Norman, D.B. 1977 *On the anatomy of the ornithischian dinosaur* Iguanodon. Ph.D. thesis, University of London.

Norman, D.B. 1980 On the anatomy of the ornithischian dinosaur *Iguanodon bernissartensis* from Bernissart, Belgium. *Mémoires de l'Institut Royale de Sciences Naturelles de Belge* **178**, 1–105.

Norman, D.B. 1984*a* On the cranial morphology and evolution of ornithopod dinosaurs. *Symposia of the Zoological Society of London* **52**, 521–547.

Norman, D.B. 1984*b* A systematic reappraisal of the reptile order Ornithischia. In *Proceedings of the Third Symposium on Mesozoic Terrestrial Ecosystems*, ed. W.-E. Reif & F. Westphal, pp. 157–162. Tübingen: Attempto Verlag.

Norman, D.B. 1985 *The Illustrated Encyclopedia of Dinosaurs*. London: Salamander Press.

Norman, D.B. & Weishampel, D.B. 1985 Ornithopod feeding mechanisms: their bearing on the evolution of herbivory. *American Naturalist* **126**, 151–164.

Norman, D.B. & Weishampel, D.B. 1987 Vegetarian dinosaurs chew it differently. *New Scientist* **114** (1159), 42–45.

Ostrom, J.H. 1964 A functional analysis of jaw mechanics in the dinosaur *Triceratops*. *Postilla* **88**, 1–35.

Paul, G.S. 1984 The archosaurs: a phylogenetic study. In *Proceedings of the Third Symposium on Mesozoic Terrestrial Ecosystems*, ed. W.-E. Reif & F. Westphal, pp. 175–180. Tübingen: Attempto Verlag.

Raath, M. 1974 Fossil vertebrate studies in Rhodesia: further evidence of gastroliths in prosauropod dinosaurs. *Arnoldia* **7**, 1–7.

Romer, A.S. 1956 *Osteology of the Reptiles*, third edition. Chicago University Press.

Santa-Luca, A.P. 1980 The postcranial skeleton of *Heterodontosaurus tucki*

(Reptilia: Ornithischia) from the Stormberg of South Africa. *Annals of the South African Museum* **79**, 159–211.

Seeley, H.G. 1888 The classification of the Dinosauria. *Report of the British Association for the Advancement of Science Meeting, Manchester* **LVII**, 698–699.

Sereno, P.C. 1984 The phylogeny of the Ornithischia: a reappraisal. In *Proceedings of the Third Symposium on Mesozoic Terrestrial Ecosystems*, ed. W.-E. Reif & F. Westphal, pp. 219–226. Tübingen: Attempto Verlag.

Throckmorton, G.S. 1976 Oral food processing in two herbivorous lizards, *Iguana iguana* (Iguanidae) and *Uromastix aegyptius* (Agamidae). *Journal of Morphology* **48**, 363–390.

Thulborn, R.A. 1970 The skull of *Fabrosaurus australis*, a Triassic ornithischian dinosaur *Fabrosaurus*. *Palaeontology* **13**, 414–432.

Thulborn, R.A. 1971 Tooth wear and jaw action in the Triassic ornithischian dinosaur *Fabrosaurus*. *Journal of Zoology, London* **164**, 165–179.

Weishampel, D.B. 1983 Hadrosaurid jaw mechanics. *Acta Palaeontologica Polonica* **28**, 271–280.

Weishampel, D.B. 1984*a* The evolution of jaw mechanism in ornithopod dinosaurs. *Advances in Anatomy, Embryology and Cell Biology* **87**, 1–110.

Weishampel, D.B. 1984*b* Interactions between Mesozoic plants and vertebrates: fructifications and seed predation. *Neues Jahrbuch für Geologie und Paläontologie* **167** , 224–250.

Weishampel, D.B. & Norman, D.B. 1989 Vertebrate herbivory in the Mesozoic: jaws, plants and evolutionary metrics. In *Paleobiology of the Dinosaurs*, ed. J.O. Farlow, pp. 87–100. *Special Paper of the Geological Society of America, Boulder, Colorado* **238**.

JEREMY M. V. RAYNER

# Avian flight evolution and the problem of *Archaeopteryx*

## Introduction

Flight is the characteristic avian adaptation of birds, and there are few features of the biology of a bird which do not reflect the demands imposed by this mode of locomotion. For this reason the problems of the evolution of birds and of the evolution of flight are closely interlinked. A biomechanical model of the processes by which flight first appeared in the avian stem group must not only explain how and why the adaptation evolved, and how bird precursors may have functioned; it must also be consistent with what the fossil record can tell us of the structure and relationships of the first birds, and ideally should clarify the function of these fossils. Equally, phylogenetic hypotheses about the evolution of bird-like characters must be plausible in the context of the morphological and mechanical developments associated with the evolution of flight. In this chapter I attempt to address biomechanical aspects of this question by defining the aerodynamic and morphological adaptations which are essential for flight, by looking for evidence of correlated features in the fossil record, and by comparing and contrasting alternative palaeobiological 'scenarios' for the behavioural context of flight.

Fossil evidence is crucial to this problem, and the reptile–bird transition is exemplified by the six specimens of *Archaeopteryx lithographica* from mid Jurassic limestone deposits in the region of Eichstätt and Solenhofen, in central Bavaria. *Archaeopteryx* is a small, bipedal theropod dinosaur, which is unique in possessing a few characters – notably feathers and wings – which are distinctly avian. Bearing as it does on the problem of the origins of a major vertebrate class and a major locomotor adaptation, *Archaeopteryx* has long been recognized as having enormous significance, and it has been the subject of much discussion since it was first described in 1860 (for references see Stephan 1987, and chapters in Hecht *et al.* 1985). Despite this attention, it is only recently that consensus has been reached on the true relationships of *Archaeopteryx*; the problem of bird flight evolution remains the subject of some debate.

One of the roots of the controversy surrounding *Archaeopteryx* has been – and remains – semantic. Precision over the definitions of words such as 'bird' and 'flight' is critical, yet much confusion has arisen because authors have failed to explain their interpretation of these terms, or have been inconsistent in the way they have used them. It is of passing interest that the difficulty of defining these terms exactly mirrors the separate phylogenetic and mechanical problems associated with the origin of birds.

Flight may be defined crudely – and perhaps too broadly – as the ability to move in air. For the discussion in this chapter it is more helpful to define 'flight' as a more refined and specialized ability, allowing generation of some useful propulsive force by flapping wings. This definition excludes passive parachuters or gliders which can generate aerodynamic forces by passive airflows over wings or other flight surfaces, but which are unable to generate additional force by moving those surfaces; it also excludes animals which travel through the air in the course of a jump. I believe this narrowing of the definition of flight is justified, since it was development of the ability to produce a horizontal thrust force by flapping the wings that enabled birds to radiate and become the dominant aerial form; gliding has evolved frequently in vertebrates, but flapping is far rarer, having evolved only three or four times in vertebrates (Rayner 1981, 1986).

A definition of modern birds would have to indicate the possession of feathers and wings (which are apomorphies of the Class Aves), and oviparity (which is plesiomorphic). It might also include thermoregulation – which is a character of uncertain status while debate continues to surround the 'hot-blooded' dinosaurs – and flight. Without further qualification flight is a poor character, but even when clearly understood it is in some ways unhelpful because several lineages and many individual species of birds have secondarily become flightless. In the evolutionary context, defining birds by their possession of flight risks confusion, because flight may not have evolved simultaneously with other bird-specific characters such as feathers; moreover, flapping flight probably followed an evolutionary progression which included forms with incipient – but ecologically valuable – aerial locomotion, but without the ability to produce thrust. This is a problem that careful study of the fossil record can help to resolve, and in this chapter I show how it is possible to place limits on the likely flight performance of *Archaeopteryx*, and from this basis to indicate how flight may have evolved in birds.

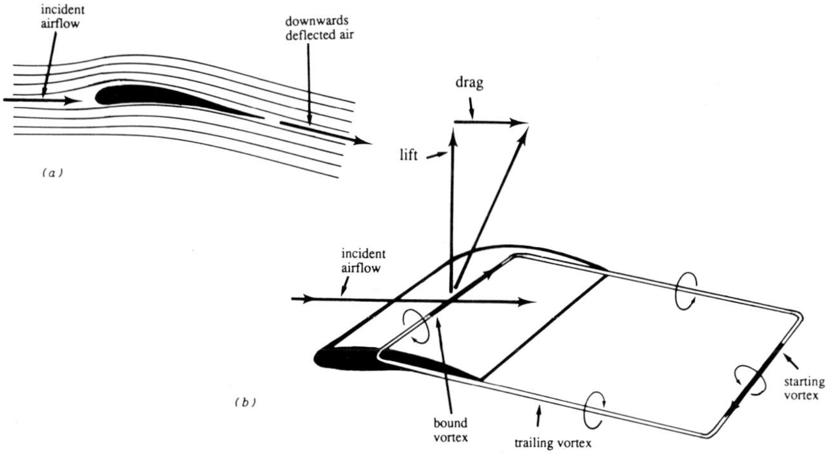


Figure 1. The section of the wings of flying vertebrates is an *aerofoil.* This profile has a rounded leading edge and a pointed trailing edge, is relatively thin, and is usually cambered so that upper and lower surfaces are asymmetrical. (*a*) The streamlined, tapered aerofoil section deflects incident air downwards, and generates a transverse lift force by a reaction to the downward momentum imparted to the air. (*b*) A steadily moving aerofoil is associated with a vortex system consisting of a bound vortex on the wing and trailing vortices behind the wingtips. A transverse starting vortex is shed whenever the magnitude of the circulation or strength of the bound vortex changes. The lift force is proportional to the product of circulation and velocity. There is also a drag force, parallel to the incident airflow, representing the energy required to generate the wake vortices and to overcome friction on the aerofoil surface. Visualization of the wake vortices clarifies the temporal patterns of force generation in flapping flight. (From Rayner 1988a.)

## Flight

The first stage in the discussion is to consider the mechanical functions of the wings in flapping flight in modern birds, and to use this as the basis of a definition of the major adaptations essential for flight.

### *Physical forces and locomotion*

Animals in locomotion use reciprocating movements of their body and limbs to generate physical forces which transmit energy into the environment. Owing to the magnitude of the forces and energy rates involved, selection may be supposed to act to favour systems in which locomotion is safe, yet economic and/or efficient, without constraining the animal's ability to exploit its environment and to survive and reproduce. The forces required are two-fold: a vertical *lift* force to support the animal's weight against gravity, and a horizontal *thrust* force to provide propulsion against resistive forces such as friction and induced (vortex) drag (Rayner 1985*a*, 1988*b*). Different physical media require different solutions for the generation of these forces. The major problem associated with flight is that flying animals must rely entirely on aerodynamic forces generated at the surface of their wings for thrust and weight support. This imposes severe mechanical constraints on the animals. The only mechanism capable of producing forces of the required magnitude in air is aerofoil action (figure 1), which demands wings that have large surface area, are relatively thin in cross-section, and have high aspect ratio (wingspan divided by wing chord or breadth), and yet which are stiff enough to transmit the aerodynamic forces they generate to the body.

An aerofoil can generate force only if the surrounding air moves relative to the wing surface. Thus, unless they are travelling relatively fast, flying animals must flap their wings. The thin, streamlined, usually cambered wing surface with rounded leading and sharp trailing edges (figure 1) causes air passing the upper surface to accelerate, and decelerates air passing under the lower surface. This imparts a downwards acceleration to air behind the wing, and as a reaction a pressure difference is experienced between the sides of the wing. This is the *lift* force which is responsible for weight support and thrust in flight; lift acts perpendicular to the relative air flow (equivalently to the movement of the aerofoil or to the local movement of the air adjacent to the wing), and is accompanied by a *drag* force, generated by friction and pressure forces, acting parallel to the line of flight. With a well-designed aerofoil drag is usually much smaller than lift: lift to drag ratios of the order of 10 to 15 are typical for animal wings. There is additional drag due to the body of the animal, and to any trailing legs, tail etc.; any lift from these other surfaces is usually small, and is negligible in terms of the overall forces, although it may contribute usefully to balance and stability.

### *The passive aerofoil and gliding*

The central limitation on the use of the aerofoil is that drag represents an energy loss, and therefore always retards the animal, so that a force must be applied to overcome it. An aerofoil can *never* generate a force with a positive component along the local direction of motion. Many organisms exploit passive aerofoil or hydrofoil action in air and/or water for self-stability, dispersal or transport (Vogel 1981): various plants and marine invertebrates control orientation by aerofoil lift, and many plant and tree seeds are designed to use aerofoil action to retard fall to exploit horizontal wind currents. Moreover, many animals are adapted for *gliding*.

Gliding requires little more than the development of a broad aerodynamic surface, has evolved in each of the five vertebrate classes, and is common in the fossil record (mainly in reptiles) (Rayner 1981, 1986; U. M. Norberg 1985*a*). The flight surface has been achieved by enlarging limb extremities (flying fish and frogs, some lizards), spreading a membrane between fore and hindlimbs (gliding mammals), and by lengthening the ribs and using them to support a membrane (*Draco*, and various Mesozoic fossil reptiles). A glider uses potential energy to maintain speed against drag, and must lose height relative to the air (the same is true of gliding and soaring birds while they are not flapping their wings). Flapping the wings so that mean lift has a horizontal component is the sole method by which a horizontal thrust force can be produced.

### *Flight and flapping wings*

I have argued above that as a phylogenetic character flight may most usefully be characterized by the ability to flap the wings to achieve thrust; this does not necessarily imply sufficient thrust to balance all drag and to allow steady, level flight. By this definition passive gliders are not capable of flight; none the less, this locomotor adaptation will be seen to be important in the context of avian evolution for it is the probable precursor of flight.

By contrast to gliding, true flight makes considerable demands; the influence of these are immediately evident in the long limbs, robust pectoral and pelvic girdles, and enlarged flight musculature of birds and bats. There is no difficulty in tracing comparable adaptations in fossil animals. The purpose of flapping is to generate a horizontal thrust while maintaining a sufficient vertical force to balance weight. In birds and bats a mosaic of constraints delimit the geometry of the wingbeat, many of which are associated with force and energy flow optimization (Rayner 1987, 1988 *a*,*b*).

Wing movements in flapping flight consist primarily of oscillatory dorso-ventral flapping, combined with varying degrees of deformation of the wing surface. At cruising speeds positive thrust appears to be generated only during the downstroke, when the wings move downwards relative to the body and forwards relative to the air: the mean lift in this phase is directed upwards and forwards. High-speed photography and aerodynamic experiments based on visualization of the vortex wakes during flapping have shown that during the downstroke the wing is more or less flat and is fully extended to maximize lift (Rayner 1979, 1980, 1987, 1988a).

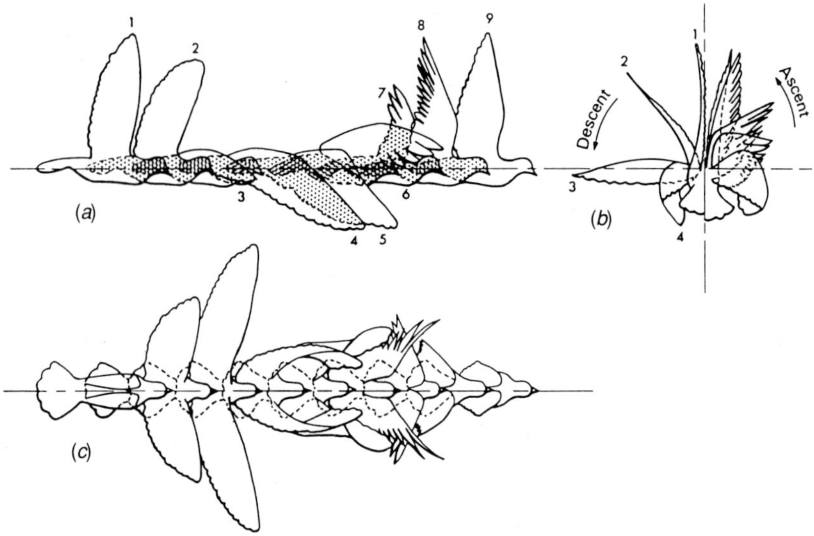


Figure 2. Wingbeat kinematics in the pigeon *Columba livia* in very slow flight at approx. 2 m/s. During the upstroke the wing is highly flexed and the wingtip is brought close to the body to minimize friction drag and wing inertia. The upstroke requires active contraction of the supracoracoideus muscle, in addition to non-planar wing flexure; I argue in the text that this form of wingbeat correlates with derived features in birds, and was inaccessible to *Archaeopteryx*. (From Guidi 1939.)

By contrast the upstroke shows more variation in kinematics and aerodynamic function; this variation is associated with both wing design (wing skeleton, wing outline size and shape) and with flight speed. At low speeds, and also at higher speeds in animals with short and rounded (low aspect ratio) wings, the upstroke is aerodynamically inactive, and generates no lift. As the wing is raised the wingtip is brought close to the body to minimize inertial forces and drag, and the wing surface is considerably deformed and flexed (figure 2). In animals with larger, more pointed wings the upstroke becomes aerodynamically active at higher (cruising) flight speeds; lift acts primarily vertically to support weight, but it also has a horizontal and backwards component which retards the animal. Flow visualization experiments have clarified the reason for change of wing shape during the upstroke even when the wing is generating lift (Spedding 1986, 1987; Spedding, Rayner & Pennycuick 1984; see also Rayner 1988a). Wingspan must be reduced somewhat during the upstroke, for, if the wings remained as fully extended as they are in the downstroke, the magnitude of the horizontal retarding component of lift would equal any thrust from the downstroke, and there would be no net horizontal force. Yet the purpose of

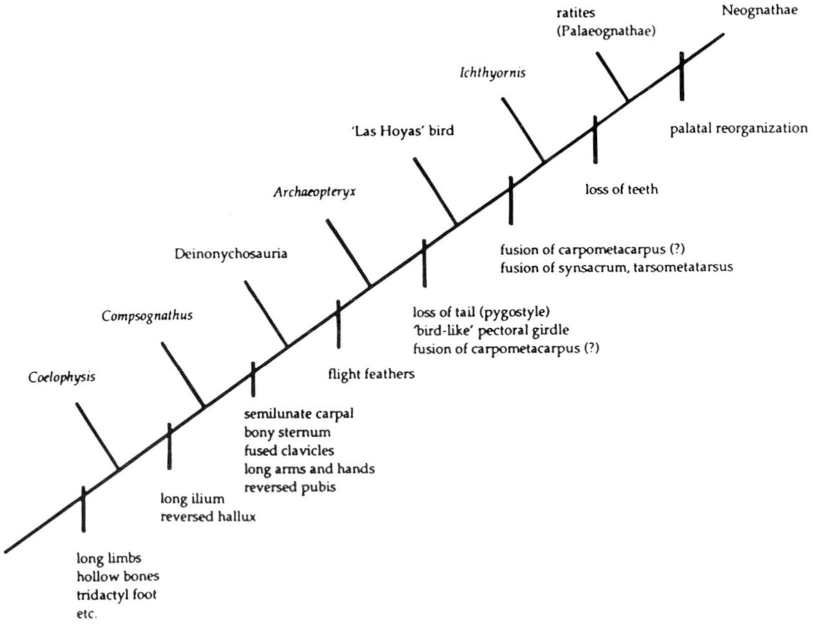


Figure 3. Cladogram showing the relationship of theropods to *Archaeopteryx* and Cretaceous birds. The emphasis is on characters regarded as avian or as 'flight-related'. (After Padian (1987), based on information from Gauthier & Padian (1985), with further information from Sanz *et al.* (1988).)

flapping is to produce a positive thrust. Most birds reduce the effective wingspan during the upstroke by flexing the wing at the carpal (wrist) joint so that the leading edge of the wingtip is swept back, and follows a circular or elliptic path when seen in lateral view; the avian hand wing can continue to function as an efficient aerofoil as the feathers slide over one another. In most cases there is little useful shortening from flexing the upper arm.

Experiments have revealed that – despite wide differences in the anatomy of the wing and in the kinematics of the wingtip – birds and bats are aerodynamically very similar. The vortex wake in both groups consists of continuous trailing vortices behind the wingtips, and the strength of the bound vortex on the wings remains constant throughout the wingbeat (Spedding 1987; Rayner 1985*a*, 1987, 1988a). This reinforces the suggestion that mechanical constraints on flight are very severe. There are strict limits on wing dimensions, reflecting deterioration of certain aspects of mechanical performance (e.g. speed, range, energy, manoeuvrability, agility) with sub-optimal wing design (Norberg & Rayner 1987; Rayner 1988*a*). These constraints would have been no less severe in the early stages of the evolution of flight, and where they reflect mechanical (rather than ecological pressures) are likely to have had a more significant effect on animals with incipient or imperfect flight adaptations. Early fliers are likely to have used wingbeats similar to those of present birds, since these represent the most efficient means of generating aerodynamic force, and it should be possible to reconstruct their design and performance with this in mind.

From the above arguments and from other studies of modern birds it is possible to identify four adaptive 'domains' which are essential for the evolution of flight.

(1) A flying animal must have wings of aerofoil profile to generate sufficient lift to support the weight.
(2) It must be able to flap its wings to generate thrust: this demands the ability to deform the wings appropriately during the upstroke, and will be accompanied by the evolution of large pectoral muscles to elevate and depress the wing.
(3) It must develop physiological systems capable of mobilizing fuel and oxygen in sufficient quantities, and sufficiently rapidly, to power the flight muscles, and must be able to radiate considerable quantities of surplus heat during flight.
(4) Behavioural, sensory and ecological adaptations must provide the animal with the means to exploit suitable habitat and food reserves efficiently; this last domain also implicates the dynamics of the ecological community to which the animal belongs, and we can only surmise this in the case of the early evolution of birds.

Of these four domains, the first two are a useful starting point for the study of avian evolution. The other two domains are no less important, but within present knowledge of archosaur palaeobiology any discussion of them must to some extent be speculative; they are not considered this chapter.

## The phylogeny of avian flight, the fossil evidence, and the flight of *Archaeopteryx*

Birds are the dominant group of flying vertebrates. Their origin and relationships and the evolution of their flight have been the subject of considerable study, and some controversy (Stephan 1987; Hecht *et al.* 1985; Padian 1986). There is overwhelming evidence that birds evolved from small bipedal theropod dinosaurs during the Lower Middle Jurassic, probably

140–160 million years ago (Huxley 1868; Abel 1911, 1912; Heilmann 1926; Ostrom 1976*a*; Thulborn 1984; Gauthier & Padian 1985; Gauthier 1986; Padian 1987; Wellnhofer 1988*a,b*). The transformations in morphology and size from dinosaur to bird appear from the limited fossil evidence to have been fairly modest (figure 3), but in the process birds acquired two features – the power of flight and the possession of feathers – which set them apart from other dinosaurs. The sister group of birds is believed to be the deinonychosaurs, exemplified by *Deinonychus*, a predatory cursorial theropod some 2–3 m in body length; some other theropods contemporary with *Archaeopteryx* such as *Compsognathus* were much smaller than *Deinonychus*, and more comparable with the size at which it seems likely that flight would have evolved.

Little is known of the first birds, and the limited fossil record may obscure the true avian origin. The most significant evidence is provided by the six fossils of *Archaeopteryx lithographica* (figure 4), which is distinguished from other theropods by its small size (mass 0.2–0.4 kg, comparable to that of a pigeon *Columba livia*; Yalden 1984, 1985), and it is the first theropod identified to have both feathers and other bird-like features which indicate its ability to fly (see below). It is unlikely that *Archaeopteryx* itself is a direct avian ancestor, but there is little doubt that it is a sister group to the stem-group birds (Gauthier & Padian 1985); it represents therefore the best available indication of the morphology of the pro-avian. Later in this chapter a mechanical argument is developed to show that a gliding ancestry for flight is inescapable. This argument will be supported – for birds at least – by study of the skeletons of *Archaeopteryx*.

The first significant question concerns not the flight but the terrestrial locomotion of *Archaeopteryx*. Like other theropods it was an active, probably fast running, biped. Whether it could climb has become central in the debate about flight evolution, for some authors demand arboreal climbing as a prerequisite for the gliding origin of flight. I do not consider climbing ability to be essential. *Archaeopteryx* apparently shares the cursorial morphology of other theropods, and its skeleton shows no features associated with climbing or arboreality. Yalden (1985) has argued that the morphology of its fore– and hindlimb claws are similar to those of arboreal birds and mammals, but the shape of these claws is also characteristic of larger theropods which were too big for climbing to have been feasible. Moreover, the recently discovered 'Solnhofen' specimen (figure 4) shows exceptionally preserved keratinous claw sheaths, which show no sign of the abrasion to be expected from climbing (J. M. V. Rayner & K. Padian, personal observations).

Figure 4. (*a*) For legend see facing page.

Compared to other archosaurs *Archaeopteryx* was small in size, and it is relatively modest in dimensions compared with many modern birds. There are strong mechanical reasons (Rayner 1985*b*; Pennycuick 1986) for expecting flight to have evolved in relatively small animals, certainly much smaller than the closest known non-feathered dinosaur *Deinonychus*. Although the skeletons of *Archaeopteryx* and *Deinonychus* are morphologically similar (Gauthier 1986), the implied correlation between osteology and behaviour may be misleading, since small animals are often more generalized in their repertoire of locomotor behaviour: the apparent absence of arboreal features in *Archaeopteryx* does not entirely rule out the possibility of climbing. More convincing, perhaps, is the palaeobiological

(*b*)

Figure 4. (*a*) The Solnhofen *Archaeopteryx*. This specimen was identified as *Archaeopteryx lithographica* in 1987, having previously been attributed to the similar-sized and contemporary dinosaur *Compsognathus*. See Wellnhofer (1988*a*,*b*) for detailed descriptions of this specimen, which is on display in the Bürgermeister-Müller-Museum, Solnhofen. Impressions of the feather shafts of the left wing are evident below the elbow joint of the right wing, running beneath the femur of the right leg. Note also the sharp and highly recurved claws of the wing digits. (*b*) Drawing of the Berlin *Archaeopteryx*, the best known and best preserved of the six known specimens. (Drawing by Nicola Hector.)

evidence pointing to an absence of trees in the vicinity of the Solenhofen lagoon (Abel 1922; Viohl 1985).

In the context of flight the most obvious question is whether *Archaeopteryx* could flap its wings, and if so what degree of useful aerodynamic force it could generate. Four separate lines of evidence may be invoked.

### 1. *Wings and wing feathers*

The most striking feature of *Archaeopteryx* is its well developed feathered wings. These wings are not significantly different in size and shape from those of modern birds such as magpies or coucals, and they give every indication that *Archaeopteryx* was a flying bird. The feathers also appear to be strong evidence of flight ability, but need to be interpreted with some caution. They are the main structure by which the flight surface achieves mechanical integrity, but in modern birds they perform other functions. It is probable that feathers evolved prior to flight for some other function, and thus the presence of feathers alone does not imply flight. Possible non-biomechanical hypotheses for feather evolution have included thermo-regulation, water-repellancy, display and fighting (Cowen & Lipps 1982; Thulborn & Hamley 1985; Regal 1975, 1985; Dyck 1985). If this is the case the avian stem group would contain animals with feathers but without the ability to fly (although its feathers may have correlated with other adaptations which pre-adapted to flight). The fact that no such animal has yet been found does not negate the hypothesis that feathers evolved prior to flight.

In *Archaeopteryx* the feathers are remarkably similar to those of modern birds. They have a stiffened central shaft to transmit aerodynamic forces generated over the feather vanes to the body, and this would not be expected if the feathers had no mechanical function. More significantly, the feather shaft is set asymmetrically against the vanes of the feather; this permits the feather to distort optimally to compensate for bending in flight due to aerodynamic loads, and is important in both gliding and flapping flight (R. Å. Norberg 1985). Vane asymmetry is characteristic of modern flying birds, but the feathers of most modern flightless birds are symmetrical (Feduccia & Tordoff 1979). Thus the feathers of *Archaeopteryx* imply at least gliding flight for this animal.

### 2. *Pectoral girdle morphology and the sternum*

The pectoral girdle of *Archaeopteryx* is similar to that of non-flying theropods; the coracoids are not the solid strut-like braces of modern birds; some specimens have a furcula, but all lack a calcified sternum (Ostrom 1974). These features suggest relatively weak specialization for flapping flight. In modern birds the sternum is the main origin for the *m. supracoracoideus*, which elevates the wing, and for some fibres of the much larger *m. pectoralis* (Olson & Feduccia 1979). The majority of pectoralis fibres originate from the furcula and from tissue surfaces below the humeral joint, and apply a positive vertical or *roll* moment to the humerus, depressing the wing; pectoralis fibres originating from the sternum below *and behind* the humeral joint attach to the forward edge of the humerus, and, as well as depressing the wing during the downstroke, apply a pitching moment which pronates the wing around its long axis (Rayner 1988*b*). The sternum may be missing through poor preservation, or may have been cartilaginous; however, it is more likely that *Archaeopteryx* had no sternum, for without it only the pronating fibres, and the supracoracoideus muscle (see below) would have been absent; on the basis of modern birds, we would expect little osteological evidence of the depressing proximal fibres of the pectoralis. Bats do not have functional sterna, although of course their pectoral anatomy is very different from that of birds; most theropods also lack calcified sterna (Ostrom 1976*b*). According to this model, *Archaeopteryx* might well have been able to depress the wing, but would have been unable to pronate it effectively. However, extreme pronation is required only in slow flight, where large changes in the angle of incidence of the wing are needed to control the varying vortex strength and to flex the wing surface during the upstroke (Rayner 1979, 1980, 1988*a,b*; see also above and figure 2); in fast (cruising) flight the wingbeat is primarily dorso-ventral, the wing remains near planar throughout, and changes in incidence are more moderate. The absence of a sternum implies that early birds (including *Archaeopteryx*) were unable to fly slowly: the keeled sternum is a derived avian character associated with later development of slow flight and of the ability to manoeuvre in flight.

### 3. *Elevating the wings: the upstroke musculature*

More definite evidence is available for the upstroke musculature. In modern birds the wing elevator is the *M. supracoracoideus*, which originates on the sternum deep under the pectoralis, and inserts on to the dorsal side of the humerus through a cavity within the humeral joint (the *foramen triosseum*). Today this arrangement is unique to birds, although it was also present in pterosaurs (Padian 1985, 1987); it was absent in *Archaeopteryx* (Ostrom 1974, 1976*b*), and must therefore have evolved subsequently. Early flying birds may have used other muscles such as the deltoid group to elevate the wing, but if this were the case it is hard to envisage the pressures that must have forced the development of a novel muscle system to form the supracoracoideus in place of the deltoid group. In flight an elevator muscle is required only at low speeds (Rayner 1985*b*, 1986, 1988*a,b*), and at higher (cruising) flight speeds the wing is elevated by aerodynamic forces. It is likely therefore that *Archaeopteryx* flew in such a way that it never needed active muscle action to elevate the wing. Again, *Archaeopteryx* could be capable of steady cruising flight, but it could not fly slowly.

### 4. *Flexing the wrist: the semi-lunate carpal*

The wing skeleton also indicates poor slow flight performance. The semi-lunate carpal forming the wrist articulation is shared with theropods such as *Deinonychus* (Ostrom 1974; Gauthier & Padian 1985; Padian 1987), but as an individual bone was lost later in the bird lineage as the bones of the manus became fused. This configuration appears to prevent the wing distorting out of its plane, but it ideally permits sweeping of the wingtip within the plane of the wing. This is precisely the movement which occurs during the upstroke in cruising flight in most modern birds (above), but is not the movement found commonly in slow flight (figure 2) or during take-off and landing. Since – like all tetrapods – *Archaeopteryx* was capable of dorso-ventral movement of the humerus, this flexure indicates that *Archaeopteryx* was capable of cruising flight, but could not manage the complex twisting, folding and flexing of the wing which must be used in slow flight.

### *Flight in* Archaeopteryx

Without elevator muscles, without longitudinal pectoralis fibres originating from a sternum, and with no ability to deform the wing out of its plane during the upstroke, *Archaeopteryx* had none of the derived features associated with slow flight in modern birds. It seems unlikely that it could have flown slowly, and slow flight and the associated specialities of take off and landing are therefore derived in birds. Fast or cruising flight is less strenuous than slow flight: the forces required from the pectoralis are less extreme, no elevator muscle is needed (Rayner 1986, 1987, 1988*a,b*), mechanical energy demands are less, and the wingbeat geometry is simpler.

It is likely that *Archaeopteryx* would have been able to flap at reasonably high speeds (of the same order of magnitude as the flight speeds of a gliding proto-bird). *Archaeopteryx* is unlikely to have been proficient in slow flight, or to have been very agile and manoeuvrable, and take-offs and landings were probably quite awkward (Rayner 1985*b*, 1988*b*).

At a later stage in the bird radiation the sternum and the supracoracoideus muscle would have evolved in response to pressures for slow flight and manoeuvrability (figure 3). Improved flight performance would also have been associated with fusing of the manus (including the semi-lunate carpal), loss of the clawed wing digits, stiffening of the pectoral girdle, and disappearance of the bony tail. Many of these features are seen in a bird reported recently from the Early Cretaceous of Los Hoyas in Spain (Sanz, Bonaparte & Lacasa 1988; Cracraft 1988); this specimen (as yet unnamed) is unfortunately incomplete, but its very small size (comparable to a sparrow) and its apparently primitive hindlimbs suggest that after the appearance of flight in the proto-avian ancestor of *Archaeopteryx*, selection was directed primarily towards improved flight performance as birds radiated into otherwise unoccupied niches.

## Theories and hypotheses of flight evolution

In the previous sections of this chapter I have discussed some of the major mechanical constraints which must be met before flight can evolve, and have considered the phylogenetic and morphological transformations associated with the appearance of birds. The demands imposed by flight are so substantial – and so unusual – that the selective pressures which encouraged flight evolution must have been very significant. Not surprisingly, many hypotheses have been advanced in attempts to explain the mechanisms which accompanied the appearance of flight in vertebrates. The problem has been complicated by occasional confusion of the significance of feathers, which as was argued above need not have evolved for flight. Some hypotheses are far-fetched, others more convincing, but few can be discounted outright; useful models may suggest behavioural correlates or precursors of flight, but must be irrelevant to this discussion if they do not address the major question of how an animal might have evolved a flapping wing capable of generating useful aerodynamic forces. As I have argued above, it is this which distinguishes birds from their non-flying relatives.

Three models attack this problem. Each has its adherents, and each has been the butt of valid (and sometimes also invalid) criticisms. Space does not permit a full review of the arguments here, but I have considered the models and their advantages and disadvantages elsewhere (Rayner 1985*b*,*c*, 1986, 1988*b*; see also U. M. Norberg 1985*b*). Similar arguments, driven by similar mechanical constraints, characterize the debate over insect flight evolution (Wootton & Ellington, this volume), where they may be seen to lead to similar conclusions.

### *The gliding model*

The oldest of the models was proposed first by Darwin (1859), who argued that bats evolved through a gliding stage, and that proto-bats were not dissimilar to contemporary gliding squirrels. (The colugo (*Cynocephalus*, Dermoptera), a relatively large glider from South-East Asia, is a close relative of megachiropteran fruit bats; J. D. Pettigrew, personal communication) Very similar phraseology was later used by Marsh (1880) in advancing the gliding model for birds. Many authors have since accepted this idea, and its implications have been considered extensively (see, for example, Bock 1986). The predominant feature of the gliding model is that proto-flappers used gravity to power their flights from trees or within vegetation (or perhaps among rocks and cliffs).

This hypothesis has been referred to by most authors as the 'arboreal' model. Animals often live or move in trees, and many are able to climb; trees are a ready source of potential energy with which to start a glide. Moreover, virtually all contemporary tetrapod gliders are associated with trees. However, this nomenclature has been the source of one of the main objections to the gliding model as applied to birds, for their is no evidence that the theropod dinosaurs related to the first birds made use of trees. The term 'arboreal' in the context of bird evolution therefore biases the discussion, and for this reason I prefer to use the term 'gliding' to emphasize the mechanical, rather than ecological, processes of the first flight. (There have been few cogent objections to the validity of the arboreal model, or of the term 'arboreal', when applied to bats, since trees and arboreal habit in mammals were widespread by the Eocene. The morphology of bats – particularly the quadrupedal support of the flight membrane – and their relationships to other arboreal and gliding mammals support this view (Rayner 1986).) In birds, as in bats, the animals' use of gravity justifies the intuitive attraction of this model, especially to the biomechanicist.

(*a*) (*b*)

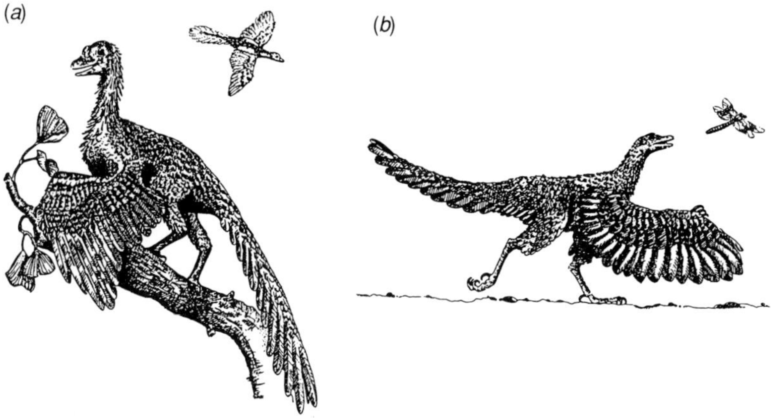

Figure 5. The cursorial and gliding models of flight evolution illustrated for the first bird, *Archaeopteryx*. (*a*) Gliding, showing the possible stance of an arboreal animal which uses the forelimb claws for support (it is in fact unproven that *Archaeopteryx* was arboreal), and (*b*) the cursorial model, illustrated by Ostrom's (1974, 1979) running insect-gatherer. (Drawings by Nicola Hector.)

*The cursorial model*

The most widely voiced alternative to gliding is the 'cursorial' model. This envisages the proto-flier as a running biped making short jumps from the ground. These jumps become extended, the wings are used for balance and propulsion, and the animal begins to fly. The hypothesis apparently originated with Williston (1879) but was formulated formally first by von Nopsca (1907, 1923). With growing understanding of flight mechanics the intuitive advantages of gliding caused the cursorial model to fall into disfavour. A major problem is that it is apparently impossible to provide thrust with the legs at the same time as gaining any useful height or distance while jumping from the ground.

The greatest omission from the theory as originally presented has been – and remains – a clear selective pressure favouring the appearance of flight in a cursor. However, the dual realization of the bipedal cursorial habits of theropods and of their relationship to birds forced reconsideration of the role of running in bird flight evolution. Ostrom (1974, 1979) proposed that catching flying insects while running and jumping could favour the development of wings functioning like tennis rackets (figure 5); Ostrom argued that the wingstroke needed to trap a flying insect would be similar to the arm movements available to *Deinonychus*, and was claimed to be similar to the flight stroke of a bird (see also Padian 1985, 1987). Disagreeing with the insect-catching hypothesis, Caple, Balda & Willis (1983) proposed that a biped would benefit from enhanced stability while running, perhaps at high speed while escaping a predator, and that this forms a strong pressure for wing enhancement. Both of these ideas are cogent models for the appearance of wings in a cursor, and both are consistent with the apparent terrestrial habits of the larger relatives of proto-birds. But both are implausible, for two reasons. First, there are no contemporary analogues of running animals feeding with their wings, or using forelimbs as lifting surfaces for stability: running birds generally fold their wings to minimize drag. Second, and more important, neither model sensibly addresses the question of *why* the wings should be flapped, and how the imprecise forms of wing waving needed for predation or stability developed into true flapping. The cursorial model as generally understood is only a first step towards the evolution of flight, and cannot be compared directly with the more completely formulated gliding model.

### *Fluttering proto-flappers*

A third possible proposal is more elusive to mechanical formulation. The 'fluttering' model proposes that flight began with animals beating their wings erratically, and finding that they could generate useful aerodynamic forces; flight would thus have evolved rapidly, almost by saltation. The fluttering scenario has been advanced most commonly for bats (see, for example, Caple *et al.* 1983), and the main pressure for this model has come from unfounded objections to the gliding model, often arising from misunderstanding of flapping flight mechanics (as in Balda, Caple & Willis 1985). The fluttering model must fail because it takes no account of the extreme morphological, physiological and behavioural specializations required for flight. Slow flight, take-off and landing are the most demanding types of flight experienced by most birds, and yet the first 'flights' of a fluttering proto-flapper would have been at low speeds, where the energetic demands of flight are at their most extreme (Clark 1977), and the wingbeat cycle is at its most complex (Rayner 1988*a*,*b*). This model need not be considered further.

## The gliding and cursorial models

Flight is a mechanical phenomenon, and a theoretical resolution of the problem of flight evolution should compare the gliding and cursorial hypotheses on terms as equal as possible. The sequence of appearance of flight adaptations and pre-adaptations must be consistent with phylogeny, and the biomechanical and physiological developments as flight improves must be gradual, consistent with morphology, as known from the fossil record.

Biomechanical analysis relies on estimation of quantities such as energy rates, speeds and forces in the muscles and limb segments, or for the whole animal. These data can be used in two ways: first, they may form the basis for comparison with what an animal of the assumed design might have been able to achieve, and second they allow the results of similar assumptions for the two models to be considered side-by-side. The first programme has the very real difficulty that it demands accurate knowledge of the dimensions of a hypothetical proto-flier and of the mechanical properties of its tissues; any absolute predictions may therefore be unreliable. The second approach can be more informative: the *relative* performance for the two models should give a fair comparison of them as alternatives, and should indicate whether one model is more or less likely (or perhaps even impossible). In this discussion I consider only the step from the presence of a wing generating static aerodynamic forces to the flapping of that wing to generate thrust. As I have argued above, it is this which distinguishes flight in the avian phylogeny; moreover several scenarios independent of biomechanics or aerial locomotion can be invoked to explain the appearance of feathered wings. I summarize the methods that I have adopted and the main conclusions of the model here, and have given the detailed arguments elsewhere (Rayner 1985*b*,*c*, 1986). The first stage in the analysis is to consider the requirements of the two models.

### *The gliding model*: *step* 1 – *climbing*

As well as being able to support its weight in air, the glider must be able to gain height. Glides are driven by gravity, and during a glide the animal loses height in the air. I assume that the proto-flier had access to a suitable habitat, and had the morphological and physiological ability to climb; for this it may be sufficient that it is relatively small. Climbing performance gives a reasonable estimate of the maximum forces and energy rates that it might be able to generate at the stage prior to the appearance of flight: the main forces in climbing are associated with the forelimbs. For instance, if the animal were to hang in a suspended position, say against a tree trunk, with its limbs at an angle of 45° to the vertical, the animal would have to impose a static moment of 0.36 $Mgb$ at the humeral joint, where $M$ is body mass, $g$ the gravitational acceleration, and $b$ the length of one forelimb. In active climbing a somewhat greater mean moment would be needed, perhaps approaching twice this value. The energy rates in climbing are hard to estimate without knowing climbing speed, but a minimum estimate is the rate of gain of potential energy.

### *The gliding model: step 2 – gliding*

The glides of extant gliding tetrapods consist of an initial leap, followed by a drop to gain speed before a sustained steady glide. This strategy is necessary to clear the take-off substrate and to attain sufficient speed for an aerodynamically stable glide. Presumably the proto-flier would have been similar. Glide speed depends on the area of the flight surface: slowing the glide reduces the risk of accident: there is pressure to increase wing area. Height loss is related to drag, which depends in part on the shape of the wing. With constant area, a longer and narrower wing gives a shallower glide. The lift generated by the wing must be responsible for weight support, and a static (isometric) moment is required at the humeral joint to maintain the wing in position; I have estimated this moment as 0.23 $Mgb$, where $b$ is now the wing semi-span. This value is less than that in hanging and vertical climbing (for a wing of the same length). A climbing animal should therefore already have the muscular capacity to hold its wing outstretched in gliding. That the acquisition of gliding by a climber is simple if the animal can develop a flight surface is seen by the frequency with which this has happened in vertebrates.

The main advantages of gliding are that it is energetically economical, and relatively fast: a gliding squirrel might travel at up to 15 m/s, while its running speed on level ground is unlikely to exceed 2 m/s (Scholey 1986). In a suitable habitat climbing and gliding becomes an attractive mode of locomotion: the energy for transport is expended in climbing rather than running, but the time expended in climbing is outweighed by the high speeds of gliding, and cost of transport can be reduced (Rayner 1986; Scholey 1986).

### *The gliding model: step 3 – flapping*

A glider loses height as it expends potential energy against drag. By flapping the wings the animal can generate thrust, and, provided there is no significant reduction in weight support (which would be accompanied by either downwards acceleration or faster flight), even a small thrust reduces the glide angle and increases the distance travelled appreciably. I have described the geometry of the wake vortex in flapping flight above; in a fully developed flapper flying at the normal glide speed the effect of incipient flapping can be predicted (Rayner 1986). Small, low amplitude wingbeats are liable to enhance drag or reduce weight support, but wingbeats with amplitude greater than about 25° and with upstroke wing flexure can generate useful thrust while still balancing weight. The increment in forces at the wing root are relatively modest, and mainly reflect wing inertia rather than additional aerodynamic thrust and drag (Rayner 1988*a*,*b*). The peak positive roll moments at high speeds remain comparable to those in climbing, as does the energy required for flight (Rayner 1985*c*, 1986, 1988*b*).

### *The cursorial model*

The cursorial model is less straightforward to analyse since existing hypotheses fail to address the mechanical environment in which the flight stroke first appeared. The problem of getting off the ground is compounded by the apparent change of action of selection forces, which would favour the hindlimbs for speed and agility in a cursor, but would switch to the forelimbs as the animal begins to fly. Did the proto-flier have to transport different muscles to give good performance in the two modes?

To make it comparable with the gliding model, I envisage a situation in which the cursor makes the best possible use of its hindlimbs to fly: it gains height while still running by jumping from the ground. Its initial flaps can then take place while descending, albeit briefly, and like the gliding model the first flights are assisted by gravity. This comparison must favour the cursorial model, for any – perhaps more realistic – alternative working *against* gravity would increase the mechanical and energetic demands of incipient flight and make the model relatively less attractive.

### *The cursorial model*: *step* 1 – *running*

The mechanics of running in bipedal birds, mammals and lizards are well understood. Without using the forelimbs the animals have good agility, and can move at relatively high sustained speeds up to 2 m/s; the energy rates required at higher speeds are extreme, and can be sustained only for brief periods.

### *The cursorial model*: *step* 2 – *jumping and flying*

If it is to jump before initiating its flights, a running proto-flier must use some of the energy available from its hindlimbs to leave the ground. At maximum exertion this must be associated with degradation of running performance, and I have calculated that for a biped of mass 0.2 kg and typical theropod morphology the drop in running speed associated with the jump is of the order of 30–40% (Rayner 1985*c*). A running speed of 2 m/s is well below that at which flight for an animal of the same mass is economical or efficient (see above). Thus the reduced speed from jumping implies very low speeds for the initial glides or flights. The animal would be unable to extend its jumps significantly by flapping to obtain thrust since the energy demanded would be considerable, and moreover the 'slow speed' wingbeat derived above is not available to an animal of theropod morphology. Acceleration to higher and more efficient flight speeds takes time, and would not be permitted by the duration and extent of short initial leaps.

None the less, this strategy of running, jumping and gliding is energetically attractive, in that cost of transport at maximum aerobic exertion is slightly below that for continuous running, even at these low speeds (Rayner 1988*b*; figure 6). A biped *could* get sufficiently off the ground to glide briefly, and *could* save energy by so doing. It *could* benefit further by flapping. But it has insufficient energy to reach speeds at which flapping is mechanically straightforward, and the costs of flight at these low speeds are so high that the demands on the forelimb musculature become extreme. It seems that mechanically the cursorial gliding model could just work; but practically it is highly unlikely. Since this model was designed to favour the cursorial hypothesis, it seems probable that flight could not have evolved unassisted from running over the ground.

### *The models compared*

The separate analysis of the two models strongly suggests that gliding is an essential precursor of the appearance of flapping flight, but cannot rule out either alternative absolutely. To compare the two models I assume that the purpose of movement is for transport, and that the animal will choose to move in such a way that minimizes the cost of transport; that is, the energy to transport unit mass through unit distance. This ignores other possible – but less readily quantifiable – benefits such as escape from predation or access to new microhabitats. I then compare two situations. (1) An animal climbs to a height, then glides to the ground or to the next height; during the glide it gradually begins flapping small wings. As the wings and the wingbeat develop, so the thrust generated increases until the flight path is level. (2) A running biped makes small jumps from the ground, simultaneously flapping its wings to extend the jumps. As the wings develop, the thrust increases and the jumps become longer. The assumptions involved in these comparisons are detailed elsewhere (Rayner 1985*b*,*c*, 1986).

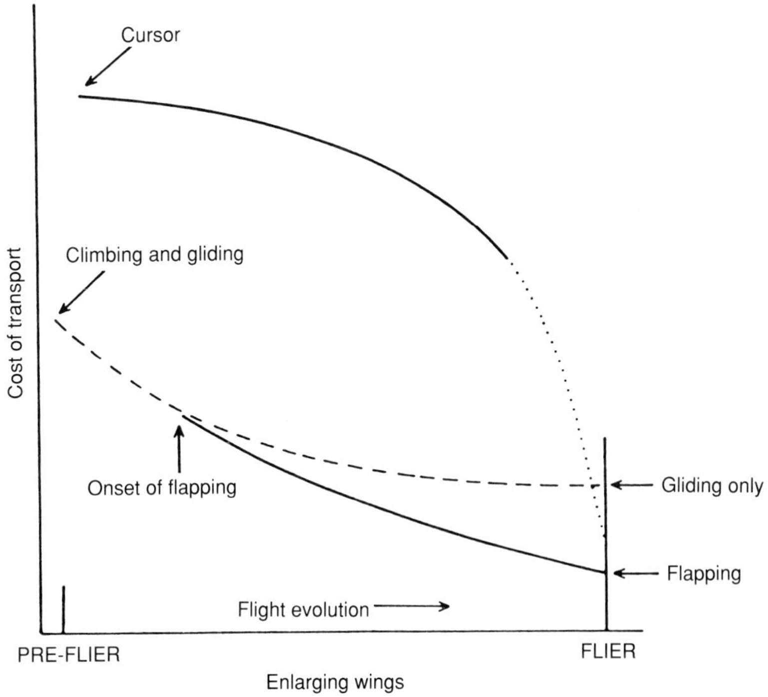


Figure 6. Comparison of predicted transport energetics for climbing and gliding and for cursorial proto-fliers. The graph shows how cost of transport changes as wings enlarge and lengthen for the gliding (lower curves) and cursorial (upper curve) models. Calculation for the gliding model explained by Rayner (1985*b*, 1986 for bats, 1985*c* for birds); cursorial model after Rayner (1985*c*). The graphs are non-dimensional, but the qualitative predictions are relevant for climbers or cursors between about 20 g and 0.5 kg in mass, and there is little change in the relative difference between the two options with size. Flapping with the short, broad wings of a passive glider is disadvantageous, but as wings lengthen slightly flapping brings considerable advantage, reducing total cost of transport by as much as 80%. For the cursorial model the initial energy saving is small; only as flapping flight becomes fully developed is there a significant reduction in cost; the final part of the curve (shown dotted) is notional and has not been modelled in detail. (From Rayner 1988*b*.)

*Cost of transport in gliding*

Climbing and gliding is the typical locomotory mode of gliding squirrels, and reduces cost of transport considerably compared to running (Scholey 1986; figure 6). This form of gliding-based movement is adaptively successful even if the wings are not flapped, but it becomes more attractive if the wings enlarge (although long non-flapping wings may be a liability in a cluttered habitat). Adding flaps as the wings develop gives a greater advantage, with a continuous and gradual reduction in cost of transport to as little as 20% of the climbing and gliding value. Provided the animal can take advantage of the reduced cost (by travelling further to forage, for instance) there is strong feedback encouraging development of the wings and other flight organs (figure 6). Moreover, if the contribution to fitness is inversely proportional to cost of transport, feedback 'encouraging' the further growth of the wings is stronger in the early stages of wing development.

### *Cost of transport for a cursor*

By comparison, the cost for running is rather higher than for climbing and gliding, and the initial savings for running and jumping are very small, since enlarging wings to reduce height loss have little effect when the bulk of the work of locomotion is done in jumping. The end result, assuming progression to a fully adapted flier, must be the same as for the gliding model, so the total reduction in cost is considerably greater than for the climber (figure 6). But feedback from the new adaptation is at first weak, and only becomes significant in the later stages when wings and flight are fully developed, and the cost reduction can be attributed more to adaptation and radiation than to initial evolution of flight. These initial cost savings are unlikely to outweigh the developmental and other costs of enhancing the forelimbs for flapping while remaining a viable running biped. Given the simultaneous problems of attaining sufficient speed and sufficiently long jumps, it is unlikely that a cursorial runner could have begun to fly. Comparison of the two models strongly implies that on mechanical grounds a gliding origin of flapping flight in birds is the more probable.

### *A compromise model: running and gliding in a wind*

Evolution and adaptation are processes embedded in compromises. It is very tempting to attempt to resolve conflicts between scientific hypotheses by a compromise model. I conjecture that this is possible – and probably essential – in this case. Both models have to cope with significant problems. The gliding model conflicts with the fossil evidence that the avian ancestors were cursorial, that *Archaeopteryx* did not climb, and that there was no suitable habitat for a climbing proto-bird in the Solenhofen vicinity. The cursorial model is more consistent with the fossil evidence, but fails on mechanical grounds. The cursorial pathway to flight is energetically too expensive, largely because the available sustained running speeds for a small bipedal dinosaur are too low, and below the range of air speeds at which flight is viable.

Phrasing the problem in this way immediately suggests a possible conjectural resolution. A cursor could maintain ground speeds in the morphologically appropriate range, and yet could achieve viable air speeds simply by running into a wind. An appropriate air speed would be of the order of 7–9 m/s (Rayner 1988*a*), so the wind speed need not exceed 5 m/s for a gliding or kite-like take-off to be viable. This speed is relatively modest, and comfortably in the range of wind speeds reliably experienced in modern littoral habitats. (In this way the wind-gliding hypothesis is more consistent with the marine fossilization site of *Archaeopteryx*.) Somewhat higher speeds would be required for an animal with smaller wings at an earlier stage of flight evolution. We may therefore envisage a scenario in which a proto-avian develops a feathered forelimb or primitive wing for behavioural or physiological reasons, and lives in a habitat with reliable steady winds. The animal begins to jump into the wind, and as the wind sustains the jumps flapping wings are able to generate thrust to lengthen the brief initial glides. This hypothesis retains the reliance on gliding as an initial phase of flight which was deduced above on mechanical grounds and on the basis of the morphology of *Archaeopteryx*. It also avoids the major objections to the 'pure' gliding and cursorial models. However, the model has one serious problem. It relies on a factor – the wind – which is purely conjectural; thus the model cannot be tested formally.

To complete the discussion, a further aspect of the problem should be considered. I have modelled the evolution of flying vertebrates as a gradual progression through adapted and adaptive levels. It is not essential that every level is 'optimal', and it is probably misleading to consider the steps as stable or as 'plateaus'. The process of flight evolution probably took place in a specialized habitat over a brief period of geological time. None the less some stages pre-adapt to following stages (as for instance does gliding to flapping), and the sequence can be driven by consistent, readily identified and realistic selective pressures. Intermediate stages might have been maladaptive compared to their precursors or successors: all that is necessary is that they were not subject to significant predation or other pressures. It is of course possible that flight evolved rapidly (by saltation), so that no intermediates would have existed. (The apparent identification of *Archaeopteryx* as a sister group to the stem-group pro-avian, and the transition to the Los Hoyas bird, apparently discount this.) Or proto-flappers might have benefited from some remarkable environmental conditions in which selective pressures were sufficiently strong and mechanical constraints sufficiently weak that the comparisons I have drawn with the aerodynamics of contemporary fliers is completely invalid. However, if flight did appear suddenly there is no mechanical – or evolutionary – problem to discuss, and further there is no hypothesis open to scientific testing. The first bird or bat would have had the same attributes as modern animals, and its success in flight would be mediated by much the same ecological pressures; questions of its evolution become meaningless.

In this survey I have not considered all the problems of flight evolution. I have ignored features such as the behavioural and ecological flexibility of an incipient flier, its likely predators, and the habitats it might have used. I have not considered the important role of control and stability. (Does *Archaeopteryx* have a long bony tail just because its ancestors did? Does the superficial similarity with magpies (*Pica*, Corvidae) indicate arboreal habit?) The biomechanical approach cannot unequivocally resolve these points, any more than it can *prove* that proto-fliers had to be gliders. Its value lies in showing that some possibilities – such as the fluttering model – are untenable, and that some hypotheses are *more likely* than others. The arguments are convincing, but as in all evolutionary debates the central difficulty lies not only in assembling the evidence but also in assessing its worth, particularly when predictions or interpretations conflict. Mechanical arguments would be devalued if they predicted the most likely behaviour or morphology of a proto-flier to be phylogenetically or developmentally untenable. But this does not occur in this case: evidence from aerodynamics and flight mechanics is reinforced by the interpretation of *Archaeopteryx* as a bird capable of forward flight but not of slow flight or of much manoeuvrability. This approach strongly predicts a gliding origin of flight in all groups of vertebrates.

## Acknowledgements

My work on flight evolution has been funded by grants from SERC and by a Research Fellowship from the Royal Society. I am indebted to Kevin Padian for many discussions and for his comments on the manuscript of this chapter.

## References

Abel, O. 1911 Die Vorfahren der Vögel und ihre Lebensweise. *Verhandlungen der Zoologisch-botanischen Gesellschaft in Wien* **61**, 144–191. Stuttgart

Abel, O. 1912 *Grundzüge der Palaeobiologie der Wirbeltiere*. Stuttgart: E. Schweizerbart'sche Verlagsbuchhandlung.

Abel, O. 1922 *Lebensbilder aus der Tierwelt der Vorzeit*. Jena: Gustav Fischer Verlag.

Balda, R.P., Caple, G. & Willis, W.R. 1985 Comparison of the gliding to flapping sequence with the flapping to gliding sequence. In *The Beginnings of Birds*, ed. M.K. Hecht, J.H. Ostrom, G. Viohl & P. Wellnhofer, pp. 267–277. Eichstätt: JuraMuseum.

Bock, W. 1986 The arboreal origin of avian flight. In *The Origin of Birds and the Evolution of Flight*, ed. K. Padian, *Memoirs of the California Academy of Sciences* **8**, pp. 57 72.

Caple, G., Balda, R.P. & Willis, W.R. 1983 The physics of leaping animals and the evolution of preflight. *American Naturalist* **121**, 455–467.

Clark, B.D. 1977 Energetics of hovering flight and the origin of bats. In *Major Patterns in Vertebrate Evolution*, ed. M.K. Hecht, P. Goody & B.M. Hecht, pp. 423–425. New York: Plenum Press.

Cowen, R. & Lipps, J.H. 1982 An adaptive scenario for the origin of birds and their flight. *Proceedings of the Third North American Paleontological Convention*, vol. 1, pp. 109–112.

Cracraft, J. 1988 Early evolution of birds. *Nature, London* **331**, 389–390.

Darwin, C. 1859.*On the Origin of Species*. London: John Murray.

Dyck, J. 1985 The evolution of feathers. *Zoologica Scripta* **14**, 137–154.

Feduccia, A. & Tordoff, M. 1979 Feathers of *Archaeopteryx*: asymmetric vanes indicate aerodynamic function. *Science* **203**, 1021–1022.

Gauthier, J.A. 1986 Saurischian monophyly and the origin of birds. In *The Origin of Birds and the Evolution of Flight*, ed. K. Padian, *Memoirs of the California Academy of Sciences* **8**, pp. 1–55.

Gauthier, J.A. & Padian, K. 1985 Phylogenetic, functional and aerodynamic analyses of the origin of birds and their flight. In *The Beginnings of Birds*, ed. M.K. Hecht, J.H. Ostrom, G. Viohl & P. Wellnhofer, pp. 185–197. Eichstätt: JuraMuseum.

Guidi, G. 1939 La battuta alare del piccione. *Aerotecnica, Roma* **19**, 121–132.

Hecht, M.K., Ostrom, J.H., Viohl, G. & Wellnhofer, P. (eds) 1985 *The Beginnings of Birds*. Eichstätt: JuraMuseum.

Heilmann, G. 1926 *The Origin of Birds*. London: Witherby; reprinted New York: Appleton 1927 and New York: Dover 1972.

Huxley, T.H. 1868 On the animals which are most nearly intermediate between birds and dinosaurs. *Annals and Magazine of Natural History* **4** (2), 66–75.

Marsh, O.C. 1880 Odontornithes: a monograph on the extinct toothed birds of North America. *Professional Papers of the Engineers Department of the U.S. Army* **18**.

Norberg, R.Å. 1985 Function of vane asymmetry and shaft curvature in bird flight feathers; inference on flight ability in Archaeopteryx. In *The Beginnings of Birds*, ed. M.K. Hecht, J.H. Ostrom, G. Viohl & P. Wellnhofer, pp. 303–318. Eichstätt: JuraMuseum

Norberg, U.M. 1985*a* Flying, gliding, soaring. In *Functional Vertebrate Morphology*, ed. M. Hildebrand, D.M. Bramble, K.F. Liem & D.B. Wake, pp. 129–158. Harvard University Press.

Norberg, U.M. 1985*b* Evolution of vertebrate flight: an aerodynamic model for the transition form gliding to active flight. *American Naturalist* **126**, 303–327.

Norberg, U.M. & Rayner, J.M.V. 1987 Ecological morphology and flight in bats (Mammalia, Chiroptera): wing adaptations, flight performance, foraging strategies and echolocation. *Philosophical Transactions of the Royal Society of London Series* B **316**, 335–427.

Olson, S.L. & Feduccia, A. 1979 Flight capacity and the pectoral girdle of *Archaeopteryx*. *Nature, London* **278**, 247–248.

Ostrom, J.H. 1974 *Archaeopteryx* and the origin of flight. *Quarterly Review of Biology* **49**, 27–47.

Ostrom, J.H. 1976*a* *Archaeopteryx* and the origin of birds. *Biological Journal of the Linnean Society* **8**, 91–182.

Ostrom, J. H. 1976*b* Some hypothetical anatomical stages in the evolution of avian flight. *Smithsonian Contributions to Paleobiology* **27**, 1–21.

Ostrom, J.H. 1979 Bird flight: how did it begin? *American Scientist* **67**, 46–56.

Padian, K. 1985 The origins and aerodynamics of flight in extinct vertebrates. *Palaeontology* **28**, 413–433.

Padian, K. (ed.) 1986 *The Origin of Birds and the Evolution of Flight. Memoirs of the California Academy of Sciences* **8**.

Padian, K. 1987 A comparative phylogenetic and functional approach to the origin of vertebrate flight. In *Recent Advances in the Study of Bats*, ed. M.B. Fenton, P.A. Racey & J.M.V. Rayner, pp. 3–19. Cambridge University Press.

Pennycuick, C.J. 1986 Mechanical constraints on the evolution of flight. In *The Origin of Birds and the Evolution of Flight*, ed. K. Padian, *Memoirs of the California Academy of Sciences* **8**, pp. 83–98.

Rayner, J.M.V. 1979 A new approach to animal flight mechanics. *Journal of Experimental Biology* **80**, 17–54.

Rayner, J.M.V. 1980 Vorticity and animal flight. In *Aspects of Animal Movement*, ed. H.Y. Elder & E.R. Trueman, *Seminar Series of the Society for Experimental Biology* **5**, pp. 177–189. Cambridge University Press.

Rayner, J.M.V. 1981 Flight adaptations in vertebrates. *Symposium of the Zoological Society of London* **48**, 137–172.

Rayner, J.M.V. 1985*a* Vorticity and propulsion mechanics in swimming and flying vertebrates. In *Konstruktionsprinzipien lebender und ausgestorbener Reptilien*, ed. J. Riess & E. Frey, *Konzepte SFB*230 **4**, pp. 89–118. Universität Tübingen.

Rayner, J.M.V. 1985*b* Mechanical and ecological constraints on flight evolution. In *The Beginnings of Birds*, ed. M.K. Hecht, J.H. Ostrom, G. Viohl & P. Wellnhofer, pp. 279–288. Eichstätt: JuraMuseum.

Rayner, J.M.V. 1985*c* Cursorial gliding in proto-birds. In *The Beginnings of*

*Birds*, ed. M.K. Hecht, J.H. Ostrom, G. Viohl & P. Wellnhofer, pp. 289–292. Eichstätt: JuraMuseum.

Rayner, J.M.V. 1986 Vertebrate flapping flight mechanics and aerodynamics, and the evolution of flight in bats. In *Biona Report* **5**, *Bat Flight-Fledermausflug*, ed. W. Nachtigall, pp. 27–74. Stuttgart: Gustav Fischer Verlag.

Rayner, J.M.V. 1987 The mechanics of flapping flight in bats. In *Recent Advances in the Study of Bats*, ed. M.B. Fenton, P.A. Racey & J.M.V. Rayner, pp. 23–42. Cambridge University Press.

Rayner, J.M.V. 1988*a* The evolution of vertebrate flight. *Biological Journal of the Linnean Society* **34**, 269–287

Rayner, J.M.V. 1988*b* Form and function in avian flight. *Current Ornithology* **5**, 1–77.

Regal, P.J. 1975 The evolutionary origin of feathers. *Quarterly Review of Biology* **50**, 35–66.

Regal, P.J. 1985 Common sense and reconstructions of the biology of fossils: *Archaeopteryx* and feathers. In *The Beginnings of Birds*, ed. M.K. Hecht, J.H. Ostrom, G. Viohl & P. Wellnhofer, pp. 67–74. Eichstätt: JuraMuseum.

Sanz, J.L., Bonaparte, J.F. & Lacasa, A. 1988 Unusual Early Cretaceous birds from Spain. *Nature, London* **331**, 433–435.

Scholey, K.D. 1986 The evolution of flight in bats. In *Biona Report* **5**, *Bat Flight-Fledermausflug*, ed. W. Nachtigall, pp. 1–12. Stuttgart: Gustav Fischer Verlag.

Spedding, G.R. 1986 The wake of a jackdaw (*Corvus monedula*) in slow flight. *Journal of Experimental Biology* **125**, 287–307.

Spedding, G.R. 1987 The wake of a kestrel (*Falco tinnunculus*) in flapping flight. *Journal of Experimental Biology* **127**, 59–78.

Spedding, G.R., Rayner, J.M.V. & Pennycuick, C.J. 1984 Momentum and energy in the wake of a pigeon (*Columba livia*) in slow flight. *Journal of Experimental Biology* **111**, 81–102.

Stephan, B. 1987 *Urvögel* (3rd edn). *Neue Brehm-Bücherei* **465**. Wittenberg-Lutherstadt: A. Ziemsen Verlag.

Thulborn, R.A. 1984 The avian relationships of *Archaeopteryx*, and the origin of birds. *Zoological Journal of the Linnean Society* **82**, 119–158.

Thulborn, R.A. & Hamley, T.L. 1985 A new palaeoecological role for *Archaeopteryx*. In *The Beginnings of Birds*, ed. M.K. Hecht, J.H. Ostrom, G. Viohl & P. Wellnhofer, pp. 81–89. Eichstätt: JuraMuseum.

Viohl, G. 1985 Geology of the Solnhofen lithographic limestone and the habitat of *Archaeopteryx*. In *The Beginnings of Birds*, ed. M.K. Hecht, J.H. Ostrom, G. Viohl & P. Wellnhofer, pp. 31–44. Eichstätt: JuraMuseum.

Vogel, S. 1981 *Life in Moving Fluids*. Boston: Willard Grant.

von Nopsca, F. 1907 Ideas on the origin of flight. *Proceedings of the Zoological Society of London* pp. 222–236.

von Nopsca, F. 1923 On the origin of flight in birds. *Proceedings of the Zoological Society of London* pp. 463–477.

Wellnhofer, P. 1988*a* A new specimen of *Archaeopteryx*. *Science* **240**, 1790–1792.

Wellnhofer, P. 1988*b* Ein neues Examplar von *Archaeopteryx*. *Archaeopteryx* **6**, 1–30.

Williston, S.W. 1879 Are birds derived from dinosaurs? *Kansas City Review of Science* **3**, 457–460.

Yalden, D.W. 1984 What size was *Archaeopteryx*? *Zoological Journal of the Linnean Society* **82**, 177–188.

Yalden, D.W. 1985 Forelimb function in *Archaeopteryx*. In *The Beginnings of Birds*, ed. M.K. Hecht, J.H. Ostrom, G. Viohl & P. Wellnhofer, pp. 91–97. Eichstätt: JuraMuseum.

JEFFREY J. THOMASON

# Functional interpretation of locomotory adaptations during equid evolution

The limbs of the horse show extreme mechanical specializations for cursorial habit which are at least as well understood as for any other living mammal. Study of the living horse can be used to make inferences about the locomotory adaptations in two fossil equids, *Mesohippus* and *Merychippus*, whose limb bones show progressive modifications towards the situation in *Equus*. Comparison of limb osteology indicates that the two fossils had many of the cursorial adaptations of *Equus*. The digital flexor musculature and the digital joints are modified in *Equus* in accordance with its unguligrade form of locomotion. Muscle attachment scars and joint shapes on the fossil bones indicate that *Merychippus* was probably also unguligrade, while *Mesohippus* probably had a digital pad as well as hooves, and its foot may have resembled superficially that of the modern tapir. The trabecular orientation and density of the metacarpals are consistent with these interpretations. The major differences in limb anatomy and function among the three equids are in the feet. The transition from a subunguligrade to an unguligrade stance probably occurred between *Mesohippus* and *Merychippus*, while two digits were lost between *Merychippus* and *Equus*. Changes in this lineage may have been related to the increase in size from approximately 25 kg body mass in *Mesohippus* to 500 kg in *Equus*. The transition to an unguligrade stance may have increased the shock absorbing capacity of the feet, while the loss of the digits may have reduced the moment of inertia of the limbs. Estimates of stresses experienced by the third metacarpals of each bone indicate that each change may have increased locomotory stress levels were it not for disproportionate thickening of the third metacarpal in relation to its length.

## Introduction

The limbs of the horse show extreme cursorial specializations. They are long, restricted to movements in a parasagittal plane and taper distally; there is only one fully functional digit in each limb. During the last one hundred years, the locomotory abilities and adaptations of the domestic horse, *Equus caballus* , have been extensively studied by a variety of techniques including dissection, morphometry, surgical interference with selected anatomical structures, cinematography, force plates and shoes, and implanted strain gauges (literature cited by Leach & Dagg 1983*a,b*). Thus the mechanical significance of many of the structural specializations is at least as well known for the horse as for any other living mammal. My aim in this review is to demonstrate the extent to which the methods of study of the living horse, and the data obtained, can be used to make inferences about the locomotory adaptations in the two fossil equids *Mesohippus* and *Merychippus*. These two genera were selected from the radiation within the family Equidae because their limb bones appear to show stages in a progressive modification towards the situation in *Equus*. Because the fossils are represented solely by bones, only osteological information is available for all three animals: shapes, proportions and numbers of bones and joints. My method is to progress through anatomical, kinematic and mechanical analyses of locomotion in *Equus*, and to extrapolate similar information for the fossils, using comparisons of osteological features among the three genera as intermediaries in each case. Data from studies of other extant mammals will be also used where appropriate. I will be dealing exclusively with the forelimb, but many of the points to be discussed also apply to the hindlimb (Hussain 1975).

## Osteology, body mass and the principles of kinetics

Mounted skeletons of the three equids differ dramatically in size. *Mesohippus* stood approximately 0.6 m at the shoulder, *Merychippus*, 0.8 to 1.3 m, while most living breeds and species of *Equus* stand between 1.5 and 2 m. Domestic horses commonly have a body mass of approximately 500 kg (with a range among the different breeds of less than half to more than double that value). The body mass of *Mesohippus* was probably in the range 25 to 40 kg, and that of *Merychippus*, 60 to 100 kg (Thomason 1985*a*; MacFadden 1986). Some of the variability in these estimates for the fossils results from individual variation, and some from the lack of a standardized procedure for making such estimates. Fortunately, even the ranges give a strong indication of the size increase from *Mesohippus* to *Equus*. My estimates were made by comparing the skeletal proportions of mounted specimens of the fossils with the skeletons of extant mammals of similar size and proportion. MacFadden (1986) has developed a method in which the relation is first established between certain cranial dimensions and body mass

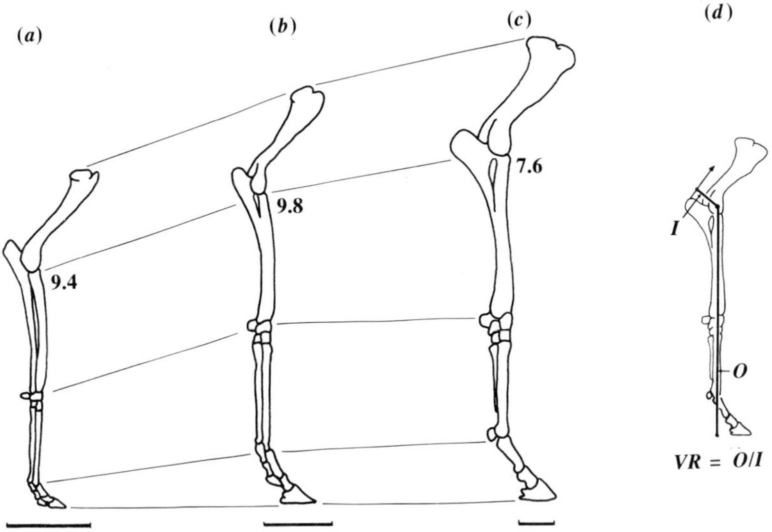


Figure 1. The forelimbs of (*a*) *Mesohippus*, (*b*) *Merychippus* and (*c*) *Equus* showing the length of the antebrachium and manus relative to the humerus (which is drawn to approximately the same length in each case). The numbers indicate the approximate velocity ratios (VR) of the hoof to the insertion of the triceps on the olecranon. Part (*d*) shows how the ratios were calculated. Scale bars are 10 cm. The length of the metacarpus relative to the humerus increases between *Mesohippus* and *Merychippus*; the relative length of the radius increases between *Merychippus* and *Equus*, while the relative length of the third digit increases in both stages.

for living *Equus* over a range of 28 to 530 kg. This relation is then used to predict body masses for extinct equids from comparable cranial measurements, on the assumption that allometric change in the fossils was the same as in *Equus*. MacFadden (1986) discusses and defends that assumption and, in future studies, this work should help to minimize procedural variation.

Despite the marked differences in body mass, the limb bones of all three equids share cursorial adaptations which are commonly found in ungulates and which have been qualitatively interpreted by using the principles of kinetics (Gregory 1912; Smith & Savage 1956; Hildebrand 1984). To begin proximally in the forelimb, all three equids are aclaviculate; this allows the scapula to rotate, thereby raising the centre of rotation of the limb and increasing the effective length of the limb (ELL). More distally, ELL is further enhanced because the antebrachium and manus are long in comparison to the brachium (figure 1), with little variation in the relative lengths of the limb bones among the three equids. (One consistent trend is that the third digit increases in relative length from *Mesohippus* to *Equus* (Sondaar 1968).) Thus the velocity ratio of hoof movement to the rate of contraction of the triceps (and shoulder musculature) is high in all three animals (figure 1). The ulna is reduced in *Mesohippus* and fused to the radius in the two later animals, so the forelimb of each moves primarily in the sagittal plane (Matthew 1926; Sondaar 1968). In the carpus of all three animals, the anterior portion of each articulation is flattened to prevent extension beyond 180° (Sondaar 1968; Yalden 1971). Thus, when fully extended, the carpus locks in a position which maximizes the combined length of antebrachium and metacarpus; that is, the axes of the two limb segments are in line (Thomason 1985*b*). In the manus, the number of digits is reduced to three in each fossil equid; the third digit in *Mesohippus* is slightly longer and more robust than digits 2 and 4, while the difference between the digits is more pronounced in *Merychippus*. In adult *Equus*, digits 2 and 4 are normally represented only by vestigial metacarpals, and all the weight is born on the third.

The combined result of all these adaptations is that *Mesohippus*, *Merychippus* and *Equus* all have long limbs in which the centres of rotation and inertia are situated proximally and the muscles of retraction have high velocity ratios. This conformation allows for a long, rapid stride, and it is quite probable that as a runner *Mesohippus* was equally as fast as *Equus*, notwithstanding the difference in size (Matthew 1926; Simpson 1951). *Mesohippus* was in all likelihood considerably more agile, which correlates with its apparent habitat of low-lying marshy forest.

## Myology and kinematics

For *Equus*, data on the soft-tissue adaptations in the limbs are available in addition to the osteological information just described. Proximally, the specializations appear to be minor: increases in thickness and extent of fasciae and intramuscular tendons. These may be important functionally to the horse, but they have not been analysed as rigourously as more distal modifications, nor do they leave distinctive marks of attachment on the bones; this makes it difficult to infer their presence in the fossils. An exception is the long tendon of the biceps (lacertosus fibrosus), which runs over the anterior aspect of the radius and carpus to insert on a prominence on the anterior aspect of the third metacarpal (MC III). It appears to be important in locking the carpus in extension during weight-bearing in *Equus*, because the carpus may not fully extend if the tendon is cut (Ivanov & Borissov 1935). The two fossils have prominences on MC III comparable to that in *Equus*, indicating the presence of a lacertosus fibrosus and the possibility of a similar function.

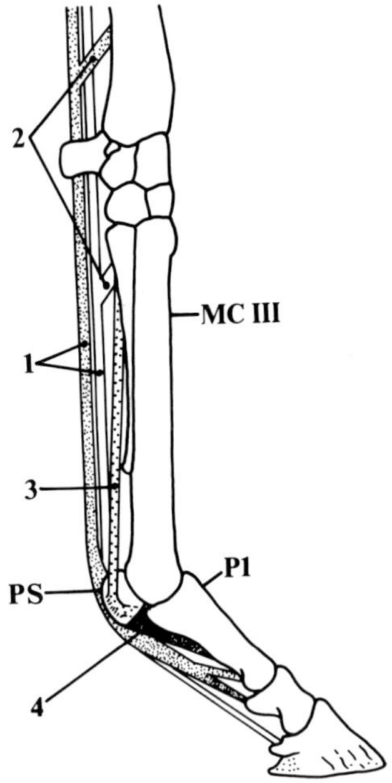


Figure 2. The main elements of the suspensory apparatus in the manus of *Equus*. Some smaller structures which also contribute to the apparatus have been omitted. The numbers correspond to descriptions in the text. MC III, third metacarpal; P1, proximal phalanx; PS, proximal sesamoid bones (one behind the other). (Adapted from Thomason, 1985*a*.)

Distally in the limb of *Equus*, the muscles and ligaments are modified to form a 'suspensory' apparatus, which has been extensively described and analysed (for example by Camp & Smith 1942; Alexander & Dimery 1985*b*; Lochner *et al.* 1980; Riemersma & De Bruyn 1986). We can now begin to add data from *in vivo* kinematic studies to anatomical information. The suspensory apparatus has four main components (figure 2): (1) the tendons of the superficial and deep digital flexor muscles; (2) the radial and carpal 'check' ligaments which allow strain in these tendons to be passively resisted in part; (3) the suspensory ligament, which is an involuted interosseous muscle of the third digit (Callegari 1968) running from MC III to a pair of proximal sesamoid bones at the metacarpo-phalangeal joint; and (4) a complex of sesamoidean ligaments connecting the sesamoids to the first phalanx. The suspensory apparatus supports the digit from the ground while standing (the unguligrade stance) and during locomotion, which adds to the effective limb length. It is easily observed, and confirmed by cinematography, that during locomotion the digit extends as weight is put on the leg and flexes towards the end of the step. As the digit extends, the suspensory apparatus

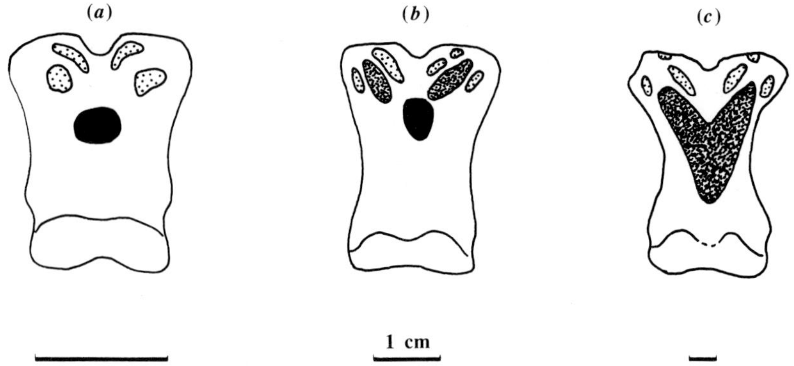


Figure 3. Volar (inferior) views of the first phalanges of (*a*) *Mesohippus*, (*b*) *Merychippus* and (*c*) *Equus* to show the attachment scars of the major sesamoidean ligaments (black, heavy and light stipple). Note the replacement of the central ligament (black) in *Mesohippus* by the increasingly large oblique ligament (heavy stipple), and the relatively large number of scars in *Merychippus*. (Adapted from Camp & Smith 1942.)

is stretched, and then elastically rebounds, reducing the requirement for metabolic energy to raise the centre of mass of the body and also maintaining forward momentum. The radius and metacarpus remain aligned during the step because of the carpal locking mechanism.

Two ways of assessing whether the fossil equids had similar soft tissue modifications and kinematic patterns are, first, to look for scars on bones comparable to those left by elements of the suspensory apparatus in *Equus* and, second, to compare the shapes of the digital joints as I have done above for the carpus. The sesamoidean ligaments have pronounced marks of attachment to P1 in all three equids (figure 3; Camp & Smith 1942). The differences in size and number of these scars led Camp & Smith to conclude that *Mesohippus* probably had a foot with a digital pad in addition to hooves (somewhat like that of the extant tapir), while *Merychippus* probably had a springing suspensory apparatus similar to that of *Equus*. Some third metacarpals of *Merychippus* have a prominence on the posterior aspect which may have been the attachment site of a check ligament (Thomason 1985*b*). This lends support to the suggestion that *Merychippus* was unguligrade.

Comparison of the joints of the digit also supports Camp & Smith (Thomason 1985*b*). The metacarpo-phalangeal (MC-P) joint in *Equus* is particularly convex to allow for a large range of digital rotation, but there is a noticeable difference in the radii of curvature of the anterior aspect, between MC III and P1, and the posterior aspect, between MC III and the proximal sesamoids (figure 4*a*). This difference in curvature may have to do with the way sesamoids affect the forces in the suspensory and sesamoidean ligaments which attach to them (Alexander & Dimery 1985*a*). What is important here is that the differential appears to correlate with an unguligrade stance; it is reversed in the tapir, which is subunguligrade and in the dog which is digitigrade (Thomason 1985*b*). In *Merychippus* (figure 4*b*) the curvatures are similar to those in *Equus*, while in *Mesohippus* (figure 4*c*) they are similar to the non-unguligrade mammals. Therefore, it may be inferred from both ligamentary attachments and the shapes of digital joints that the transition from a subunguligrade to an unguligrade limb preceded the transition from being tridactyl to being monodactyl.

## Mechanics

Recently there have been a number of studies of the forces acting on the limbs of *Equus* and of other mammals, and also of the strains and stresses experienced by the material of individual limb bones (see references in Leach & Dagg 1983*a*; Thomason 1985*a*). I have used the patterns evident from the results of such studies to estimate the locomotory forces and stresses acting on the MC III of *Mesohippus* and *Merychippus*, and also to evaluate the results (Thomason 1985*a*). That paper contains more detailed discussion of the points made below. The equid manus appears to be suitable for a free-body type of analysis, because the similar specializations among the three equids either simplify or allow easier rationalization of the assumptions necessary to estimate forces and stresses. For example, the vector of the ground reaction force (**G**) acting on the hoof of *Equus* during one step remains nearly parallel to the axis of the long segment of the limb formed by the radius and metacarpus. At the midstep, both the vector **G** and the limb segment are close to vertical. This appears to be simply a result of the geometry of the limb. If the carpus could extend further, as it does in non-ungulate mammals, then either or both of the radius or metacarpus would become inclined to vector **G**. For most extinct animals, the exact inclination of each bone to the force vector at the midstep would be impossible to determine. However, it can be assumed that vector **G**, the radius and the metacarpus were probably all close to the vertical at the midstep of *Mesohippus* and *Merychippus* because of the similarity in the geometry of their limbs to that in *Equus*.

The mechanical situation in the manus of *Equus* at the midstep (figure 5*a*) can be simplified thus: the ground force vector exerts a turning moment ($\mathbf{G} \times \mathbf{g}$) about the metacarpo-phalangeal joint which is resisted by an equal

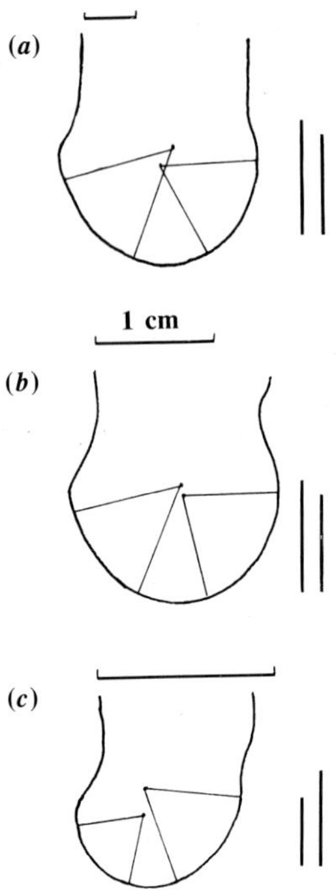


Figure 4. Diagrammatic sections through the distal ends of the third metacarpals of (*a*) *Equus*, (*b*) *Merychippus* and (*c*) *Mesohippus* showing the relative radii of curvature of the anterior aspect, which articulates with the proximal phalanx, and the posterior aspect, which articulates with the proximal sesamoid bones. The left vertical line in each pair illustrates the posterior radius; the right, the anterior radius.

and opposite moment of the tension in the suspensory apparatus ($\mathbf{T} \times \mathbf{t}$). Line **t** runs from the approximate centre of rotation of the MC-P joint to a point estimated to be on the line of action of the combined tensile forces **T** in the elements of the suspensory apparatus.

In fact, the digital flexor tendons and suspensory ligament probably experience quite different tensions (Forssell 1915) and should be treated separately because of the different ways in which they transmit forces about the joint (Alexander & Dimery 1985*a*). The tensile forces in the digital flexor tendons may be assumed to be essentially equal above and below the MC-P joint because they pass over the joint in a groove between the proximal sesamoid bones, a situation which approximates to a frictionless pulley. The sesamoidean ligament, however, attaches to the proximal sesamoid bones, and is continued distally by extensor branches of the ligament and by the proximal sesamoidean ligaments which were studied by Camp & Smith

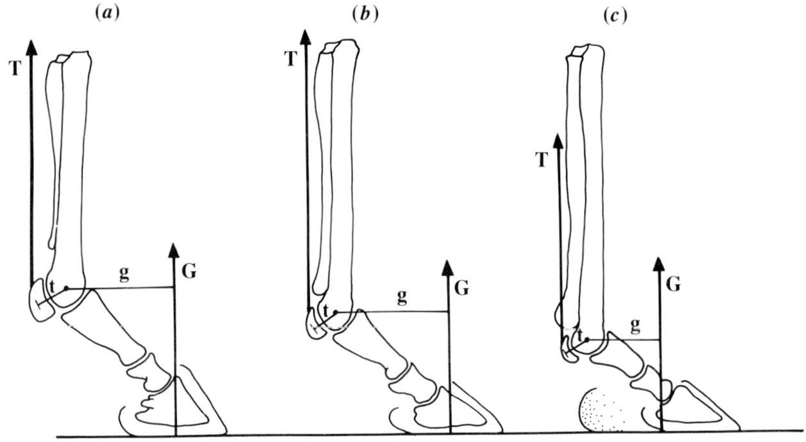


Figure 5. The bones of the manus in (*a*) *Equus*, (*b*) *Merychippus* and (*c*) *Mesohippus* to show the magnification of ground force by the leverage of the digit about the metacarpo-phalangeal (MC-P) joint. In (*b*) and ( *c*) the small side digits have been omitted for clarity. **G**, vector of the ground reaction force on the foot; **g**, moment arm of force **G** about the MC-P joint; **T**, tension in the flexor elements of the suspensory apparatus; **t**, effective moment arm of force **T** about the MC-P joint. (Adapted from Thomason 1985*a*.)

(1942). In this case the tension in the elements above the joint may differ from that below it.

A detailed analysis of the forces across the metacarpo-phalangeal joint has not yet been attempted for *Equus* – although strains in the elements of the suspensory apparatus have been recorded *in vivo* (Lochner *et al.* 1980) – and is probably not feasible for the fossils. For individuals of *Equus*, both strain measurements and data on the cross-sectional areas and material properties of the elements of the suspensory apparatus would be necessary to calculate accurately forces across the MC-P joint. None of this information is available for the fossil equids and it is not readily estimable, so although the model of the forces in the digit of each equid is an approximation, it is probably almost as precise as can be constructed.

Even allowing for simplifications in figure 5*a*, the ground force **G** has a considerable mechanical advantage over the suspensory apparatus, equal to the ratio $g/t$ (where $g=|\mathbf{g}|$ and $t=|\mathbf{t}|$). This ratio is approximately 3.5, depending on the length and inclination of the digit. If allowances are made for the extra digits in the two fossils and the probable presence of a digital pad in *Mesohippus*, then the mechanical situation in the forefoot of each can be similarly simplified (figure 5*b*, *c*). The ratio $g/t$ appears to have been highest in *Merychippus* at about 5.8, and lowest in *Mesohippus* at 3.1 because of the pad. If these data are combined with measurements of the dimensions and cross-sectional geometry of MC III for each animal, the stresses acting at the midshaft of this bone at the mid-step of a medium-paced trot may be estimated (Thomason 1985*a*). (At the specified gait, it can be assumed that the magnitude of **G** approximates the body weight.) The estimates of peak (compressive) stress vary more-or-less in accordance with the ratio $g/t$, being highest for *Merychippus* (−40 to −75 MN/m²), intermediate for *Equus* (−24 to −48 MN/m²) and lowest for *Mesohippus* (−10 to −33 MN/m²). In each case the estimates fall within the range of values (−5 to −100 MN/m²) derived from experiments on a variety of living mammals of different sizes and moving at different speeds.

The estimates of stress may be used as a basis for interpreting the differences in foot morphology. An area that is of particular interest to me is the internal architecture of the bones. Approximate values of pressure between MC III and the first phalanx and between MC III and the proximal sesamoid bones can be determined from the data that were used for the stresses. The density and orientation of trabeculae within the third metacarpals correlates well with these pressures (Thomason 1985*b*,*c*), which once more supports the ideas on the differences in foot morphology between the two fossils. The stress data are also useful in speculating on the possible reasons for the changes in limb morphology.

## Limb morphology and size

Two points arise from the preceding discussion. First, the major detectable differences in limb morphology are in the manus; more proximally the cursorial adaptations are shared. There appears to have been a transition from a padded, subunguligrade foot in *Mesohippus* to an unguligrade, but still tridactyl, foot in *Merychippus*, with a later transition to monodactyly in *Equus*. Second, there was a marked size increase between *Mesohippus* and *Equus*. Matthew (1926) and Simpson (1951) suggested that the changes in limb structure and function were necessary to enable *Equus* to preserve the cursorial abilities of *Mesohippus*. Certainly the 'design' of the manus in *Equus* appears to be something of a compromise. On the one hand, the unguligrade stance has several interrelated benefits. It adds to the effective length of the limb and hence to stride length. The downward deflection of the digit as weight is put on the limb spreads the forces of impact over a relatively long time period and reduces their magnitude. The energy stored in the suspensory apparatus as the digit deflects is recovered later in the step, thus increasing the energetic efficiency of locomotion (Alexander & Dimery 1985*b*). Finally, the tapering of the limb distally reduces the magnitudes of the fluctuations in its kinetic energy (Smith & Savage 1956; Hildebrand & Hurley 1985). On the other hand, the cost of these benefits is that the forces experienced by the musculo-skeletal elements of the manus are magnified with respect to **G** by the leverage of the digit: the ratio $g/t$. Presumably the advantages weigh more heavily, though the frequency of mechanical failure of limb elements is quite high in both domestic and wild populations of *Equus*.

At this point, extrapolation to the fossils becomes difficult because of the combined effects of changes in size with those of foot morphology. To quantify the advantages of a given foot type at a given size both variables have to be taken into account. We need to be able to compare, for example, the absorption of impact by padded and unguligrade feet, in tridactyl and monodactyl animals, at each of the three sizes under discussion. Or we need to compare the kinetic energy of oscillation for limbs ending in each type of foot at each size. One problem is that animals of the right sizes and foot types do not necessarily exist. Another is that the experimental procedures to determine accurately such quantities as kinetic energy of the limbs, energy storage and impact absorption are time-consuming and have been performed for few mammals (Fedak, Heglund & Taylor 1982) and incompletely for *Equus* (Sprigings & Leach 1986).

The data from the mechanical analyses in the previous section are of some use here, in conjunction with information on the allometric changes in MC III, because they indicate that the changes in foot morphology occurred despite potentially detrimental effects with regard to loading of the limbs. In principle, both the change to an unguligrade foot and the increase in size should increase the levels of stress in the bones of the manus. However, stress levels appear to have been essentially independent of size, as has been observed for mammals in general (Biewener 1982), although the change in morphology appears to have affected these stresses. Biewener (1983) observed that one of the mechanisms for compensating for the potential effects of size on bone stresses in large mammals is to align the longer limb bones with the vector **G**. This mechanism was unavailable to the equids because the long bones of their antebrachium and manus were already so aligned in *Mesohippus*. One important method of compensation for the increase in size in the equids appears to have been by increase in the cross-sectional area of MC III (Thomason 1985*a*).

Changes in the proportions of limb bones solely with size are unusual in mammals, which tend to scale approximately geometrically (Alexander *et al.* 1979). An exception is the ungulate family Bovidae, the limbs of which appear to scale elastically (McMahon 1975). Figure 6 shows log–log plots of

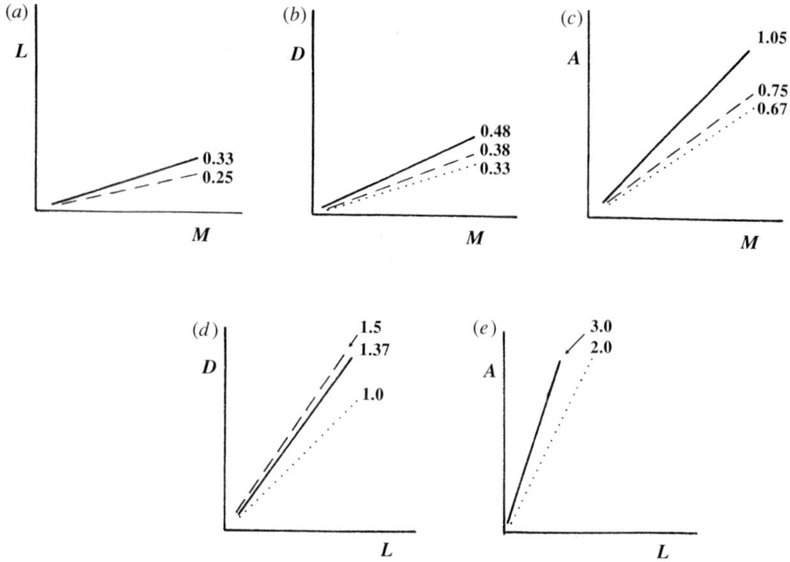


Figure 6. Log–log graphs to show the scaling pattern of MC III in the three equids (continuous line) compared to the predictions of geometric (dotted line) and elastic scaling (dashed line). In (*a*) continuous and dotted lines overlap, while in (*e*) continuous and dashed lines overlap. Only the slopes of each line are plotted, with no vertical displacement. *A*, log(cross-sectional area of bone); *D*, log(antero-posterior diameter of bone); *L*, log(length of bone); *M*, log(mass of animal). Data for equid metacarpals are from Thomason (1985a).

the scaling of various dimensions of the equid MC III compared with the predictions of geometric and elastic scaling. The length of MC III scales geometrically with body mass (*a*) while both (antero-posterior) diameter (*b*) and cross-sectional area (*c*) increase more rapidly with mass than would be predicted by either geometric or elastic scaling. The lower two graphs in figure 6 show how the measured dimensions of the metacarpals scale with respect to each other. The plot of diameter against length (*d*) gives an exponent slightly less than that predicted by elastic scaling, while for area against length (*e*) the exponent is the same as for elastic scaling. (The apparent discrepancy here arises from the change in latero-medial diameter – not shown – which has a higher exponent than for elastic scaling.) Given that the lower graphs show that the dimensions of the equid MC III scale very closely to elastically, is it possible that the upper graphs show errors in the estimates for the body masses of the two fossils, rather than a different relation to those predicted? I think that to be unlikely because, if lines of slope 0.67 and 0.75 are drawn through the point representing the known area and body mass for *Equus* and masses for the fossils are read off from known areas, the body masses are implausibly low (Thomason 1985*a*). Therefore the metacarpus of the equids does not seem to scale in accordance with the predictions of elastic and geometric scaling.

Let us return to the question: why change the morphology of the foot? Although no definitive answer may be given, the discussion above does permit some speculation. The acquisition of an unguligrade foot may have been related to the greater need to absorb the forces of impact of the foot with the ground on increase in size, coupled with the obvious advantages of conserving energy, and also with a change in habitat (Simpson 1951; Thomason 1986). *Merychippus* was a plains- rather than a forest-dweller. The later loss of the second and fourth digits may have been an attempt at reducing the increase in the moment of inertia of the limb during the size increase between *Merychippus* and *Equus*.

## Conclusions and suggestions for future study

The family Equidae is particularly suitable for studies of the functional morphology of its members. There is a living representative, the locomotory adaptations of which have been extensively analysed, and there is a well preserved phylogeny for which the functional significance of osteological differences in limb structure may be inferred by comparison with *Equus*. I have concentrated here on two extinct equids to show some of the methods by which their locomotory adaptations may be analysed, and also to identify areas which remain impracticable to pursue. There is considerable scope for further work solely within the Equidae, not to mention the possibilities for applying these methods to artiodactyls. There are more than twenty genera of extinct equids, and several lineages increase in size with time. Camp & Smith (1942) suggested that some large forms (e.g. *Anchitherium*) retained a padded foot, on the basis of the pattern of their sesamoidean ligaments. However, the large size of the proximal sesamoid bones and the curvature of the metacarpo-phalangeal joint suggests to me that *Anchitherium* may have been unguligrade (Thomason, personal observations). This is just one example of a question to which the methods outlined above may be applicable.

## References

Alexander, R.McN. & Dimery, N.J. 1985*a* The significance of sesamoids and retro-articular processes for the mechanics of joints. *Journal of Zoology, London* A **205**, 357–371.

Alexander, R.McN. & Dimery, N.J. 1985*b* Elastic properties of the forefoot of the donkey, *Equus asinus*. *Journal of Zoology, London* A **205**, 511–524.

Alexander, R.McN., Hayes, A.S, Maloiy, G.M.O. & Wathuta, E.M. 1979 Allometry of the limb bones of mammals from shrews (*Sorex*) to elephants (*Loxodonta*). *Journal of Zoology, London* **189**, 305–314.

Biewener, A.A. 1982 Bone strength in small mammals and bipedal birds: do safety factors change with body size? *Journal of Experimental Biology* **98**, 289–301.

Biewener, A.A. 1983 Allometry of quadrupedal locomotion: the scaling of duty factor, bone curvature and limb orientation to body size. *Journal of Experimental Biology* **105**, 147–171.

Callegari, E. 1968 The involution of the muscle tissue of the proximal sesamoidean ligament of the horse. *Lo Sperimentale* **118**, 57–69.

Camp, C.L. & Smith, N. 1942 Phylogeny and functions of the digital ligaments of the horse. *Memoirs of the University of California, Berkeley* **13**, 69–124.

Fedak, M.A., Heglund, N.C. & Taylor, C.R. 1982 Energetics and mechanics of terrestrial locomotion. II. Kinetic energy changes of the limbs and body as a function of speed and body size in birds and mammals. *Journal of Experimental Biology* **79**, 23–40.

Forssell, C.E.G. 1915 Untersuchungen über die Wirkungsweise der Beugesehnen am Vorderfuss des Pferdes. *Zeitschrift für Tiermedizin* **18**, 184–188.

Gregory, W.K. 1912 Notes on the principles of quadrupedal locomotion and on the mechanics of the limb in hooved mammals. *Annals of the New York Academy of Sciences* **22**, 267–294.

Hildebrand, M. 1984 Walking and running. In *Functional Vertebrate Morphology*, ed. M. Hildebrand, D.M. Bramble, K.F. Liem & D.B. Wake, pp. 38–57. Cambridge, Massachusetts: The Belknap Press at Harvard University Press.

Hildebrand, M. & Hurley, J.P. 1985 Energy of the oscillating legs of a fast-moving cheetah, pronghorn, jackrabbit and elephant. *Journal of Morphology* **184**, 23–31.

Hussain, S.T. 1975 Evolutionary and functional anatomy of the pelvic limb in fossil and Recent Equidae. *Anatomia Histologia Embryologia* **4**, 179–222.

Ivanov, S.V & Borissov, V.M. 1935 [Importance of the lacertosus fibrosus in the standing and motion of the horse.] *Arkhiv Anatomii, Gistologii i Embryologii* 14, 51–55. [In Russian.]

Leach, D.H. & Dagg, A.I. 1983*a* Evolution of equine locomotion research. *Equine Veterinary Journal* **15**, 87–92.

Leach, D.H. & Dagg, A.I. 1983*b* A review of research on equine locomotion and biomechanics. *Equine Veterinary Journal* **15**, 93–102.

Lochner, F.K., Milne, D.W., Mills, E.J. & Groom, J.J. 1980 In vivo and in vitro measurement of tendon strain in the horse. *American Journal of Veterinary Research* **41**, 1929–1937.

MacFadden, B.J. 1986 Fossil horses from 'Eohippus' (*Hyracotherium*) to

*Equus*: scaling, Cope's Law and the evolution of body size. *Paleobiology* **12**, 355–369.

Matthew, W.D. 1926 The evolution of the horse: a record and its interpretation. *Quarterly Review of Biology* **1**, 139–185.

McMahon, T.A. 1975 Allometry and biomechanics: limb bones of adult ungulates. *American Naturalist* **109**, 547–563.

Riemersma, D.J. & De Bruyn, P. 1986 Variations in cross-sectional area and composition of equine tendons with regard to their mechanical function. *Research in Veterinary Science* **41**, 7–13.

Simpson, G.G. 1951 *Horses*. Oxford University Press.

Smith, J.M. & Savage, R.J.G. 1956 Some locomotory adaptations in mammals. *Zoological Journal of the Linnean Society* **42**, 603–622.

Sondaar, P.Y. 1968 The osteology of the manus of fossil and Recent Equidae. *Verhandelingen der Koninklijke Nederlandse Akademie van Wetenschappen Afdeling Natuurkunde Tweede Reeks* **25**, 1–76.

Sprigings, E. & Leach, D. 1986 Standardised technique for determining the centre of gravity of body and limb segments of horses. *Equine Veterinary Journal* **18**, 43–49.

Thomason, J.J. 1985*a* Estimation of locomotory forces and stresses in the limb bones of Recent and extinct equids. *Paleobiology* **11**, 209–220.

Thomason, J.J. 1985*b* The relationship of trabecular architecture to inferred loading patterns in the third metacarpals of the extinct equids *Merychippus* and *Mesohippus*. *Paleobiology* **11**, 323–335.

Thomason, J.J. 1985*c* The relationship of structure to mechanical function in the third metacarpal of the horse. *Canadian Journal of Zoology* **63**, 1420–1428.

Thomason, J.J. 1986 The functional morphology of the manus in the tridactyl equids *Merychippus* and *Mesohippus*: paleontological inferences from neontological models. *Journal of Vertebrate Paleontology* **6**, 143–161.

Yalden, D.W. 1971 The functional morphology of the carpus in ungulate mammals. *Acta Anatomica* **78**, 461–487.

ANDREW C. KITCHENER

# The evolution and mechanical design of horns and antlers

## Introduction

Living horned ruminants (Bovidae and Cervidae) offer a unique opportunity to investigate the co-evolution of morphology, behaviour and ecology within a diverse group of organisms. Differences in horn and antler morphology, locomotor adaptations and dentition may readily be correlated with differences in social behaviour, reproductive strategies, diet and other ecological requirements (Geist 1966; Schaffer & Reed 1972; Jarman 1974; Leuthold 1977; Janis 1982; Kingdon 1982*a*,*b*; Walther, Mungall & Grau 1983; Spinage 1986; Gordon & Ilius 1988). The comparatively recent evolution and continuing radiation of bovids and cervids means that the different morphological grades apparent in their evolutionary histories have living equivalents which provide useful models to help to analyse the major trends and selection pressures which have produced today's diverse assemblage of species (Gentry 1978; Janis 1982; Kingdon 1982*a*; Savage & Long 1986; Spinage 1986).

Horns and antlers are used as weapons in intraspecific combat between males to compete for access to females in oestrus (Geist 1966; Walther 1966; Leuthold 1977; Clutton-Brock 1982; Goss 1983). This chapter looks at the factors which affect the mechanical design of horns and antlers of living animals and their interaction with developmental, behavioural and ecological factors. This approach should provide a better understanding of the importance of mechanical constraints in the evolution of horns and antlers.

However, in order to investigate the mechanical design of any structure, it is essential to know how the structure is used, the forces acting on it and the mechanical properties of the materials from which it is made. Therefore, the aims and 'rules' of the different types of fighting are reviewed to ascertain the most important design criteria and to model how horns and antlers function in fighting.

The adaptation of horn and antler structure to the predictable forces of apparently ritualized fighting (Geist 1978) are examined under the assumption that these structures are optimized for fighting; that is, that they are minimum-weight structures.

Finally, the relative merits of horns and antlers as structures for fighting are discussed in relation to behaviour and ecology.

## Horns and antlers: structure, growth and morphology

Horn-like organs have evolved independently many times within the mammals, and show a great diversity of structure, size and shape (Geist 1966; Halstead 1978; Janis 1982; Savage & Long 1986). Today horns are confined to five families of mammals, but only the keratinous horns of the Bovidae and the bony antlers of the Cervidae are considered here.

The horns of bovids are made up of a sheath of tightly packed, flattened cells containing α-keratin which overlies a core of spongy bone. There is a germinative epithelium continuous with the skin and covering the bony core, the cells of which divide periodically to produce more horn sheath. Horns grow from their bases. Each year a new layer of horn is produced which pushes the old growth away from the skull (figure 1*a*; Goss 1983). Bovid horns are never shed, although the horn sheath may be worn away in fighting and the old layers may exfoliate (Geist 1971; O'Gara & Matson 1975; Spinage 1982; Petocz & Shank 1983).

The antlers of deer are analogous to the horns of bovids. The outer layer of compact bone in antlers bears most of the forces produced in fighting so that it has a function comparable to that of the keratinous sheath of the horns of bovids. Both antlers and horns have a core of spongy bone. However, in many ways antlers differ from horns. Antlers grow as branched structures covered in skin (velvet) such that the growing points always move away from the skull (figure 1*b*; Goss 1983). When they are fully grown, the blood supply is cut off, the velvet is shed and the antlers die before they are used in fighting. Antlers are cast and regrown annually. The growth of both horn sheath and antlers is mainly under the control of testosterone (Chapman 1975; Chaplin 1977; Goss 1983).

Horns and antlers show a wide variety of sizes and shapes. The smallest cervid and bovid species have simple, spiky structures which have increased in size and complexity with the increasing body size of the different species (Whitehead 1972; Goss 1983; Spinage 1986). Some of the diversity in horn and antler shape is shown in figure 2.

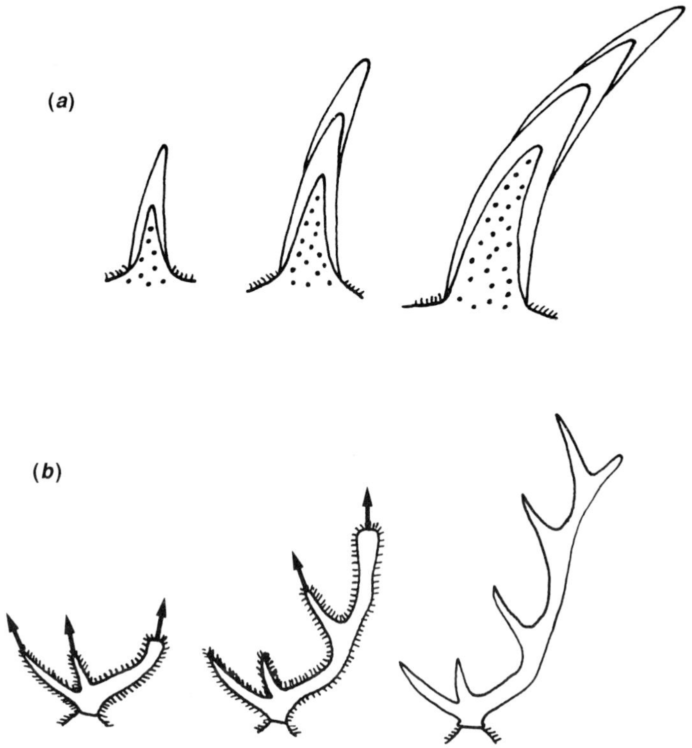


Figure 1. (*a*) The annual incremental growth of the keratin sheath of bovid horns. A new layer of keratin grows from the germinative epithelium covering the bony core each year so that the old growth is pushed away from the skull. (*b*) The seasonal incremental growth of antlers. The growing points are at the tips and move continually away from the skull. (After Goss 1983.)

## The evolution of horns and antlers

The first bovids and cervids appeared during the early Miocene about 20 million years ago (Halstead 1978; Janis 1982; Savage & Long 1986; R. J. G. Savage, personal communication). The first bovids were represented by *Walangania* sp., *Protragocerus* sp., *Gazella* sp. and *Eotragus* sp. in Africa and *Eotragus artenensis* in Europe (Gentry 1978). The most primitive of these is *Eotragus*, but it is too large to be ancestral to today's bovids, and indicates a much earlier origin for the Bovidae, perhaps in Central Asia during the Oligocene (Gentry 1978; Kingdon 1982*a*). *Eotragus* had simple, spiky, upright horns similar to the horns of the dwarf antelopes (Neotraginae; figure 3; Kingdon 1982*a*). The first cervid, *Euprox* from the early Miocene of Europe, is thought to be directly ancestral to living muntjacs (*Muntiacus*

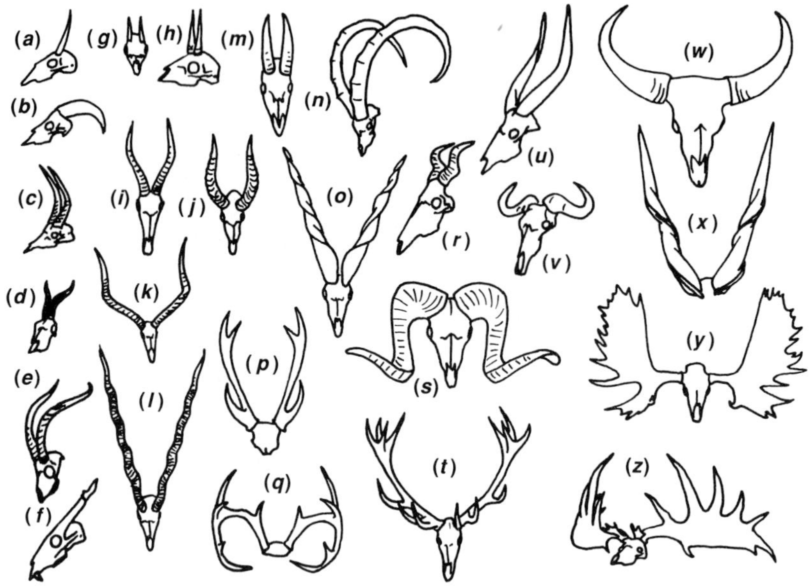


Figure 2. The diversity of horns and antlers: (*a*) beira (*Dorcatragus*), (*b*) tahr (*Hemitragus*), (*c*) dibatag (*Ammodorcas*), (*d*) gazelle (*Procapra*), (*e*) gazelle (*Gazella*), (*f*) muntjac (*Muntiacus*), (*g*) dik-dik (*Madoqua*), (*h*) klipspringer (*Oreotragus*), (*i*) blesbok (*Damaliscus*), (*j*) puku (*Kobus*), (*k*) impala (*Aepyceros*), (*l*) blackbuck (*Antilope*), (*m*) serow (*Capricornis*), (*n*) wild goat (*Capra*), (*o*) markhor (*Capra*), (*p*) sambar (*Cervus*), (*q*) white-tail deer (*Odocoileus*), (*r*) hartebeest (*Alcelaphus*), (*s*) argali (*Ovis*), (*t*) red deer (*Cervus*), (*u*) bongo (*Boöcercus*), (*v*) wildebeest (*Connochaetes*), (*w*) gaur (*Bos*), (*x*) eland (*Taurotragus*), (*y*) moose (*Alces*), (*z*) Irish elk (*Megaloceros*). Not to scale, but shown roughly in increasing body size from left to right. (After Ward 1903; Groves & Grubb 1982.)

spp.; Savage & Long 1986; R. J. G. Savage, personal communication). *Euprox* had simple spiky antlers which, like those of the muntjac, were mounted on long bony pedicles projecting from the frontal bone of the skull (figure 3).

The early cervids and bovids probably inhabited forests or woodland where they fed on scattered fruits and shoots which are high in energy and digestibility (Janis 1982). This diet would have been necessary for small animals with a high metabolic rate. They would have been monogamous and have defended their territory against intruders. There would have been little sexual dimorphism and no ritualized fighting. It is quite likely that both *Eotragus* and *Euprox* would have used their horns and antlers in direct attacks against the body of an opponent rather than interlocking them.

The major evolutionary trends within the Bovidae and Cervidae include an

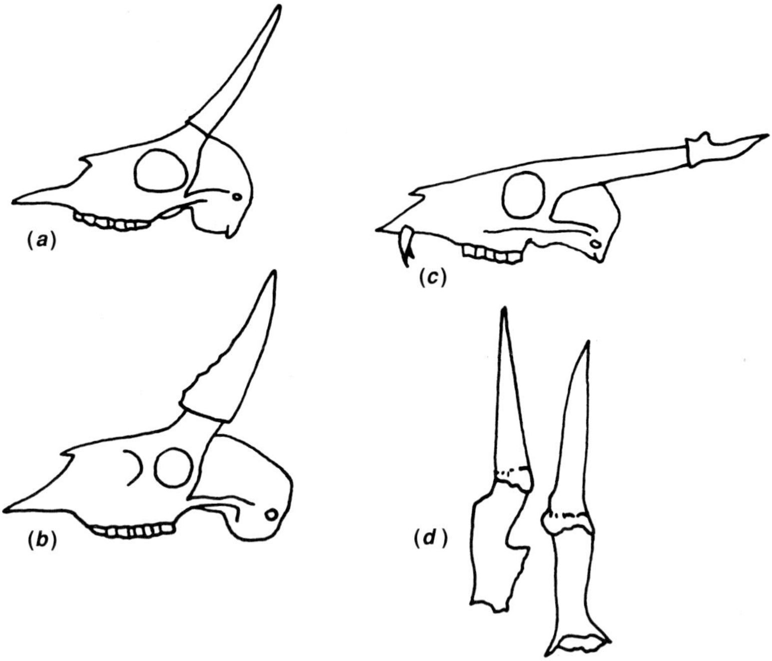


Figure 3. A comparison between the earliest bovids and cervids and similar living forms. (*a*) Beira (*Dorcatragus*), (*b*) *Eotragus*, (*c*) muntjac (*Muntiacus*), (*d*) *Euprox*. (After Groves & Grubb 1982; Kingdon 1982*a*; Savage & Long 1986.)

increase in body size and an increase in size and complexity of horns and antlers (Savage & Long 1986). The horns and antlers of large species of bovids and cervids are relatively bigger than those of small species (figure 4; Huxley 1932; Gould 1974; Clutton-Brock, Albon & Harvey 1980). Large species live in open habitats and are usually polygynous, so that they fight more frequently than do small species which live in closed habitats and are monogamous. Consequently, large species risk more injury and invest more material in horns and antlers which act as defensive guards as well as offensive weapons (Clutton-Brock *et al.* 1980). Polygyny resulted from the need for animals to herd together for protection against predators in the new, open grasslands of the late Miocene and Pliocene (Clutton-Brock, Guinness & Albon 1982; Kingdon 1982*a*). This gave males the opportunity to increase their own reproductive success by monopolizing breeding, which would have led to direct competition between males for access to females in oestrus. Selection would favour traits such as increased body size and larger, more complex horns and antlers which improved fighting success (Clutton-Brock *et al.* 1982).

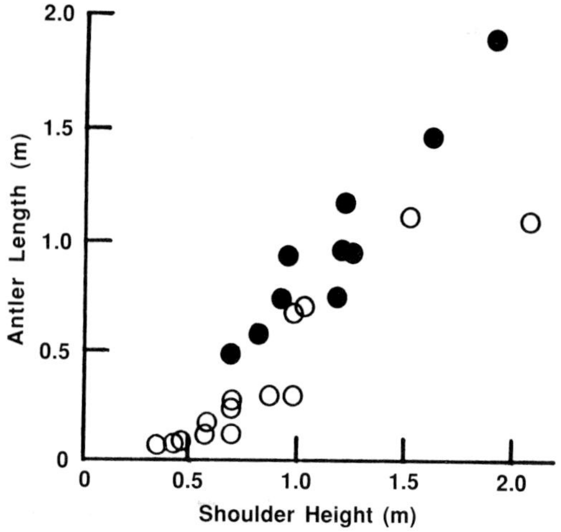


Figure 4. The antlers of deer with large breeding groups (> 6; filled circles) have relatively larger antlers than do deer with small breeding groups (⩽ 6; open circles). (After Clutton-Brock *et al.* 1980.)

However, in the early Miocene forests the density of highly digestible and energy rich foods would be too low to support herding and the total energy requirements of large animals, even though they would have lower metabolic rates. Large, complex horns and antlers would also be a hindrance to movement through a closed habitat. The major radiation of both cervids and bovids occurred later in the Miocene when grasslands began to replace forests. Grass represents an abundant but low energy food source which can be utilized only by an efficient food processing system necessitating improved dentition, jaw mechanics and rumen function (Kingdon 1982*a*; Norman & Weishampel, this volume). However, a change from high to low energy diet would require an increase in body size so that metabolic rates would be reduced. Therefore, the evolution of a larger body size was accompanied by a change in diet which promoted success in fighting.

There was also a fundamental change in the function and method of fighting (Geist 1966). Formerly fighting would be used to defend the resources of the territory of a monogamous pair, but in open grassland food was so abundant that fighting was directed towards increasing reproductive success alone. In forests fighting was unritualized and attacks were directed against the body, but in the grasslands attacks were focused on the horns and antlers, which were used to catch the blows of the greater forces being produced (Geist 1966). Polygyny increased the frequency of fighting so that a complex series of displays have developed in order to assess the fighting ability of an opponent so that unnecessary fights are avoided (see below).

Therefore, there was a fundamental change in the mechanical function of horns from stabbing organs in the ancestral forest forms to structures which resist and transmit the forces of fighting in their descendants of open habitats.

## The aim of fighting

Bovids and cervids interlock the bases of their horns and antlers in the most forceful encounters. They try to lever their opponents off balance so that they can use the tips of their horns or the terminal branches (tines) of their antlers to stab the neck, flank or belly of the opponent. The twists and ridges on horns and the lower tines of antlers ensure that these structures interlock securely and so avoid accidental injury caused by slippage (Walther 1966; Leuthold 1977).

Fighting may deter an opponent by maximizing superficial damage to the skin (Geist 1978). However, waterbuck (*Kobus ellipsiprymnus*) do not hesitate in inflicting deep and fatal wounds if the opportunity arises (Spinage 1982, 1986). Injuries and deaths from fighting are by no means rare (table 1). A recent study of the tanned hides of deer suggests that the incidence of injury in fighting may be underestimated by as much as an order of magnitude (Geist 1986).

Therefore, the aim of fighting appears to be to deter rivals from gaining access to females in oestrus by injuring them while avoiding being injured oneself.

## The rules of fighting

All larger bovids and cervids enter into a sometimes very complex assessment procedure prior to fighting; this ensures that only animals of approximately equal body size fight (Schaller 1967, 1977; Geist 1971; Clutton-Brock *et al.* 1979; Spinage 1982; Walther *et al.* 1983). Small animals avoid fighting larger animals because they will inevitably lose and risk serious injury. Large animals avoid fighting small animals because they will inevitably win, and would waste considerable energy doing so. During the rut or breeding season male cervids and bovids feed very little, and this puts a great demand on their fat stores. For example, red deer (*Cervus elaphus*) on the Island of Rhum can lose up to 20% of their body mass during the six week rut (Clutton-Brock *et al.* 1982). The high energy demands of the rut mean that stags enter winter in poor condition and are subject to a higher winter mortality than are females, so that conservation of energy is essential (Clutton-Brock & Albon 1985).

Table 1. *Annual rates of injury and mortality associated with intraspecific fighting (as percentage of the adult male population per annum) in various cervids and bovids*

+ indicates that mortality or injury has been recorded, but not quantified. Sources: (*a*) Barette (1977), (*b*) Clark (1981), (*c*) Clutton-Brock *et al.* (1979, 1982), (*d*) Leslie & Jenkins (1985), (*e*) Geist (1971), (*f*) Mloszewski (1983), (*g*) R. Cindery, personal communication, (*h*) Spinage (1986), (*i*) Gosling (1974), (*j*) Nowak & Paradiso (1983), (*k*) BBC TV film *The passing of the buck*.

| Species | Mortality | Injury | Source |
|---|---|---|---|
| **Cervidae** | | | |
| *Muntiacus muntjak* | | + | *a* |
| *M. reevesi* | | + | *b* |
| *Cervus elaphus* | 5–29 | 6–23 | *c* |
| *C. canadensis* | + | + | *d* |
| *Alces alces* | 4 | + | *e* |
| *Odocoileus hemionus* | | 19 | *c* |
| *Rangifer tarandus* | + | | *c* |
| **Bovidae** | | | |
| *Synceros caffer* | + | + | *f* |
| *Boselaphus tragocamelus* | + | | *g* |
| *Kobus* spp. | + | + | *h* |
| *Oryx gazella* | + | | *g* |
| *Alcelaphus buselaphus cokii* | + | | *i* |
| *Pelea capreolus* | + | | *j* |
| *Antilope cervicapra* | | + | *k* |
| *Saiga tatarica* | + | | *j* |
| *Oreamnos americanus* | | + | *e* |
| *Ovibos moschatus* | 5–10 | | *c* |
| *Ovis canadensis* | | + | *e* |

The assessment procedure involves visual, vocal and olfactory displays designed to indicate the body condition and body size of a potential combatant. Red deer begin with a roaring contest which indicates their readiness to fight (Clutton-Brock *et al.* 1982). Next the two animals enter into a parallel walk which emphasizes body size in a lateral display. Fighting usually occurs only after this procedure. The assessment procedures of most cervids and bovids involve some form of lateral display to emphasize body size and, hence, potential force of fighting (Schaller 1967, 1977; Geist 1971; Leuthold 1977; Spinage 1982; Walther *et al.* 1983).

It is commonly believed that cervids and bovids use the size of antlers and horns as an indirect measure of body size and potential fighting ability.

However, in red deer fighting success was not correlated with antler size (Clutton-Brock *et al.* 1979). Instead, in the only detailed study of its kind (Clutton-Brock *et al.* 1979), body size was the best indicator of fighting success. Spinage (1982) found that the horns of waterbuck were largest in males that were too young to fight or who fought infrequently. Mature males who fought frequently had relatively smaller horns because of the wear caused by fighting. It is unlikely, therefore, that waterbuck use horn size as an indicator of body size and potential fighting success. The complex series of displays used in thc assessment procedure provides a considerably more sensitive measure of the body condition and body size of a potential opponent than would display of a fixed length of antler or horn (Clutton-Brock *et al.* 1982).

Therefore, assessment ensures that the force of fighting is predictable in a way related to body mass and to the type of fighting behaviour (see below).

The fighting of cervids and bovids often appears to be ritualized. This may be a consequence of the fact that particular horn shapes can be moved defensively and offensively only in a limited number of ways (Geist 1978). In their most forceful encounters horns and antlers interlock in a predictable way dictated by their overall morphology and the orientation of their twists, ridges and tines. In virtually all cases horns and antlers interlock most securely at their bases. Therefore, the direction and position of the maximum force are predictable.

To summarize, the rules of fighting ensure that the force of fighting is predictable in magnitude, direction and its position on antlers and horns.

## The fighting behaviour of bovids and cervids

Bovids exhibit different fighting behaviours which are classified here in order of increasing force.

(1) The dwarf antelopes or neotragines, thin-horned duikers (*Cephalophus* spp. and *Sylvicapra grimmia*), the rhebok (*Pelea capreolus*) and the goat-antelopes (Rupicaprini) have simple spiky horns. They have little or no ritualized assessment procedure or fighting. There is no horn-to-horn contact in fighting because the horns are directed towards the body in a whirl-around lateral fight (Geist 1966; Leuthold 1977; Dittrich & Böer 1980; Kingdon 1982*a*; Spinage 1986).

(2) The fighting behaviour of thick-horned duikers (*Cephalophus* spp.) is unknown, but it probably involves some sort of combined butting

involving the frontal bone and the horns. The frontal bone is massively thickened and separated from the cranium by a large frontal sinus (Kingdon 1982*a*). Similar sinuses are found in sheep and goats which use their horns in forceful clashes (Schaffer 1968; Geist 1971; Schaffer & Reed 1972). Maxwell's duiker (*Cephalophus maxwellii*) engages its horns frontally after a ritualized mutual rubbing of the preorbital glands (Ralls 1975).

(3) The majority of antelopes clash their horns together and interlock them or cross them over at their bases. In the ensuing wrestling match they try to knock each other off balance to stab the opponent with their horns (Leuthold 1977; Kingdon 1982*a*; Spinage 1982, 1986; Walther *et al.* 1983).

(4) Wildebeest (*Connochaetes* spp.), hartebeest (*Alcelaphus* spp. and *Damaliscus* spp.), oryx (*Oryx* spp.) and the sable antelope (*Hippotragus niger*) drop down on their metacarpal joints to fight. Fighting usually involves clashing and wrestling (Estes 1969; Gosling 1974; Leuthold 1977; Walther 1978; Kingdon 1982*a*).

(5) Cattle (*Bos* spp., *Bubalus* spp. and *Bison* spp.) clash and wrestle more forcefully than antelopes (Sinclair 1977).

(6) Tahrs (*Hemitragus* spp.), goats, ibex and turs (*Capra* spp.) rear up on their hind legs and crash down on an opponent (Schaller 1977).

(7) Cape buffalo (*Synceros caffer*) charge at each other like sheep, but wrestle also (Sinclair 1977; Mloszewski 1983).

(8) Sheep (*Ovis* spp.) charge at each other before clashing (Geist 1971; Schaller 1977).

Cervids fight like the majority of antelopes (group 3), although there are some variations. For example, muntjac (*Muntiacus* spp.) interlock their antlers, but use their large canine teeth for inflicting wounds (Barette 1977).

**The mechanical design of horns and antlers**

From the review of the aims, rules and types of fighting it is clear that a horn or an antler can be modelled as a short, straight cylindrical beam which is fixed at one end (namely to the skull) and which is free at the other end, where the force of fighting is applied (figure 5).

This model horn could fail in one of five different ways: (1) the bending stress could exceed the bending strength of the structure; (2) the shear stress could exceed the shear strength of the structure; (3) the horn could fail by deflecting too much; (4) the energy of fighting may exceed the maximum

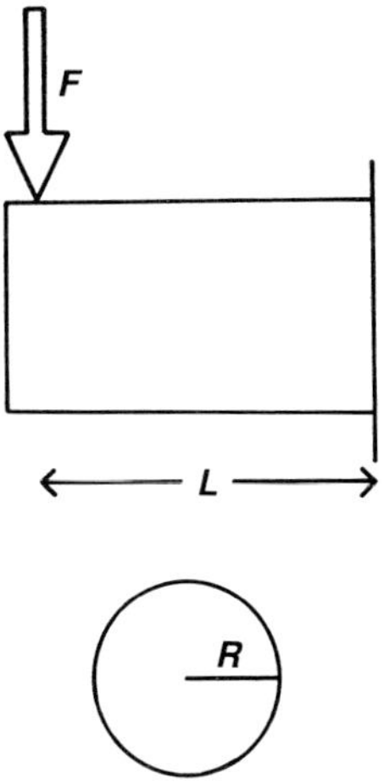


Figure 5. The model horn. A force $F$ acts on a cylindrical beam with a moment arm $L$. The beam is fixed at one end (to the skull) and is free at the other where the force acts. The beam is modelled as a tube of radius $R$. Stress in, and deflexion of, the beam are calculated by using equations (1)–(3).

strain energy that the horn can store in bending, and the horn will fail if this energy is not dissipated in other ways; (5) scratches and cracks may act as stress concentrations, which could propagate through a structure at normal stresses resulting from fighting. The sensitivity of structures to cracks is indicated by the critical crack length, which should be as large as possible if catastrophic failure is not to occur (Gordon 1976).

The forces and energies involved have been measured from an analysis of films of fighting in blackbuck (*Antilope cervicapra*) and bighorn sheep (*Ovis canadensis*; Kitchener 1988). The latter is considered to be the most forceful of all fighting bovids (Schaffer 1968; Geist 1971). When these forces and energies were applied to average-sized horns for both species in the typical fighting positions, it became apparent that horns were likely to fail only in a limited number of ways (Kitchener 1988).

The results of this analysis can be summarized as follows (Kitchener 1988). Shear stresses were negligible compared to bending stresses. Less than 1% (0.2–0.7 J) of the energy of fighting could be stored as strain energy in the horns as they bent, so that it was assumed that the body musculature was responsible for absorbing all the energy of fighting. The critical crack lengths of both blackbuck and bighorn sheep horns were more than 60% of their basal transverse dimensions. It is unlikely that animals would have fought with such extensive damage to their horns so that catastrophic failure is unlikely. Therefore, horns and antlers must be designed to resist failure in bending.

The bending stress $\sigma$ (Pa) in the model horn is given by

$$\sigma = FLR/I, \tag{1}$$

where $F$ is the force of fighting (N), $L$ is the moment arm (m), that is the distance between the base of the horn and the point where the force is applied, and $R$ is the radius of the horn (m) (Gordon 1976). $I$ is the second moment of area of the horn base ($m^4$), and is given by

$$I = \pi R^4 \tag{2}$$

for a horn with circular cross-section. The second moment of area is a shape factor which describes the distribution of material in the cross-section of a beam with respect to its neutral axis. The neutral axis is the point in the beam where there is no net tensile or compressive stress in bending, and in this case it is located at the centre of the circular cross-section of the beam (Gordon 1976). The deflexion $\delta$ (m) of this model horn is

$$\delta = FL^3/(3EI) \tag{3}$$

(Gordon 1976), where $E$ is the stiffness of horn.

From the rules of fighting it is possible to make a number of assumptions about the model horn. The force $F$ is predictable and is a function of body mass. The stiffness $E$ of horn and antler is a material property. At present it will be assumed that the stiffnesses of all horn keratins and antler bones are the same. Horns and antlers are always interlocked at their bases so that the moment arm $L$ is minimized and will be assumed to be constant even for large changes in body size. Therefore, the bending stress $\sigma$ and the deflexion $\delta$ of loaded horns and antlers can be limited by the radius, and more importantly by the second moment of area, of the base of the horn. During growth or during the evolution of larger body size, small changes in radius will produce large changes in second moment of area which balance any increases in force of fighting.

The relation between the mean second moment of area of the base of the horn and mean body mass of antelopes is linear (figure 6; Kitchener 1985). This suggests that horns are designed to deflect a constant and limited amount (equation (3)) because increases in body mass increase bending stresses in proportion to the radius of the horn (equation (1)). This means that larger animals are more likely to break their horns than are smaller animals because the safety factor in bending decreases with increasing body size.

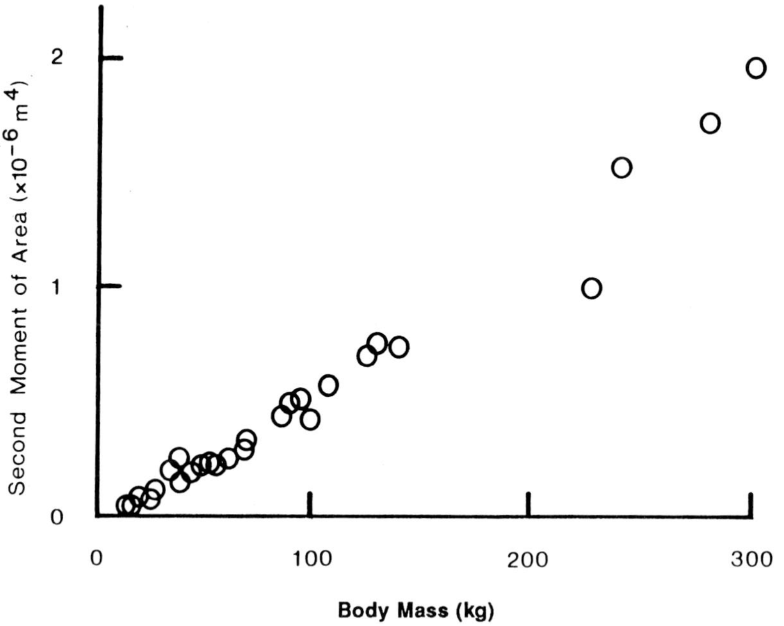


Figure 6. The linear relation between the mean second moment of area of the base of the horns and mean body mass of antelopes. Horn data are from Ward (1903, 1910), Dollman & Burlace (1935) and Best & Raw (1971). For other data see Kitchener (1985).

The safety factor is the ratio of the intended strength of a structure to the maximum load it is expected to bear (Alexander 1982). Engineering structures are typically designed to have safety factors of 10 or 20 or more depending on the material used (Roark & Young 1975). However, millions of years of evolution and the facility of *in situ* remodelling have honed down safety factors in biological structures to 10 or less (Alexander 1982).

Consider two structures which experience the same frequencies and variations in the maximum forces acting on them, including abnormal accidental loads. One structure has a low safety factor and is more likely to break than the structure with a high safety factor because there is a greater cumulative frequency or probability that a force acting on that structure will exceed its strength (figure 7). Consequently, the low safety factors expected in horns and antlers of large bovids and cervids should be associated with high incidences of breakage, assuming that the frequency and type of fighting are similar. Unfortunately, there are few data available to test this. Packer (1983) gave incidences of breakage in horns of African bovids which seem to support the predictions. The eland (*Taurotragus oryx*; mean body mass 500

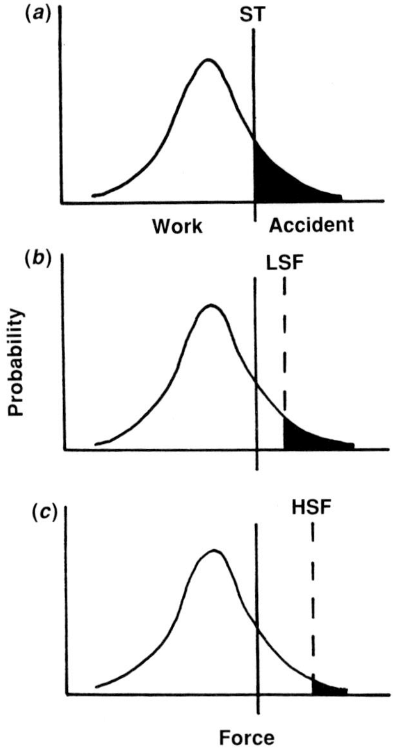


Figure 7. (*a*) A theoretical probability distribution of the maximum forces acting on a structure. The structure is intended to work up to a maximum force (ST). Forces above ST are rare and are due to accidents. (*b*) The theoretical force probability distribution for a structure designed with a low safety factor (LSF) to show that the cumulative probability of breakage (i.e. the area to the right of LSF) is greater than for a structure designed with a high safety factor (HSF) (*c*). Therefore, for given load, the horn in (*b*) is more likely to fail than that in (*c*).

kg) has a higher incidence of horn breakage (6%) than the smaller waterbuck (225 kg; 3.5%). Both these species have a higher incidence of horn breakage than the Thomson's gazelle (*Gazella thomsonii*; 20 kg; 0.8%). However, the sample sizes for Packer's (1983) data are very variable, and some are too small to be reliable: many more field data need to be collected.

Comparison of second moment of area and body mass reveals nine separate linear relations relating them, which differ in their gradients and seem to correspond broadly with the categories of fighting described above (figure 8; Kitchener 1985). In general terms, the more forceful the fighting the greater the structural stiffness (the gradient of the line) of the horn or antler.

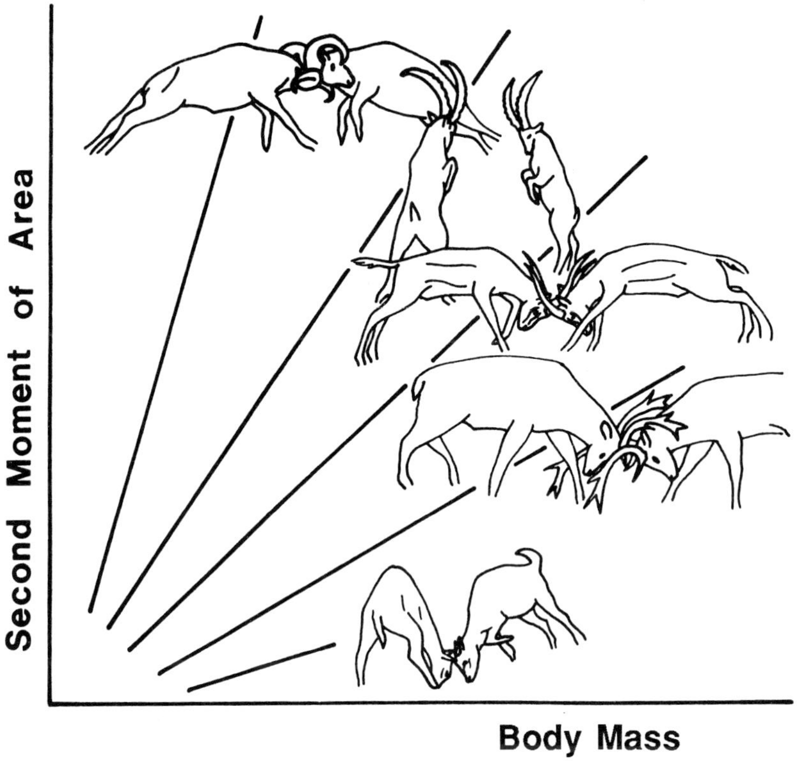


Figure 8. A diagrammatic representation of the relation between the second moment of area of horn/antler base and body mass for different types of fighting. The greater the gradient, the more forceful the fighting. In order of increasing force: duikers (*Cephalophus*), deer (*Cervus*), antelopes (*Gazella*), goats (*Capra*) and sheep (*Ovis*). (After Walther 1966; Geist 1971; Ralls 1975; Kurt 1977.)

The characteristics and generic composition of these nine different mechanical designs of horn are given in table 2.

Until now it has been assumed that the stiffnesses of horn keratin and antler bone are similar. However, although deer fight in a manner similar to that of antelopes, they have antlers with a second moment of area $I$ approximately three times less than that of the horns of bovids of similar body mass (figure 8; Kitchener 1985). This would mean that if the stiffness $E$ of antler bone is similar to that of horn keratin, antlers would deflect three times as much for the same force. However, the compact bone of antlers is about three times stiffer than horn keratin (table 3) so that antlers and horns have the same structural stiffness $EI$. Consequently, horns and antlers deflect by the same amount (see equation (3)). Also, antlers are deciduous structures

Table 2. *The characteristic behaviour, ecology and herd composition associated with animals belonging to the nine classes of mechanical design of horns and antlers. The relative structural stiffness of the horns or antlers increases from group* 1 *to group* 9

| | Sexual dimorphism | Breeding system | Fighting method | Feeding style | Territoriality | Habitat | Social grouping |
|---|---|---|---|---|---|---|---|
| 1 | – | M | S | *B/(*G) | T | C/OC | BP |
| 2 | – | M | S/(C)? | *B | T | C/OC | BP |
| 3*a* | –/+ | M/P | S/W | *B/*G/G | *T/T/HR | C/OC/O | BP/H/X |
| 3*b* | – | M | S | *B/*G | HR/*T | OC/O | BP/H |
| 3*c* | – | P | W/(C) | *B/*G | HR/*T | O | X/HSM |
| 4 | –/+ | P | W/(C) | *B/B/*G/G | HR/T/*T | OC/O | H/X/SM |
| 5 | + | P | C/(W) | G | HR/*T/T | O | H/X/SM |
| 6 | + | P | C/(W) | B/G | T/HR | C/OC/O | X/H/SM |
| 7 | + | P | C | *G/*B | HR | O | H |
| 8 | + | P | C/(W) | G | HR | C/OC/O | X/H/SM |
| 9 | + | P | C | *G/*B | HR | O | H |

*Sexual dimorphism*: +, present; –, absent.
*Breeding system*: M, monogamous; P, polygynous.
*Fighting method*: S, stabber; C, clasher; W, wrestler.
*Feeding style*: B, browser; *B, selective browser; G, grazer; *G, selective grazer.
*Territoriality*: T, territorial; *T, seasonally territorial; HR, social hierarchy.
*Habitat*: O, open; OC, open with cover; C, closed.
*Social grouping*: BP, breeding pair; H, separate herds; X, mixed herds; SM, single males.
*Composition of groups*: (1) *Madoqua*, *Ourebia*, *Oreotragus*, *Cephalophus* spp., *Raphicerus*, *Sylvicapra*, *Pelea*, *Tetracerus*, *Dorcatragus*; (2) *Cephalophus* spp. (thick-horned); (3*a*) Cervidae; (3*b*) *Oreamnos*, *Rupicapra*, *Nemorhaedus*, *Capricornis*; (3*c*) *Oryx*, *Addax*; (4) *Ammodorcas*, *Antidorcas*, *Litocranius*, *Redunca*, *Kobus*, *Hippotragus equinus*, *Gazella*, *Procapra*, *Saiga*, *Pantholops*, *Aepyceros*, *Antilope*, *Boöcercus*, *Taurotragus*, *Tragelaphus*, *Damaliscus* spp.; (5) *Alcelaphus*, *Damaliscus* spp.; (6) *Bos*, *Bubalus*; (7) *Capra*, *Ammotragus*, *Hemitragus*; (8) *Synceros caffer*; (9) *Ovis*, *Pseudois* spp.
*Sources*: Geist (1971), Jarman (1974), Ralls (1975), Leuthold (1977), Schaller (1977), Sinclair (1977), Groves (1978), Clutton-Brock *et al.* (1982), Kingdon (1982*a*), Mloszewski (1983), Spinage (1986).

so that they can afford to have a lower safety factor. Alexander's (1982) model of safety factors takes into account the cost of maintenance, breakage and replacement. The higher cost of the loss of a permanent horn would necessitate a safety factor higher than that for an annually obsolescent antler (see below).

Table 3. *The mechanical properties of horn and antler*
Work of fracture measured by using Currey's method (1979). All results from Kitchener (1987*b*, 1988) except Cervidae (M. Watkins, personal communication) and other antler results (A. C. Kitchener, C. Clayton & D. Owen, unpublished results).

| | Bending stiffness (GPa) | Work of fracture ($kJ/m^2$) | Bending strength (MPa) | Density ($kg/m^3$) |
|---|---|---|---|---|
| **Horn keratin** | | | | |
| *Ovis* | 4.1 | 22.0 | 228 | 1300 |
| *Kobus* | 3.3 | 20.0 | 245 | 1300 |
| *Oryx* | 4.3 | 19.0 | 212 | 1300 |
| **Horn core** | | | | |
| *Kobus* | 1.3 | 1.5 | 42 | 1300 |
| **Antler** | | | | |
| Cervidae | $\approx$12.0 | $\approx$10.0 | $\approx$250 | 1900 |
| *Cervus porcinus* | 12.7 | 10.0 | 246 | 2000 |
| *Cervus axis* | 11.6 | 11.5 | 223 | 1900 |
| *Cervus nippon* | 13.7 | 9.0 | 239 | 1900 |

Therefore, horns and antlers seem to be designed to resist deflexions so that they are effective in transmitting and resisting the forces of fighting. Bovids with the most forceful fighting have horns most resistant to bending.

## Structural adaptations to the forces of fighting

The ritualization of the fighting of most bovids and cervids means that the direction and magnitude of the forces acting on horns and antlers are predictable. Therefore, the mass of the cylindrical model horn (figure 5) can be reduced to save materials by optimizing the second moment of area in the direction of the predictable force of fighting.

Circular cross-sections have maximum structural stiffness in all directions. They are ideal for the unpredictable fighting of dwarf antelopes and goat-antelopes (figure 9). However, the horns of many antelopes, sheep and goats have cross-sectional shapes which are not circular, but which have a second moment of area maximized in the direction of the force of fighting and minimized in all other 'non-fighting directions' (figure 9). Therefore, by concentrating material in this way the mass of horns and antlers can be minimized. The horns of bighorn sheep and the antlers of the extinct Irish elk (*Megaloceros giganteus giganteus*) may be up to 10% of their total body

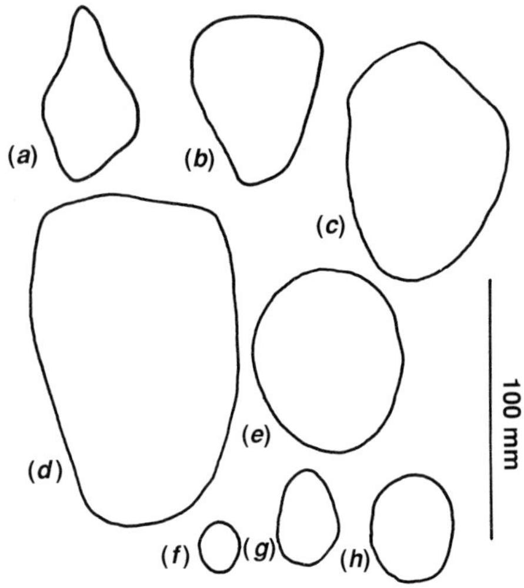


Figure 9. Cross-sections through the bases of the horns of bovids showing optimization of the second moment of area in the direction of the maximum force of fighting. The force is directed as from the top of the page. The greatest degrees of optimization are found in large bovids or those that fight forcefully. (*a*) *Capra aegagrus*, (*b*) *Ovis musimon*, (*c*) *O. orientalis gmelini*, (*d*) *O. canadensis*, (*e*) *Kobus ellipsiprymnus*, (*f*) *Cephalophus* sp., (*g*) *Gazella thomsonii*, (*h*) *Antilope cervicapra*.

mass, and yet the second moment of area of the bases of the horns and antlers of both these animals is maximized in the direction of the force of fighting (Geist 1971; Gould 1974; Kitchener 1987*a*). In the Irish elk, the second moment of area of the elliptical base of the antler is 15% greater in the direction of the expected force of fighting than is a circular antler of the same cross-sectional area (Kitchener 1987*a*).

Therefore, it should be possible to infer the degree of ritualization in the fighting behaviour of extinct bovids and cervids by looking to see if these structures are optimized to take the uniaxial bending forces of ritualized fighting.

### The best solution: horns or antlers?

The phenomenon of the casting of antlers in deer has provoked many theories based mainly on behavioural, physiological and ecological factors (Stonehouse 1968; Henshaw 1971; Chapman 1975; Chaplin 1977; Geist & Bromley 1978; Clutton-Brock 1982). However, it would seem that there is a simple mechanical factor coupled with a limitation in the growth of antlers which makes casting necessary (Kitchener 1985, 1986).

For example, male red deer increase in body mass throughout life until they no longer take part in the rut and their body mass declines (figure 10; Huxley 1932; Whitehead 1972; Goss 1983). The size of successive sets of antlers shows a similar pattern of increase until they 'grow back' in old age (figure 10; Whitehead 1972). The model of horn and antler design shows that antlers must increase in their basal dimensions if they are not to break or deflect too much, since fighting force increases with body mass. Unfortunately, antlers are dead when functional, and they grow only from their tips (figure 1*b*) so that the only way that deer can grow a larger set of antlers is to cast the old set and grow a new pair (Kitchener 1986). Figure 11 shows that the second moment of area of the base of antlers increases linearly with body mass for the red deer of Rhum so that constant deflexion in fighting is maintained.

There would seem to be a great disadvantage in growing antlers annually especially when the source of some of the calcium required for their growth is skeletal (Chaplin 1977). Most cervids are confined to the northern hemisphere where they have access to a superabundance of food during the spring and summer sufficient to support the growth of a new pair of antlers. Seasonal fluctuations in the tropics are not so great, and bovids do not have such large food resources, but their horns are permanent structures so additional annual growth is relatively much less.

The breeding season or rut of cervids is restricted to the autumn so that by casting their antlers immediately after the rut, deer can save the energy used in carrying them around, provided they have no other use for them. Winter casting is found in solitary species or in those with small breeding groups (e.g. moose *Alces alces*, roe deer *Capreolus capreolus*), but gregarious species such as red deer which use their antlers to maintain rank order within herds or to compete for limited winter food, retain their antlers until the spring (Whitehead 1972; Chapman 1975).

Bovids may not breed seasonally and may have to fight at any time during the year. Gregarious species also use their horns to maintain rank order within bachelor herds. Therefore, horns must be made of a light, durable material because they are permanent structures.

The density of horn keratin is much less than that of antler bone (table 3), but keratins are probably more energetically expensive to grow than bone. Horn keratin is also tougher than antler bone, reflecting its continued long term use (table 3). It is possible that bovids have maintenance behaviours such as horning of mud and vegetation which prevent horn keratin from drying out and becoming brittle (Leuthold 1977; Kitchener 1987*b*). Tropical deer such as hog deer (*Cervus porcinus*) and chital (*C. axis*) do not cast antlers or breed seasonally, and may have to fight at any time of the year (Schaller

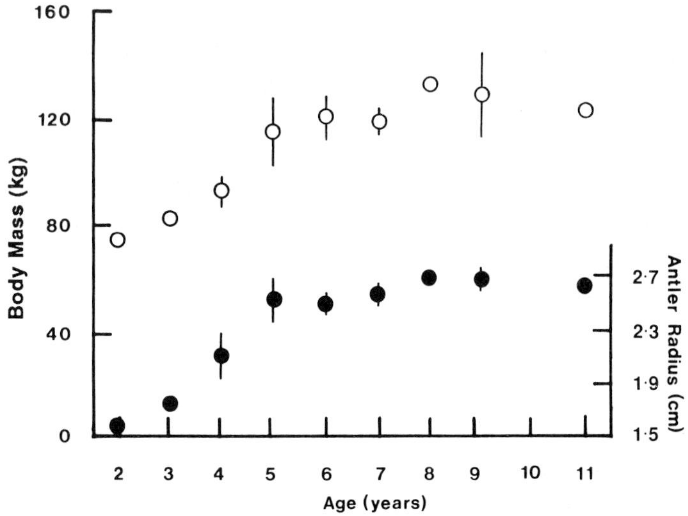


Figure 10. The change with age in body mass (open circles) and basal radius of the antlers (filled circles) of red deer (*Cervus elaphus*). The bars show standard errors of the means.

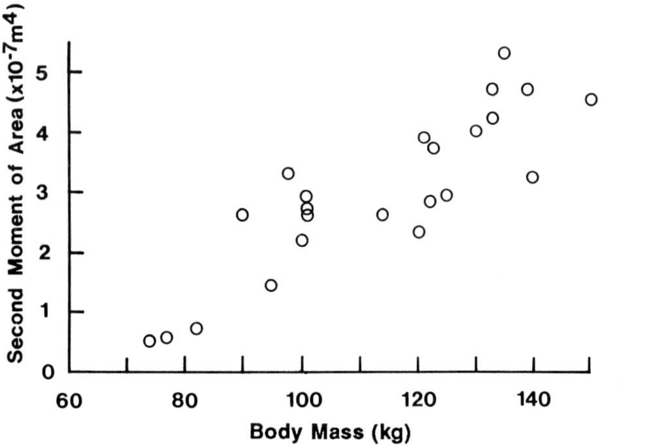


Figure 11. The linear relation between the second moment of area of the antler base and body mass of red deer (*Cervus elaphus*).

1967). Their antler bones have mechanical properties very similar to those of other deer, but the proportion of spongy bone in their antlers is considerably smaller (table 3). Consequently, tropical deer have much stiffer and stronger, but also much heavier, antlers than do deer or bovids of similar body size.

Antelopes kept in zoos in the northern hemisphere have been known to suffer frostbite in their horns. No such problem exists for dead antlers. This may explain why bovids have been relatively unsuccessful in spreading from warm climates into colder climates. In fact only bovids with very small horns (goat-antelopes) or with large frontal and horn core sinuses have achieved this (cattle, goats and sheep), but the significance of the latter feature is unknown.

Therefore, antlers seem to be ideal structures for fighting in temperate climates which have an annual superabundance of food and a strictly seasonal breeding season. Horns seem to be better suited to warmer, less seasonal climates where they may be used at any time of the year.

## The evolution and mechanical design of horns and antlers

These results demonstrate a strong correlation between the mechanical design of the horns and antlers of bovids and cervids and their fighting behaviour, reproductive strategies, diet and other ecological requirements. This provides a good model for reconstructing the behaviour and ecology of fossil species. It should be possible to determine the main selection pressures influencing the evolution of horns and antlers. For example, it seems likely that a change of habitat and diet at the end of the Miocene prompted the radiation of bovids and cervids by releasing ecological constraints on body size and social behaviour; this in turn led to a change in horn function and to diversification in horn mechanical design and fighting behaviour.

Often only horn cores and antlers are preserved in the fossil record, but these can provide much useful information to test hypotheses about the co-evolution of behaviour, ecology and morphology in bovids and cervids. Much work remains to be done. Unfortunately, keratin is rarely preserved, but horn cores do reflect the shape of the keratin sheath. Also, the roughness on the surface of a horn core appears to be directly correlated with the thickness of keratin sheath so that a reasonably accurate reconstruction of the size and shape of horn sheaths of fossil species could be attempted.

The rules of fighting and the optimization of horn and antler shape to the predictable forces of fighting can be used to reconstruct fighting behaviours and predict the magnitude of the forces acting on these structures. This has been achieved with the enormous palmate antlers of the Irish elk, which show both structural and material adaptations to the predicted forces of fighting (Kitchener 1987*a*). A biomechanical study was used to test the hypothesis that the Irish elk never used its antlers for fighting; it was shown instead that they were optimally designed for taking the forces expected from a reconstructed fighting behaviour (Kitchener 1987*a*).

Other areas of future study could include a more detailed investigation of cranial adaptations to fighting including the presence of frontal sinuses and the orientation of horns and antlers with respect to the skull (Schaffer 1968; Schaffer & Reed 1972; Kitchener 1987*a*). The role of the nuchal ligament in storing the energy of fighting has also been suggested (Geist 1971). In the long term it may be possible to reconstruct fighting in ceratopsian and pachycephalosaurian dinosaurs (Farlow & Dodson 1975; Norman 1985).

The continued investigation of the interrelation of horn and antler mechanical design, fighting behaviour, reproductive strategies and ecology in the evolution of bovids and cervids should provide a fruitful area of study for many years to come.

## Acknowledgements

I thank Dr T. H. Clutton-Brock of Cambridge University for supplying the data on red deer and Michael Kendall of the BBC Natural History Library, Bristol, who loaned the video from which fighting deer were taken in figure 8. I wish also to thank Dr Iain Bishop, Dr Juliet Clutton-Brock and Dr Kim Dennis-Bryan of the British Museum (Natural History) for the loan of horns and Mr Richard Cindery of Whipsnade Zoo for supplying horns. I am very grateful to Dr Alan McCaffery for giving up his desk so that I could type out this manuscript, to Dr Alick Jones for giving me refuge within the Zoology Department at Reading and to Charles Deeming, Dave Unwin and Dr Julian Vincent who all made comments on an earlier version of this manuscript. Mechanical tests were carried out on an Instron testing machine which was largely funded by the Royal Society. Finally, I thank Dr Jeremy Rayner for inviting me to submit a paper to the symposium.

## References

Alexander, R.McN. 1982 *Optima for Animals*. London: Edward Arnold.

Barette, C. 1977 Fighting behaviour of muntjac and the evolution of antlers. *Evolution* **31**, 169–176.

Best, A.A. & Raw, W.G. 1971 *Rowland Ward's Records of Big Game*, 15th edition. London: Rowland Ward.

Chaplin, R.E. 1977 *Deer*. Poole: Blanford.

Chapman, D.I. 1975 Antlers: bones of contention. *Mammal Review* **5**, 121–172.

Clark, M. 1981 *Mammal Watching*. London: Hamlyn.

Clutton-Brock, T.H. 1982 The function of antlers. *Behaviour* **79**, 108–125.

Clutton-Brock, T.H. & Albon, S.D. 1985 Competition and population regulation in social mammals. In *Behavioural Ecology: Ecological*

*Consequences of Adaptive Behaviour*, ed. R.M. Sibly & R.H. Smith, pp. 557–575. Oxford: Blackwell Scientific.

Clutton-Brock, T.H., Albon, S.D., Gibson, R.M. & Guinness, F.E. 1979 The logical stag: adaptive aspects of fighting in red deer (*Cervus elaphus* L.). *Animal Behaviour* **27**, 211–225

Clutton-Brock, T.H., Albon, S.D. & Harvey, P.H. 1980 Antlers, body size and breeding group size in the Cervidae. *Nature, London* **285**, 565–567.

Clutton-Brock, T.H., Guinness, F.E. & Albon S.D. 1982 *Red Deer. Behaviour and Ecology of Two Sexes*. Edinburgh University Press.

Currey, J.D. 1979 Mechanical properties of bone tissues with greatly differing functions. *Journal of Biomechanics* **12**, 313–319.

Dittrich, L. & Böer, M. 1980 *Verhalten und Fortpflanzung von Kirks Rüssell-Dikdiks im Zoologischer Garten*. Hanover: Freimann & Fuchs.

Dollman, J.G. & Burlace, J.B. 1935 *Rowland Ward's Records of Big Game, African and Asian sections*, 10th edition. London: Rowland Ward.

Estes, R.D. 1969 Territorial behaviour of the wildebeest (*Connochaetes taurinus taurinus* Burchell, 1823). *Zeitschrift für Tierpsychologie* **26**, 284–370.

Farlow, J.O. & Dodson, P. 1975 The behavioural significance of frill and horn morphology in ceratopsian dinosaurs. *Evolution* **29**, 353–361.

Geist, V. 1966 The evolution of horn-like organs. *Behaviour* **27**, 175–214.

Geist, V. 1971 *Mountain Sheep*. Chicago University Press.

Geist, V. 1978 On weapons, combat and ecology. In *Aggression, Dominance and Spacing*, ed. L. Krames, P. Pliner & T. Alloway, pp. 1–30. London: Plenum Press.

Geist, V. 1986 New evidence of high frequency of antler wounding in cervids. *Canadian Journal of Zoology* **64**, 380–384.

Geist, V. & Bromley, P.T. 1978 Why deer shed antlers. *Zeitschrift für Säugetierkunde* **43**, 223–231.

Gentry, A.W. 1978 Bovidae. In *The Evolution of African Mammals*, ed. V.J. Maglio & H.B.S. Cooke, pp. 540–572. Harvard University Press.

Gordon, I.J. & Ilius, A.W. 1988 Incisor arcade structure and diet selection in ruminants. *Functional Ecology* **2**, 15–22.

Gordon, J.E. 1976 *The New Science of Strong Materials*, 2nd edition. Harmondsworth: Penguin.

Gosling, L.M. 1974 The social behaviour of Coke's hartebeest. In *The Social Behaviour of Ungulates and its Relation to Management*, ed. V. Geist & F. Walther, vol. 2, pp. 488–511. Morges, Switzerland: IUCN.

Goss, R.J. 1983 *Deer Antlers*. London: Academic Press.

Gould, S.J. 1974 The evolutionary significance of 'bizarre' structures: antler size and skull size in the 'Irish elk' *Megaloceros giganteus*. *Evolution* **28**, 191–220.

Groves, C. P. 1978 The taxonomic status of the dwarf blue sheep. *Säugetierkundliche Mitteilungen* **26**, 177–183.

Groves, C.P. & Grubb, P. 1982 The species of muntjac (genus *Muntiacus*) in Borneo: unrecognised sympatry in tropical deer. *Zoologische Mededelingen, Leiden* **56**, 203–216.

Halstead, L.B. 1978 *The Evolution of the Mammals.* London: Peter Lowe.

Henshaw, J. 1971 Antlers: the unbrittle bones of contention. *Nature, London* **231**, 469.

Huxley, J.S. 1932 *Problems of Relative Growth.* London: Methuen.

Janis, C. 1982 Evolution of horns in ungulates: ecology and palaeoecology. *Biological Reviews* **57**, 261–318.

Jarman, P.J. 1974 The social organisation of antelope in relation to their ecology. *Behavior* **48**, 215–267.

Kingdon, J. 1982*a* *East African Mammals. An Atlas of Evolution in Africa*, vol. IIIC. London: Academic Press.

Kingdon, J. 1982*b* *East African Mammals. An Atlas of Evolution in Africa*, vol. IIID. London: Academic Press.

Kitchener, A. 1985 The effect of behaviour and body weight on the mechanical design of horns. *Journal of Zoology, London* A **205**, 191–203.

Kitchener, A. 1986 Why deer shed their antlers. *Deer* **6**, 403–406.

Kitchener, A. 1987*a* The fighting behaviour of the extinct Irish elk. *Modern Geology* **11**, 1–28.

Kitchener, A. 1987*b* Fracture toughness of horns and a reinterpretation of the horning behaviour of bovids. *Journal of Zoology, London* **213**, 621–639.

Kitchener, A. 1988 An analysis of the forces of fighting of the blackbuck (*Antilope cervicapra*) and the bighorn sheep (*Ovis canadensis*) and the mechanical design of the horns of bovids. *Journal of Zoology, London* **214**, 1–20.

Kurt, F. 1977 In *The Book of Hunting*, ed. N. Gelb, pp. 174–175. London: Paddington Press.

Leslie, D.M. & Jenkins, K.J. 1985 Rutting mortality among male Roosevelt elk. *Journal of Mammalogy* **66**, 163–164.

Leuthold, W. 1977 *African Ungulates.* Heidelberg: Springer Verlag.

Mloszewski, M.J. 1983 *The Behaviour and Ecology of the African Buffalo.* Cambridge University Press.

Norman, D.B. 1985 *The Illustrated Encyclopaedia of Dinosaurs.* London: Salamander Press.

Nowak, R.M. & Paradiso, J.L. 1983 *Walker's Mammals of the World*, 4th edition. Johns Hopkins University Press.

O'Gara, B.W. & Matson, G. 1975 Growth and casting of horns by pronghorns and exfoliation of horns by bovids. *Journal of Mammalogy* **56**, 829–846.

Packer, C. 1983 Sexual dimorphism: the horns of African antelopes. *Science* **221**, 1191–1193.

Petocz, R.G. & Shank, C.C. 1983 Horn exfoliation in Marco Polo sheep *Ovis ammon poli* in the Afghan Pamir. *Journal of Mammalogy* **64**, 136–138.

Ralls, K. 1975 Agonistic behaviour in Maxwell's duiker, *Cephalophus maxwelli. Mammalia* **39**, 241–249.

Roark, R.J. & Young, W.C. 1975 *Formulas for Stress and Strain*, 5th edition. London: McGraw-Hill.

Savage, R.J.G. & Long, M.R. 1986 *Mammal Evolution*. London: British Museum (Natural History).
Schaffer, W.M. 1968 Intraspecific combat and the evolution of the Caprini. *Evolution* **22**, 817–825.
Schaffer, W.M. & Reed, C.A. 1972 The coevolution of social behaviour and cranial morphology in sheep and goats. *Fieldiana Zoology* **61**, 1–88.
Schaller, G.B. 1967 *The Deer and the Tiger*. Chicago University Press.
Schaller, G.B. 1977 *Mountain Monarchs*. Chicago University Press.
Sinclair, A.R.E. 1977 *The African Buffalo*. Chicago University Press.
Spinage, C.A. 1982 *A Territorial Antelope: the Uganda Waterbuck*. London: Academic Press.
Spinage, C.A. 1986 *The Natural History of Antelopes*. Beckenham: Croom Helm.
Stonehouse, B. 1968 Thermoregulatory function of growing antlers. *Nature, London* **218**, 870.
Walther, F. 1966 *Mit Horn und Huf*. Berlin: Paul Parey.
Walther, F. 1978 Behavioral observations on oryx antelope (*Oryx beisa*) invading the Serengeti National Park. *Journal of Mammalogy* **59**, 243–260.
Walther, F., Mungall, E.C. & Grau, G.A. 1983 *Gazelles and their Relatives*. New Jersey: Noyes.
Ward, R. 1903 *Records of Big Game*. 4th edition. London: Rowland Ward.
Ward, R. 1910 *Records of Big Game*. 6th edition. London: Rowland Ward.
Whitehead, G.K. 1972 *Deer of the World*. London: Constable.

R. McNEILL ALEXANDER

# Characteristics and advantages of human bipedalism

Some non-human primates walk bipedally on occasions, but people are the only mammals that use bipedal walking and running as their usual gaits. Our stance and gait are also unique in two other major features: we keep our knees straight while standing and in the stance phase of walking, and we hold our trunks erect. Many other peculiarities of our stance and gait seem to be related to these two. Our manners of standing and walking seem to be adapted to minimize metabolic energy costs, but there is no satisfactory direct evidence that the costs are low. Also, our ability to run bipedally may depend on our keeping our knees fairly straight and our trunks erect. Non-human primates seem unable to run bipedally.

## Bipedalism

People and birds are the only modern animals that use bipedal walking as their usual gait. Kangaroos move bipedally, but hop instead of running and use all four feet at low speeds. Bipedal rodents also hop. Some non-human primates walk bipedally, but for them bipedal walking is an occasional rather than a usual gait. Chimpanzees (*Pan*) usually walk and run on all fours but occasionally rise to walk on their hind legs, especially when carrying objects (Napier & Napier 1967). Gorillas (*Gorilla*) often stand bipedally, but seldom walk bipedally for more than a few steps (Schaller 1963). Gibbons (*Hylobates*) occasionally walk bipedally on horizontal branches (Fleagle 1980). A population of macaques (*Macaca fuscata*) in Japan has learned to walk bipedally, carrying food (Napier & Napier 1967). A major advantage of bipedalism for people, chimpanzees and macaques is that it frees the hands for uses other than locomotion.

The higher primates have at least two characteristics that seem to be preadaptations for bipedalism. First, when standing quadrupedally, they carry an unusually large proportion of their weight on their hind feet (Kimura 1985). Second, they are plantigrade, placing the heel as well as the toe on the ground when they stand or walk. Many small mammals are plantigrade, but the only large plantigrade mammals are primates and bears. The significance of plantigrady for bipedalism is that plantigrade feet have long soles, making balancing on two feet easier. Our feet are small in comparison to other primates (Lessertisseur & Jouffroy 1973) but still seem to be rather larger than is necessary for balance. Healthy people standing normally allow their centres of mass to sway backwards and forwards through a range of only about 25 mm, and could easily stand on feet a little longer than that (Debrunner 1985). Ostriches (*Struthio*) have longer legs than people but balance satisfactorily on slightly shorter (digitigrade) feet.

In addition to bipedalism, human stance and gait show many peculiarities. Most of them can be interpreted as consequences of two basic peculiarities: we keep our knees remarkably straight while our feet are on the ground, and we carry our trunks remarkably vertical. This is the theme of the next two sections.

## Straight knees

We can extend our knees slightly beyond the straight position, to an angle of about 190°. However, force is needed to do this: passive resistance due to skin, fascia and ligaments is felt, when the knee is extended beyond about 170° (Smith 1956). We usually stand with our knees extended to about 185°. When we walk, the knee angle is 160°–180° for most of the stance phase and decreases only at the end of the phase, after the heel has been lifted (Ishida, Kimura & Okada 1974; Debrunner 1985). Other mammals stand and walk on bent knees. For example, raccoons and cats bend their knees to about 110° and 100°, respectively, in the middle of the stance phase of walking (figure 2*a*; Goslow, Reinking & Stuart 1973). Chimpanzees and gorillas stand with knees bent at 120–140° (Tuttle, Basmajian & Ishida 1979). Various non-human primates walking bipedally set the foot down with the knee almost straight, but then immediately bend the knee (Ishida *et al.* 1974). Figures 1*b* and 2*b* show the marked difference between chimpanzee and human knee angles in the stance phase.

The straightness of the knees in human standing relieves the knee muscles of the need to be active. The centre of pressure of the foot on the ground is normally about halfway along the sole, and the line of action of the ground force passes anterior to the axis of the knee (figure 1*c*). The moment of the ground force about the knee is balanced largely by the passive tissues which limit its extension (Smith 1956). The quadriceps muscles (extensors of the knee) and the hamstring muscles (flexors) show little or no electrical activity

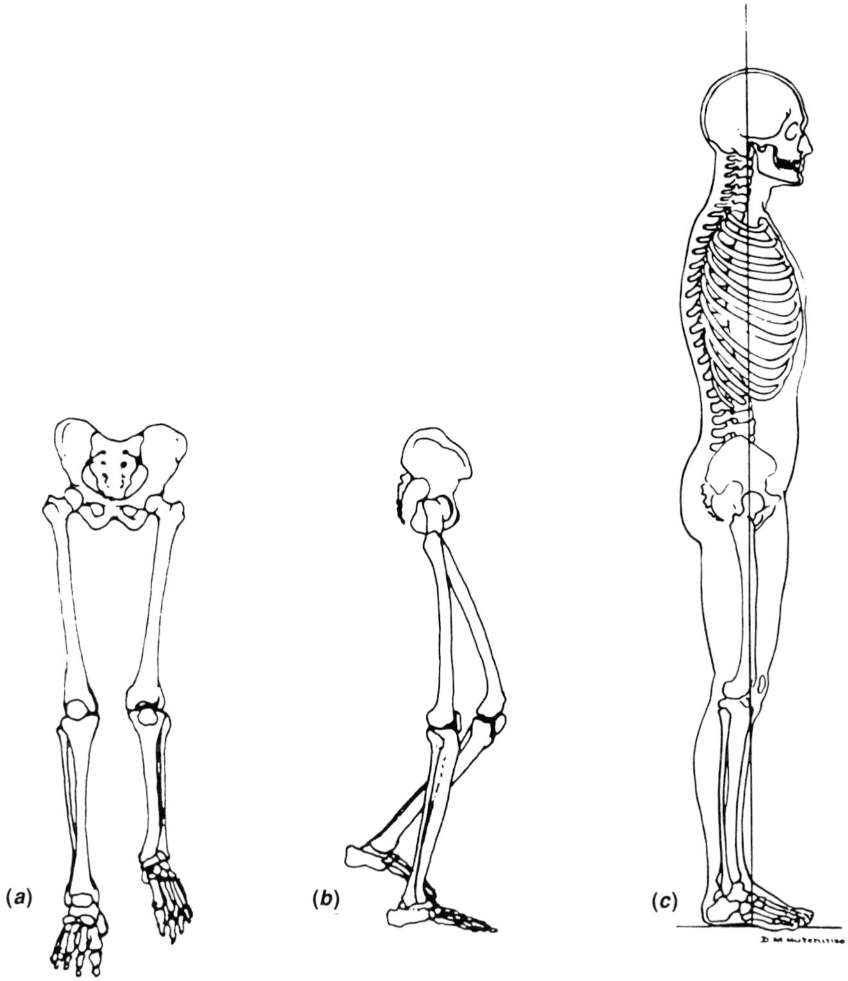


Figure 1. (*a*) and (*b*) Anterior and lateral views of the skeleton of human legs in the position of the middle of the stance phase of walking. (From Stern & Susman 1981.) (*c*) Human skeleton in a relaxed standing posture. The vertical line passes through the centre of mass of the body. (From Basmajian & De Luca 1985.)

(Basmajian & De Luca 1985). In contrast, gorillas and chimpanzees show activity in the quadriceps muscles and at least one of the hamstrings, when standing bipedally (Tuttle *et al.* 1979). The same muscles become active in people when they stand on bent knees like apes.

The straightness of the knee in the stance phase allows walking with little activity in knee muscles. The moment of the ground force tends to flex the knee during part of the phase and to extend it during another, but it is always

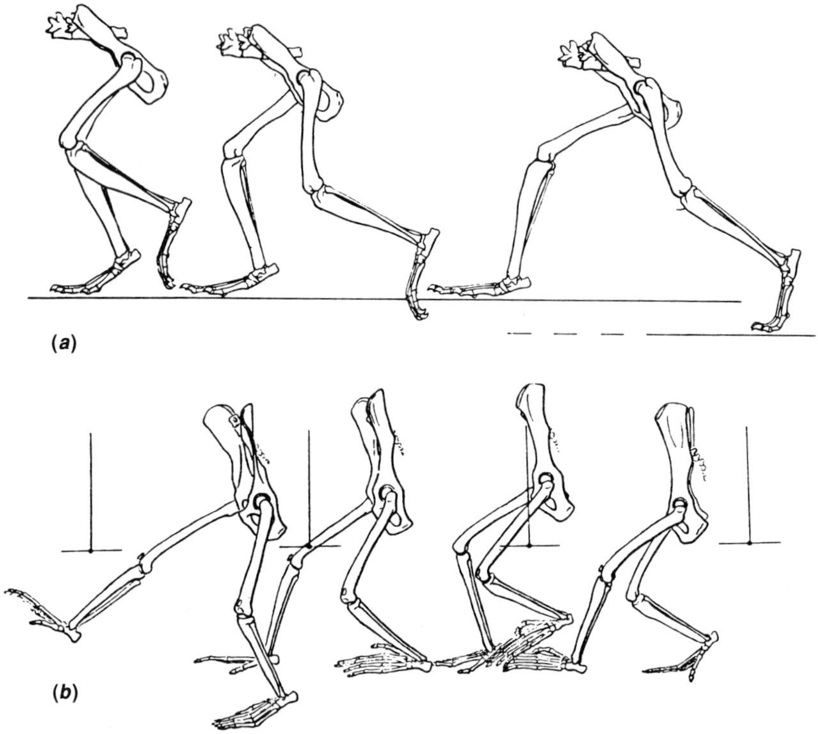

Figure 2. Drawings from cineradiographs of (*a*) a raccoon (*Procyon*) and (*b*) a chimpanzee (*Pan*) walking. ((*a*) from Jenkins & Camazine 1977; (*b*) from Jenkins 1972.)

small (Debrunner 1985). If we walked on bent knees, as do apes, large moments tending to flex the knee would act during much of the stance phase and would have to be balanced by tension in the quadriceps muscles. Electromyographic studies of humans have shown that the quadriceps and hamstring muscles are active during at least part of the stance phase (Basmajian & De Luca 1985). However, the level of activity seems to be low. Tuttle *et al.* (1979) formed the impression that the same muscles are much more active in the bipedal walking of apes than that of people. Brandell's (1973) records of human gaits indicate that activity in these muscles increases far more when the change is made from walking to running than does activity in calf muscles.

Another consequence of the straight knees in human walking is that the hip must rise and fall, describing an arc of a circle, during the part of the stance phase in which the foot is flat on the ground. This tends to make the potential energy fluctuations during the step balance the kinetic energy fluctuations, for the following reason. The force exerted on the ground, by people and other animals, tends to remain approximately in line with the leg, throughout the step. If the force is aligned with the leg and the length of the leg remains constant, no work is done, and the potential and kinetic energy changes must balance. Cavagna, Heglund & Taylor (1977) used a force platform to measure, indirectly, changes in potential and external kinetic energy. They found the energy changes to be fairly well balanced in humans throughout the normal range of walking speeds. In rheas and turkeys (*Rhea* and *Meleagris*, the only other bipeds investigated) potential and kinetic energy changes were well balanced in slow walking, but not in fast walking.

During part of the walking cycle both feet are on the ground simultaneously. If each hip then described an arc of a circle about its foot, one hip would fall while the other rose and the trunk would make a marked rolling movement. This is avoided by setting each foot down initially on its heel, and later letting the heel rise. In the diagrams in figure 3 the legs are treated as plane linkages with the left and right hips required to follow the same path. In figure 3*a* the knees as well as the hips and ankles are movable, but the feet are kept flat on the ground. The linkage has two degrees of freedom of movement, enough for the hips to be moved along any desired path within certain limits. In figure 3*b* the knees are locked and the linkage is rigid. However, in figure 3*c* the joints lost by locking the knees have been replaced by joints further down the leg: the point of contact of the heel with the ground serves as an extra joint for the forward leg, and the metatarso-phalangeal joints for the other. The mobility of the linkage has been restored.

Plantigrade mammals in general lift the heel before the toes (figure 2). However, the habit of setting the heel down well before the rest of the sole while walking seems to be peculiar to humans. Ishida *et al.* (1974) found that non-human primates walking bipedally set down the heel and the more anterior parts of the sole almost simultaneously. The heel has become much more prominent in the course of human evolution, giving rise to the characteristic arch of the human foot.

Relative stride length is defined as the ratio of stride length to hip height. Non-human primates use longer relative stride lengths than other mammals at corresponding speeds (i.e. at equal Froude numbers; Alexander & Maloiy 1984). A chimpanzee behaved like other non-human primates, whether it was walking bipedally or quadrupedally. People, however, use relative stride lengths about 30% shorter than those of other primates at any given Froude number. This may be associated with the habit of walking with straight knees, which increase hip height without necessarily changing stride length. If the chimpanzee in figure 2*b* walked on straight knees, its hip height would

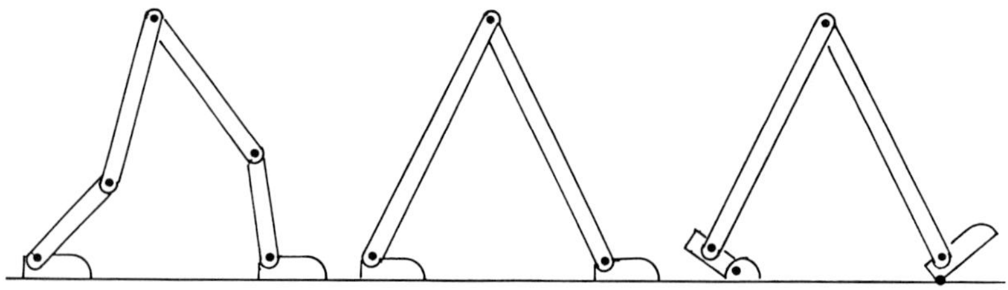


Figure 3. Diagrams of the double support phase of walking. These diagrams are explained in the text.

be increased by a factor of 1.3 and its relative stride length (for the same absolute stride length) would be reduced by 25%.

We have seen how in human walking the straightness of the knees makes the centre of mass of the body rise and fall. The body has a downward acceleration in the single-support phase (while only one foot is on the ground) and an upward acceleration in the double-support phase. The vertical component of the force exerted by the foot on the ground during the single-support phase must be less than body weight. The total of the vertical components exerted by the two feet during the double-support phase must be greater than the body weight. Figure 4 shows how this is achieved. The force exerted by each foot rises and falls in a two-humped pattern. At higher walking speeds the vertical accelerations are greater, larger fluctuations of force are required, and the pattern becomes more markedly two-humped. The change can be described quantitatively by the Fourier analysis method of Alexander & Jayes (1980): the ratio of coefficients $a_3/a_1$ measures the degree of the double hump, and takes increasingly negative values as walking speed increases (but becomes positive in running, when the force pattern is essentially single humped).

The value of $a_3/a_1$ for human walking ranges from about $-0.2$ at low speeds to $-0.7$ at high speeds (data from Alexander & Jayes 1980). Force records of quadrupedal mammals walking at various speeds nearly all gave $a_3/a_1$ between 0 and $-0.25$, though two records of dogs' hind feet gave values of about $-0.4$ (data from Alexander & Jayes 1983). Force records of non-human primates walking both quadrupedally and bipedally are only mildly (if at all) two-humped: none of the records published by Kimura *et al.* (1983) or by Kimura (1985) gives values of $a_3/a_1$ more negative than $-0.3$. The markedly two-humped records obtained for fast human walking (figure 4*c*) do not seem to be matched by any other species.

There is another difference between bipedal walking of chimpanzees and that of people, which may be related to the different angles of the knees. In

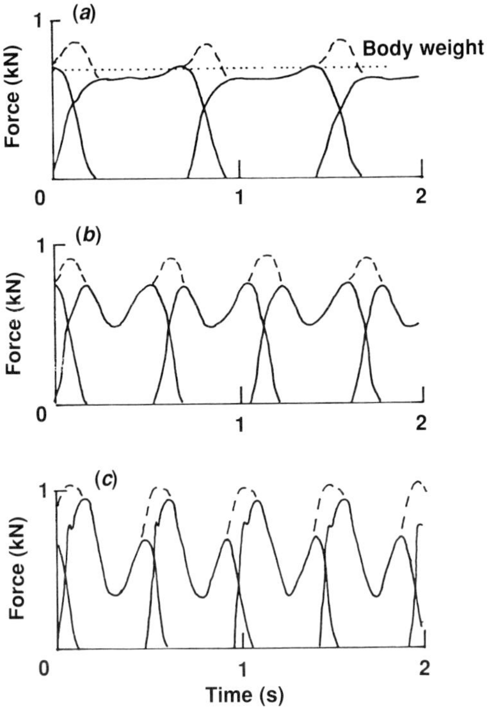


Figure 4. Graphs of the vertical components of the forces on the feet, against time, for human walking at (*a*) 0.9 m/s, (*b*) 1.5 m/s and (*c*) 2.1 m/s. These graphs are based on the data of Alexander & Jayes (1980). Only one footfall was recorded in each case, and it has been assumed that identical patterns of force were exerted in successive footfalls. Values of $a_3/a_1$ are: (*a*) $-0.22$, (*b*) $-0.36$, (*c*) $-0.55$.

the single-support phase people allow the hip of the swinging leg (left in figure 1*a*) to sink a little below that of the supporting leg. Chimpanzees raise the hip of the swinging leg above that of the supporting leg (Jenkins 1972). The straighter knee of the supporting leg in people gives more clearance for the swinging leg, possibly making it unnecessary to raise that leg's hip. The shortness of the foot may also have an effect.

### Erect trunk

Quadrupeds stand and move with trunks horizontal and people with their trunks vertical. Apes sometimes bring their trunks to a vertical position when standing bipedally, especially when reaching upwards (photographs given by Tuttle *et al.* (1979)), but their trunks usually slope forward in bipedal walking (Jenkins 1972).

The angle of the trunk depends partly on the orientation of the pelvic girdle and partly on the curvature of the vertebral column. In this paragraph, pelvic angle means the angle to the horizontal of a line along the dorsal edge of the pelvic girdle, just touching the ilium and ischium. This angle can only be estimated from ordinary films, but can be measured from the X-ray ciné films which have been taken of a few species. It is about 40° in walking raccoons (figure 2*a*), and this angle may perhaps be taken as typical of quadrupedal mammals. Chimpanzees have relatively long arms which are effectively lengthened further by the habit of walking on the knuckles rather than on the palms of their hands. Consequently the pelvic angle in quadrupedal walking is usually a little more than in raccoons, averaging about 50° (Jenkins 1972). In bipedal walking it is increased to about 80° (figure 2*b*). The more erect stance of humans increases the pelvic angle (defined as above) to about 100° (figure 1), but the shape of the girdle is so different from that of other mammals that it seems doubtful whether the comparison of angles is meaningful. Other human peculiarities related to the erectness of the trunk are the abrupt change of orientation between the sacrum and the lumbar vertebrae, and the concave curve of the lumbar region.

One consequence of the more erect trunk is that smaller moments act at the hip joints. Consider the position of the centre of mass of the body, excluding the legs. In apes standing with the trunk sloping it is well in front of the hip joints, requiring tension in hip extensor muscles. Electromyograms show activity in biceps femoris and sometimes in other hip extensor muscles (Tuttle *et al.* 1979). In people standing with the trunk erect, this centre of mass seems to be slightly behind the hip joints. Any activity in hip extensor muscles is slight, but iliopsoas (a flexor of the hip) is consequently active (Basmajian & De Luca 1985). The inactivity of the knee extensor muscles has already been noted: human standing requires remarkably little activity in leg muscles.

Similarly in bipedal walking, the erect position of the human trunk reduces the activity needed in hip extensor muscles (Tuttle *et al.* 1979).

The feet move relative to the trunk during the stance phase of walking. Quadrupeds set the feet down well forward of the hip joints and keep them on the ground until they are well behind the hip joints (figure 2*a*). In apes walking bipedally the mean position of the feet must be under the centre of mass of the whole body which (because of the sloping trunk) is well anterior to the hips. Consequently the foot is lifted while under, rather than behind, the hip joint (figure 2*b*). Although the trunk is more erect than in quadrupeds, the range of movement of the femur relative to the pelvic girdle is not grossly different. In human walking the pelvic angle is a little larger, as already noted. The erect trunk brings the centre of mass close to the hips. Also, the straightness of the knees helps to make the average position of the femur more extended, relative to the pelvis. If the pelvic girdle had the same general shape as in apes and other mammals, the hamstring muscles would lose their moment arm and could not function as extensor muscles of the hip (see references given by Zuckerman *et al.* 1973).

The pelvic girdle has become relatively short in humans, and the ilium has become very broad. These changes may perhaps be interpreted as providing support for the viscera, at the base of the erect trunk. However, the broadened ilium also provides an origin for the greatly enlarged gluteus maximus muscle, which is discussed in the next section (Stern 1972). The shortened ilium does not seem to have resulted in a shortened trunk: apes and humans of equal mass have trunks of about equal length (Aiello 1981). The changes that occurred as the human pelvis evolved have made its shape grossly different from those of all other living primates (Zuckerman *et al.* 1973).

**Advantages**

The peculiarities of human stance and gait suggest that we should be remarkably economical of energy. Our erect trunk and straight knees enable us to stand with little or no activity in the large muscles of our thighs. These muscles should therefore use little metabolic energy. In walking, kinetic and potential energy changes are largely balanced, over a wide range of speeds. Therefore the muscles have little positive and negative work to do and might be expected to use little metabolic energy.

Rates of oxygen consumption while standing and while lying have been measured, for people and some other species (Blaxter 1962). The difference between standing and lying was equivalent to 0.10 W/(kg body mass) for cattle, 0.14 W/kg for sheep and 0.21 W/kg for people. Standing seems to be less economical for us than for these other mammals, which may not, however, be typical of mammals in general.

Rates of oxygen consumption in walking and running have been measured for many species of mammals and birds. The slope of a graph of oxygen consumption against speed is often used as a measure of the energy cost of travelling unit distance. This slope tends to be less for larger species (proportional to (body mass)$^{-0.3}$) so comparisons will be made with predicted values for the same body mass, based on all available data. For humans it is 1.4 times the predicted value and for chimpanzees (whether moving bipedally or quadrupedally) 1.2 times the predicted value (Taylor, Heglund & Maloiy 1982). This makes humans seem less economical than chimpanzees and most other mammals. However, the value for humans in this comparison refers to running. It is human *walking* that seems likely to be economical.

Graphs of oxygen consumption against speed for human walking curve markedly, and so have no unique slope. Instead of comparing slopes, we will compare total costs of transport (total metabolic energy consumption divided by distance travelled and by body mass). Both for people and for horses this cost has a minimum value at a moderate walking speed. For adult men this minimum cost is 3.0–3.5 J/(kg m) (calculated from fig. 2 of Zarrugh & Radcliffe, 1978). For small horses it is 3.0 J/(kg m) (Hoyt & Taylor 1981). The horses were about twice as heavy as the people: correction of the horse value to human body mass, assuming proportionality to $(\text{mass})^{-0.3}$, gives 3.6 J/kg m. The comparison of slopes for running made people seem considerably less economical than most other mammals (including horses). The comparison of minimum total costs of transport makes people seem marginally more economical than horses, but even it does not make human walking seem remarkably economical.

Perhaps we should look for some other major advantage for the human bipedal gait. One possibility is speed. Other primates seem to be limited to low speeds, when travelling bipedally. The non-human primates in the experiments of Kimura *et al.* (1983) all moved bipedally at speeds of 1.8 m/s or less, but the human subjects ran at up to 2.8 m/s and could presumably have run much faster. I can find no published record of any non-human primate travelling bipedally faster than 2 m/s, but most people can probably attain 7 m/s, and good athletes sprint at 10 m/s.

The speeds of the non-human bipeds may be limited by the moments that their muscles can exert, about hip and knee. The peak forces on the feet are almost doubled when a person changes from a fast walk to a run (Alexander & Jayes 1980). In running, higher speeds require larger forces. The human style of running, with trunk more erect than in apes and knees less bent, enables us to exert given forces on the ground with smaller moments about hip and knee. The argument is the same as that already presented for standing and walking. We bend our knees more in the stance phase in running than in walking (to about 130°; Debrunner 1985), but not as much as other primates walking bipedally (Ishida *et al.* 1974).

The straight knees and erect trunk reduce moments about the hip in the sagittal plane, but do nothing to reduce moments in the transverse plane. It seems possible that our greatly enlarged gluteus maximus may be, as Stern (1972) suggests, an adaptation for bipedal running. The enlarged part is an abductor, not an extensor of the hip (Stern 1972).

It seems easy to believe that an ability to run bipedally was important in guiding human evolution. The ability to carry things is useful in many circumstances. Bipedalism frees the hands for carrying. Plant gatherers need to carry food, and migrants have to carry children (Sinclair, Leakey & Norton-Griffiths 1986). Other life styles might require the carrying of weapons or tools. I do not wish to join the controversies about early hominid life styles, which raise problems in anthropology more than biomechanics. However, if the ability to carry things is important, the ability to run while carrying them will generally be advantageous. Running may be useful in escaping predators, in pursuing prey and in travelling fast for other reasons.

## References

Aiello, L.C. 1981 The allometry of body proportions. *Symposia of the Zoological Society of London* **48**, 331–358.

Alexander, R.McN. & Jayes, A.S. 1980 Fourier analysis of forces exerted in walking and running. *Journal of Biomechanics* **13**, 383–390.

Alexander, R.McN. & Jayes, A.S. 1983 A dynamic similarity hypothesis for the gaits of quadrupedal mammals. *Journal of Zoology, London* **201**, 135–152.

Alexander, R.McN. & Maloiy, G.M.O. 1984 Stride lengths and stride frequencies of primates. *Journal of Zoology, London* **202**, 577–582.

Basmajian, J.V. & De Luca, C.J. 1985 *Muscles Alive*, 5th edition. Baltimore: Williams & Wilkins.

Blaxter, K.L. 1962 *The Energy Metabolism of Ruminants.* London: Hutchinson.

Brandell, B.R. 1973 An analysis of muscle coordination in walking and running gaits. *Medicine and Sport* **8**, 278–287.

Cavagna, G.A., Heglund, N.C. & Taylor, C.R. 1977 Mechanical work in terrestrial locomotion, two basic mechanisms for minimizing energy expenditure. *American Journal of Physiology* **233**, R243 – R261.

Debrunner, H.U. 1985 *Biomechanik des Fusses.* Stuttgart: Enke.

Fleagle, J.G. 1980 Locomotion and posture. In *Malayan Forest Primates*, ed. D.J. Chivers, pp. 191–207. New York: Plenum Press.

Goslow, G.E., Reinking, R.M. & Stuart, D.G. 1973 The cat step cycle, hind limb joint angles and muscle lengths during unrestrained locomotion. *Journal of Morphology* **141**, 1–42.

Hoyt, D.F. & Taylor, C.R. 1981 Gait and the energetics of locomotion in horses. *Nature, London* **292**, 239–240.

Ishida, H., Kimura, T. & Okada, M. 1974 Patterns of bipedal walking in anthropoid primates. In *Proceedings of the Symposium of the 5th Congress of the International Primatological Society*, ed. S. Kondo, pp. 287–301. Tokyo: Japan Science Press.

Jenkins, F.A. 1972 Chimpanzee bipedalism: cineradiographic analysis and implications for the evolution of gait. *Science* **178**, 877–879.

Jenkins, F.A. & Camazine, S.M. 1977 Hip structure and locomotion in ambulatory and cursorial carnivores. *Journal of Zoology, London* **181**, 351–370.

Kimura, T. 1985 Bipedal and quadrupedal walking of primates, comparative dynamics. In *Primate Morphophysiology: Locomotor Analyses and Human Bipedalism*, ed. S. Kondo, pp. 81–104. University of Tokyo Press.

Kimura, T., Okada, M., Yamazaki, N. & Ishida, H. 1983 Speed of the bipedal gaits of man and nonhuman primates. *Annales des Sciences Naturelles Série Zoologie* (13) **5**, 145–158.

Lessertisseur, J. & Jouffroy, F.K. 1973 Tendances locomotrices des primates traduites par les proportions du pied. *Folia Primatologica* **20**, 125–160.

Napier, J.R. & Napier, P.H. 1967 *A Handbook of Living Primates*. London: Academic Press.

Schaller, G.B. 1963 *The Mountain Gorilla: Ecology and Behaviour*. University of Chicago Press.

Sinclair, A.R.E, Leakey, M.D. & Norton-Griffiths, M. 1986 Migration and hominid bipedalism. *Nature, London* **324**, 307–308.

Smith, J.W. 1956 Observations on the postural mechanism of the human knee joint. *Journal of Anatomy* **90**, 236–260.

Stern, J.T. 1972 Anatomical and functional specializations of the human gluteus maximus. *American Journal of Physical Anthropology* **36**, 315–340.

Stern, J.T. & Susman, R.L. 1981 Electromyography of the gluteal muscles in *Hylobates*, *Pongo* and *Pan*: implications for the evolution of hominid bipedality. *American Journal of Physical Anthropology* **55**, 153–166.

Taylor, C.R., Heglund, N.C. & Maloiy, G.M.O. 1982 Energetics and mechanics of terrestrial locomotion. I. *Journal of Experimental Biology* **97**, 1–21.

Tuttle, R.H., Basmajian, J.V. & Ishida, H. 1979 Activities of pongid thigh muscles during bipedal behavior. *American Journal of Physical Anthropology* **50**, 123–136.

Zarrugh, M.Y. & Radcliffe, C.W. 1978 Predicting metabolic cost of level walking. *European Journal of Applied Physiology* **38**, 215–223.

Zuckerman, S., Ashton, E.H., Flinn, R.M., Oxnard, C.E. & Spence, T.F. 1973 Some locomotor features of the pelvic girdle in primates. *Symposia of the Zoological Society of London* **33**, 71–165.

# Index


acanthodians, 88, 90
acanthopterygians, 90–1
*Acipenser*, 66
*Acorus*, 32
*Addax*, 244
*Aepyceros*, 232, 244
aerodynamics, 101–3, 110, 116, 141, 149, 153–5, 183–90, 194, 196–7, 200–3, 208
aerofoil, 99, 108, 150–1, 155, 185–7, 189–90
air space, 149
*Alca*, 135
*Alcelaphus*, 232, 236, 238, 244
*Alces*, 232, 236, 247
Alcidae, 131, 135
*Ambystoma*, 13–14
*Amia*, 66
*Ammodorcas*, 232, 244
ammonoids, 57, 84–5, 89
*Ammotragus*, 244
Anatidae, 131
*Anchisaurus*, 164–5, 167, 169, 171, 177
*Anhinga*, 132
ankylosaurs, 163
antelopes, 231, 237–45, 248–9
*Antidorcas*, 244
*Antilope*, 232, 236, 239, 244, 246
antlers, 229–50
Anura, 131, 132
apes, 257–8, 261–4
arachnids, 40
*Archaeopteryx*, 99, 151, 183–208
*Archiacoceras*, 60
Archosauria, 145–59, 161–78, 183–208
argali, 232
Arran, 43–4
*Arthropleura*, 43–6, 53
Arthropoda, 37–54
Artiodactyla, 225, 229–50
auks, 131, 135
autopodia, 138–9
balance, 58, 61–2, 64, 79, 115–16, 119, 121, 157, 164, 167–8, 186–7, 199, 235, 238, 240, 256, 258–9, 263
*Baltoeurypterus*, 38–42, 51–3
bass, 12
bats, 145, 147–51, 153–6, 187, 189, 195, 198, 200, 205, 208
beams, 25–6, 33–4, 126, 238–40
behaviour, 1, 6–16, 23, 60, 69, 80, 105, 118, 122, 135, 154, 183, 190, 192, 197, 200, 207–8, 229–30, 237, 244, 246–7, 249–50
beira, 232–3
bending stress, 22, 238–40
bighorn sheep, 239, 245
biomechanical analysis, 6, 14, 48, 53, 113, 131, 146–7, 151–2, 158, 201
bipedality, 147, 155, 157, 162, 167–9, 173, 176–7, 183, 190, 199, 203, 207, 255–65
birds, 7, 8, 134, 145, 147, 149–51, 154–6, 183–208, 255, 263
bison, 238
blackbuck, 232, 239
blesbok, 232
bones, 4, 5, 7–8, 10–11, 15, 17, 62, 135–6, 139, 141, 149, 150, 154, 156–7, 163, 168–9, 171–3, 196, 213–25, 230, 232, 238, 240, 243, 247–8
bongo, 232
*Boöcercus*, 232, 244
*Bos*, 232, 238, 244
*Boselaphus*, 236
boundary layers, 65, 67, 118, 124
Bovidae, 223, 229–39, 241, 243, 245–50
bowfin, 66
Bryozoa, 88
*Bubalus*, 238, 244
buffalo, 238
buoyancy, 52, 57–64, 72–3, 78–81, 83, 90, 115, 119, 121–2, 126–7
Burgess Shale, 41, 57

Camerate shells, 58–61, 64, 80, 85
*Campylognathoides*, 150
*Capra*, 232, 236–9, 243–4, 246
*Capricornis*, 232, 244
*Carassius*, 74, 78
*Carcharhinus*, 77, 88–90
*Caretta*, 138
*Carettochelys*, 135, 138–9, 143
carpus, 139, 145, 215–16, 218–19, 225
cats, 256
cell walls, 22–5, 28–33
cellulose, 21–4, 30, 34
Centrarchidae, 12
cephalaspids, 113–28
*Cephalophus*, 237–8, 243–4, 246
cephalopods, 57–92
ceratopians, 161, 163, 173–4, 177–8, 250
*Ceresiosaurus*, 139
Cervidae, 229–50
*Cervus*, 232, 235–6, 243, 245, 247–8
Cetacea, 72, 131
chaetodonts, 82, 87–8, 91
Chelicerata, 38
Chelonia, 131, 135–6, 139
chimpanzees, 255–63
chital, 247
chondrichthians, 88, 90–1
circulation, 131, 185
co-evolution, 161, 178
cockroaches, 46
coefficient of pressure, 27
collenchyma, 34
colugo, 198
*Columba*, 151, 188, 191
combat, 229, 236
complex system, 1–3, 6–7, 9
composite, 22, 25, 34
*Compsognathus*, 151, 191, 193
*Connochaetes*, 232, 238
coots, 132
coracoid, 154, 156, 188, 195–7
corals, 88
cost of transport, 58, 81, 123–4, 126, 202, 204–6
crack length, 239
crayfish, 50
crinoids, 88
crocodiles, 145
Crustacea, 37–8, 47, 50, 53
cuticle, 37, 41, 45–6, 48
Cuvier, 2–3
*Cynocephalus*, 198
cyrtocone, 85–7, 89, 90

*Damaliscus*, 232, 238, 244
dandelion, 23, 32
Darwin, Charles, 1–2, 17, 147, 198
Decapoda, 50
deer, 229–50
*Deinonychus*, 191–2, 196, 199
dentition, 176–8, 229, 234
Dermochelidae, 131–2
*Dermochelys*, 132, 135
Dermoptera, 198
dibatag, 232
dicynodonts, 155
diet, 161–5, 169, 176–7, 229, 232, 234, 249
digitigrade, 219, 256
dik-dik, 232
*Dimorphodon*, 146, 150–1, 155, 156
dinosaurs, 145, 147, 151, 154, 156–7, 161–78, 183, 184, 190–3, 198, 207, 250
dispersal, 104, 186
display, 123, 176, 193, 194, 234, 236–7
diversity, 2, 17, 57–8, 84, 86–91, 161–2, 176, 178, 230, 232, 249
dogfish, 66, 115, 118, 124
dolphin, 140
*Dorcatragus*, 232–3, 244
*Dorygnathus*, 150
drag coefficient, 66–71, 118
drag, 22, 32, 47–53, 57–8, 61–73, 77, 81–3, 86, 101–3, 108, 110, 115–19, 122–7, 131, 133–4, 137–9, 142, 185–8, 200, 202–3
drifting, 104
ducks, 131
duiker, 169, 237–8, 243

*Echidna*, 155
ecology, 6, 47, 84, 87, 91, 142, 154, 157, 171, 184, 190, 198, 208, 229–30, 244, 246, 249–50
ectocochleates, 58–67, 69–74, 76–87, 89–92
Ediacara, 57
efficiency, 2, 15–16, 52, 77, 83, 125–6, 133, 155, 166, 177, 222
Eichstätt, 183
eland, 232, 241
*Eledone*, 74, 77
Emydidae, 132, 135, 137
endocochleates, 58, 66, 69, 74–80, 82–4
energy, 46, 64, 69, 124, 126, 133, 154, 169, 185–7, 189, 196, 198, 201–5, 218, 222–3, 225, 232, 234–5, 238–9, 247, 250, 255, 258–9, 263–4
*Eotragus*, 231–3
Ephemeroptera, 99–100
Equidae, 213–25, 264
equilibrium, 57–8, 61–4, 83, 102, 105, 115, 119, 122
*Equus*, 213–25
erect trunk, 261–4
*Esox*, 66, 77

eumalacostracans, 38
*Euparkeria*, 163
*Euproops*, 50
*Euprox*, 231–3
eurypterids, 38–41, 43, 45, 47–8, 50–3, 139
evolution, of flight, *see* flight
exoskeleton, 37, 47, 121
extinction, 58, 84, 86–9, 91, 140

fabrosaurs, 167–8, 171, 175–7
feathers, 100, 150, 155, 183–4, 189, 191–4, 197, 201, 207
feeding, 7, 12–14, 38, 89, 91–2, 161–3, 172–4, 178, 200, 244
feet, 168, 213, 223, 255–6, 259–64
fighting, 194, 229–50
fish, 12–13, 49–50, 57–63, 66–7, 70–4, 76–83, 86–92, 113–28, 149, 187
flight, 4, 7, 99–111, 134–5, 141, 145–59, 183–208 (*see also* rowing flight, underwater flight)
  evolution of, 99–111, 147, 183–208
  flapping, 111, 184–5, 187, 194–5, 200, 202, 204–6
  gliding, 100–5, 107–11, 148, 153, 184, 186–7, 191, 194, 197–208
  parachuting, 100–4, 107, 109–11, 184
flotation, 62, 80
food-processing, 166
Fourier, 260
frog, 131, 187
Froude number, 259
*Fulica*, 132

gait, 38–43, 45, 47–8, 53, 154–5, 163–4, 167, 169, 222, 255–6, 258, 263–4
gastroliths, 142, 164, 169, 174, 177
gaur, 232
*Gazella*, 231–2, 236, 242–4, 246
gibbon, 255
glide angle, 101, 104–5, 107–9, 111, 202
glide speed, 107–9, 202
glide-and-swim, 58
gliding, *see under* flight
goat, 232, 238, 243, 245, 249
goat-antelope, 237, 245
*Gorilla*, 255–7
growth, 23, 33, 46, 60, 157, 206, 230–1, 240, 246–7
gut processors, 176, 178
gut, 164, 166, 168, 176–8

hang-gliding, 149–51, 155
hartebeest, 232, 238
*Hemitragus*, 232, 238, 244
*Heracleum*, 23
herbivory, 161–6, 168–9, 171, 174, 176, 178–9
heterodontosaurs, 167, 169–74, 177
*Heterodontosaurus*, 167, 169–72
hindlimb, 133, 149–51, 154–7, 168, 191, 197, 214
*Hippotragus*, 238, 244
historical biology, 1, 3–4
hog deer, 247
hogweed, 23
horns, 169, 229–50
horses, 213–25, 264
humans, 255–65
hydrodynamics, 14, 69, 72–3, 113–16, 120, 127, 133, 141
hydrofoil, 52, 116, 131, 134, 138–40, 142–3, 186
Hydropheidae, 132
hydroplane, 58, 90
hydrostatic stability, 57, 61–4, 80, 83
*Hylobates*, 255
*Hypsilophodon*, 167, 171–3, 177

ibex, 238
ichthyosaurs, 140–1
*Iguana*, 165
impala, 232
*Inia*, 140–1
injury, 233, 235–6
insects, 99–111, 133, 198–9
*Iris*, 32–3
Irish elk, 232, 245–6, 249

*Jaekelopterus*, 51
jaws, 13–14, 59, 89–91, 161–77, 179, 234
jet propulsion, 57, 79, 83
Joggins, 48
jumping, 101, 104, 110–11, 199, 203–4, 206

kangaroos, 255
keratin, 150, 191, 230–1, 240, 243, 245, 247, 249
kinematics, 75, 148, 152, 155–6, 188–9, 216
klipspringer, 232
knee, 255–64
*Kobus*, 232, 235–6, 244–6

laminate, 25, 105
*Laticauda*, 132
leaves, 32, 35, 168, 173
legs, 41, 46, 99, 102, 104, 149–50, 154, 157, 168, 186, 193, 199, 217, 238, 255–7, 259, 261–2
*Lepomis*, 12
*Lesothosaurus*, 167–71
*Leuciscus*, 74

lift, 49–53, 58, 72–3, 101–4, 107–8, 110, 115–20, 122–5, 127, 131, 138–9, 141–2, 155, 185–8, 190, 200, 202, 256, 259, 262
lift coefficient, 119–24
ligaments, 4–7, 217–20, 225, 250, 256
lignin, 22–3
limbs, 16–17, 37–46, 51, 127–8, 131, 133, 135–43, 146, 148–57, 168–9, 171, 185, 187, 191–2, 197–207, 213–17, 219, 222–3, 225
*Limulus*, 45–6, 49, 50
*Litocranius*, 244
lizard, 145, 148–9, 161, 165–6, 187, 203
lobster, 50
locomotion, 37, 47, 50, 52, 58, 61, 63–4, 70, 72, 78, 81, 83, 85, 88, 123, 125–6, 131, 133–5, 137–43, 146, 154–5, 183–5, 191, 201–2, 206, 213–14, 217, 222, 255 (*see also* flight, running, swimming, walking, etc.)
*Loligo*, 74, 76–9, 82
Los Hoyas, 197, 207
lycopods, 104

*Macaca*, 255
macaque, 255
mackerel, 66, 126
macroevolution, 161, 178
*Madoqua*, 232, 244
magpie, 194, 208
mammals, 3, 149, 161–2, 165–6, 168, 171, 173–5, 178, 187, 191, 198, 203, 213–14, 219, 222–3, 230, 255–6, 259–60, 262–4
manoeuvrability, 80, 82, 85, 88, 91, 126, 138, 189, 197, 208
manus, 196–7, 215–17, 219, 221–3
markhor, 232
*Massospondylus*, 164–5
mayflies, 99–100, 107, 109
mechanical advantage, 221
*Megaloceros*, 232, 245
*Meleagris*, 259
*Merychippus*, 213–16, 218–22, 225
*Mesohippus*, 213–16, 218–23
*Mesolimulus*, 49–50
*Micropterus*, 12
millipedes, 39
*Mixopterus*, 41, 43
monogamy, 232–4, 244
Montceau-les-Mines, 43
Monte St Georgio, 139
moose, 232, 247
morphology, 1–17, 28, 31, 34, 37–8, 40–7, 50–3, 57–8, 67, 85, 89, 131, 135, 146, 148–50, 158, 161–2, 178, 183, 191–2, 195, 197–8, 200–1, 203–4, 207–8, 222–3, 225, 229–30, 237, 249
mortality, 235–6
motor pattern, 7–14
mouth processor, 176–8
*Muntiacus*, 231–3, 236, 238
muntjac, 231–3, 236, 238
Murchison District, 47
muscles, 4–8, 10–14, 37–8, 46, 52–3, 57, 73–6, 78–80, 83, 113, 121, 125–6, 135–8, 140–1, 156, 163, 166, 175, 188, 190, 195–7, 201, 203, 213, 216–17, 256–8, 262–4
myology, 216
myriapod, 38–9, 41, 43–4, 104

nautiloids, 57, 60, 81, 84–5
*Nautilus*, 58–63, 66, 69–83
nectridea, 115, 127
*Nemorhaedus*, 244
Neotraginae, 231
New Brunswick, 43–5
newts, 131–2
*Nitella*, 25
nothosaurs, 135, 139, 142–3

*Octopus*, 74, 76–8
*Odaraia*, 50
*Odocoileus*, 232, 236
odontocetes, 61, 132
*Ogygiocarella*, 49
*Olenoides*, 41
ontogeny, 10–14, 17
*Oreamnos*, 236, 244
*Oreotragus*, 232, 244
organismal design, 1–4, 6, 11
Ornithischia, 162–3, 166, 168, 178
ornithopods, 161, 163, 166, 168–9, 173–4, 177
ornithopters, 151
orthocone, 64, 66–7, 83, 85–7, 89–91
*Oryx*, 236, 238, 241, 244–5
osteichthians, 88
osteology, 192, 213–14
ostracoderms, 88, 90
ostrich, 256
otariids, 52, 132, 138, 143
*Ourebia*, 244
*Ovibos*, 236
oviparity, 184
*Ovis*, 232, 236, 238–9, 243–6
Owen, Richard, 147–9
oxygen, 19, 75, 83, 125–6, 190, 263–4

pachycephalosaurs, 163, 250
*Pan*, 255, 258
*Pantholops*, 244

parachuting, *see under* flight
*Parastylonurus*, 41, 50
pectoralis muscle, 195–6
*Pelea*, 236–7, 244
pelvic girdle, 187, 262–3
pelvis, 136, 157, 165, 167–8, 263
penguin, 131–2
*Perca*, 74, 77
*Periplaneta*, 104
*Phoca*, 132
phragmocone, 59, 63–4, 71–3, 83
phylogenetic analysis, 1, 4, 11–12, 158
phylogeny, 4, 6, 10–12, 14, 16, 113, 143, 152, 158, 190, 200–1, 225
*Physeter*, 132
physical biology, 3–4
*Pica*, 208
pike, 66
pinnipeds, 52
placoderms, 88, 90
plants, 1, 21–35, 100, 104, 161–2, 164, 166, 171, 176–9, 186, 265
plantigrady, 255–6, 259
*Plateosaurus*, 164–5
plesiosaurs, 52, 131–43
*Pleurodira*, 135
pleurokinesis, 172–4, 177
pliosaurs, 136
polar bears, 131–3
polygyny, 233–4, 244
power output, 53, 79, 113, 125
pre-stress, 21, 23, 29, 32–3
predation, 50, 87, 90–1, 128, 140, 162, 164, 191, 200, 204, 207–8, 233, 265
prey capture, 11–13
primates, 255–65
*Procapra*, 232, 244
*Procyon*, 258
progymnosperms, 104
prosauropods, 162–8, 175, 177–8
proto-bats, 198
protopterygotes, 99–105, 108–10
*Protragocerus*, 231
*Pseudemys*, 132
*Pseudois*, 244
*Psilotum*, 24–6, 30, 31, 32
psittacosaurs, 173–5, 177–8
*Psittacosaurus*, 174–5, 177
*Pteranodon*, 152–4, 157–8
*Pteraspis*, 115
pterodactyloids, 145–6, 152
pterosaurs, 145–59, 196
pterygotids, 50
puku, 232

quadrupeds, 164, 173, 177–8, 198, 255, 259–63
*Quetzalcoatlus*, 145, 155

raccoons, 256, 258, 262
Rallidae, 132
*Rangifer*, 236
*Raphicerus*, 244
red deer, 232, 235–7, 246–8
*Redunca*, 244
reptiles, 145, 147, 161–3, 166, 176, 178, 183, 187
respiration, 47
Reynolds number, 51, 103, 105, 107–8, 110, 114, 116, 118, 140
rhamphorhynchoids, 145–6
*Rhamphorhynchus*, 145, 148, 152
rhea, 259
rhebok, 237
*Rhynia*, 24–6, 30–3
Rhynie Chert, 30
rodent, 255
rowing flight, 138–9, 143 (*see also* underwater flight)
rowing, 52–3, 131–5, 137–40, 142–3, 184, 199, 230–1, 247
running, 39, 101, 104, 191, 199–207, 255, 258, 260, 263–5 (*see also* walking)
Rupicaprini, 237

safety factor, 32, 153, 240–2, 244
*Saiga*, 236, 244
salamander, 12–13
*Salmo*, 66, 74, 77–8
Salt Range, 41
sambar, 232
sarcopterygians, 90
Saurischia, 162–3
sauropods, 161–2, 177–8
Sauropterygia, 135
scaling, 45, 81, 126, 224–5
sclerenchyma, 33–4
*Scomber*, 66, 71
scombroids, 58, 79
sea snakes, 131–2
sealions, 132
seals, 132
Seeley, Henry Glover, 149
*Selaginella*, 34
semi-lunate carpal, 196–7
*Sepia*, 74, 77–8, 82
serow, 232
shape, 12, 23, 42, 46, 52, 66, 81–2, 89–90, 105, 108–9, 115–16, 124, 126, 138–9, 145–6, 155–7, 161, 168–9, 173, 175, 188, 191, 194, 202, 213–14, 218–19, 230, 237, 240, 245, 249, 262–3
sharks, 58, 90–1
shear stress, 238–9

sheep, 238–9, 243, 245, 249, 263
shell, 57–73, 80–91, 105, 116, 135
shell geometry, 63, 66, 71, 81
sinking speed, 104, 107–8, 110
siphuncule, 59, 60, 85
size, 12, 22, 30–1, 43, 45–7, 50, 64, 73–6, 79–82, 90, 101, 103, 124, 131, 133, 145, 157, 164, 166–9, 191–4, 197–8, 205, 213–14, 216, 218, 222–3, 225, 230, 232–40, 242, 247–9
skeleton, 57, 62, 80, 121, 135–6, 145, 149, 154, 163, 165, 168–9, 171, 174, 188, 191–2, 196, 214, 257
skull, 115, 146, 149, 157, 161–75, 230–2, 238–9, 250
Solenhofen, 183, 191, 193, 194, 206
speed, 15, 40, 49, 52–3, 58, 69, 72–4, 77–8, 80–2, 85–6, 91, 104–5, 107–11, 116, 122–8, 133, 150, 153–5, 163, 187–9, 196–7, 200–4, 206–7, 222, 255, 259–60, 263–4
Spheniscidae, 131–2
*Spheniscus*, 132
sphenopsids, 104
*Sphyraena*, 77, 79
*Spirula*, 58
squid, 58, 61, 66, 70–1, 73, 76
squirrels, 198, 202, 205
stability, 22, 34, 40–2, 50, 57, 61–4, 80, 82–6, 102–10, 119–20, 125, 157, 186, 200, 208
standing, 1, 3, 6, 17, 34, 37, 84, 87–9, 102, 146–7, 199–200, 216–17, 229, 255–7, 261–4
stegosaurs, 163
stems, of plants, 21–34
sternum, 141, 149, 156, 195, 196, 197
stiffness, 23, 25, 29, 32, 142, 150, 240, 242–5
straight knee, 256, 258–9, 263–4
stress, 21–3, 28–34, 45–6, 80, 104, 163, 213, 219, 222–3, 238–40
stride length, 43, 45, 222, 259–60
*Struthio*, 256
sturgeon, 66
sub-unguligrade, 213, 219, 222
sunfish, 12
supracoracoideus muscle, 156, 188, 195–7
suspensory apparatus, 217–18, 220–2
swimbladder, 60, 62, 72–3
swimming, 38, 48–53, 57–8, 62–4, 70, 72–4, 76–7, 80–1, 88, 91, 113–28, 131–43 (*see also* rowing, rowing flight, underwater flight)
  axial, 50, 131–2, 137
  paraxial, 131–2, 137
*Synceros*, 236, 238, 244

tahr, 232, 238
tails, 48, 50, 76, 89–90, 100, 103, 106–7, 109, 115–16, 118, 120–2, 124–5, 127, 145, 157, 164, 167–8, 186, 197, 208, 232
tapir, 213, 218–19
*Taraxacum*, 23–34
*Taurotragus*, 232, 241, 244
teeth, 89–90, 161–75, 177, 238
teleosts, 58, 88–91
terrestrial locomotion, 138, 154–5, 191 (*see also* running, walking)
*Tetracerus*, 244
*Thalarctos*, 131–2
thermoregulation, 184, 194
theropod, 183, 189–91, 195–6, 198–9, 203–4
thrust, 61–3, 74–5, 77–8, 80–2, 103, 123, 131, 133, 139–43, 155, 184, 186–90, 199, 201–4, 207
thunniforms, 77, 82, 88
tope, 124
trace fossils, 38–41, 43, 47, 53
trackways, 38, 41, 43–5, 47, 48
*Tragelaphus*, 244
trilobites, 39, 41, 43, 48–9, 53
trionychids, 135, 138–9, 143
*Triturus*, 132
trout, 66, 74, 76–8, 126
tur, 238
turgor, 21, 23, 25–34
turkey, 259
turtle, 131–2, 135–7, 142–3

underwater flight, 131, 134–5, 137–40, 142–3 (*see also* rowing flight)
unguligrady, 213, 217–19, 222–5
uniramia, 38, 40
Urodela, 132
Ursidae, 132

vane asymmetry, 194
viscosity, 105, 114
vorticity, 49, 72, 128, 131, 185–7, 189, 195, 202

wake, 49, 66, 71, 185, 187, 189, 202
*Walangania*, 231
walking, 38, 40–7, 50, 52, 138, 255–64 (*see also* running)
water beetle, 52
waterbuck, 235, 237, 242
whale, 132
white-tail deer, 232
wildebeest, 232, 238
wind-gliding, 207
wing area, 109, 145, 153, 154, 155, 157, 202
winglets, 99–111

wings, 7, 8, 75, 99, 101–2, 107, 109–10, 121, 125, 131, 136, 139–41, 143, 145, 147–50, 152–7, 183–90, 193–7, 199–207
wingspan, 145, 148, 152–4, 186, 188–9
wolf-spider, 46

*Xenopus*, 132

Young modulus, 23–4, 29

*Zalophus*, 132
zeugopodia, 139